RADAR SIGNAL PROCESSING AND ITS APPLICATIONS

edited by

Jian Li
University of Florida

Petre Stoica
Uppsala University

Robert Hummel
Defense Advance Research Projects Agency

Edmund G. Zelnio
AFRL/SNA

A Special Issue of
MULTIDIMENSIONAL SYSTEMS AND SIGNAL PROCESSING
An International Journal
Volume 14: Nos. 1-3 (2003)

KLUWER ACADEMIC PUBLISHERS
Boston / Dordrecht / London

MULTIDIMENSIONAL SYSTEMS AND SIGNAL PROCESSING

An International Journal

Volume 14, Nos. 1/2/3, January–July 2003

Special Issue: Radar Signal Processing and Its Applications
Guest Editors: Jian Li, Robert Hummel, Petre Stoica and Edmund G. Zelnio

Distributors for North, Central and South America:
Kluwer Academic Publishers
101 Philip Drive
Assinippi Park
Norwell, Massachusetts 02061 USA
Telephone (781) 871-6600
Fax (781) 681-9045
E-Mail <kluwer@wkap.com>

Distributors for all other countries:
Kluwer Academic Publishers Group
Post Office Box 322
3300 AH Dordrecht, THE NETHERLANDS
Telephone 31 78 6576 000
Fax 31 78 6576 254
E-Mail <services@wkap.nl>

Electronic Services <http://www.wkap.nl>

Library of Congress Cataloging-in-Publication Data

Radar Signal Processing and Its Applications / edited by Jian Li...[et al.].
p.cm.
Reprinted from a special issue of MULTIDIMENSIONAL SYSTEMS AND SIGNAL PROCESSING, an international journal, Volume 14: Nos. 1-3; January-July 2003.
ISBN: 1-4020-7397-6 (alk. paper)

Printed on acid-free paper.

Printed in the United States of America.

Multidimensional Systems and Signal Processing, 14, 5, 2003

Editorial

On behalf of the readers, editors, associate editors and editorial board members of this journal, I wish to express my appreciation and thanks to the four guest editors who organized this comprehensive special issue on the challenging problems in radar that are being tackled using technical devices originating in one and multidimensional signal processing. The topics spanned are extensive and pertinent as summarized in the guest editorial and detailed in the papers by the authors. We look forward to receiving comments by the readers following perusal.

There was considerable emphasis on multidimensional systems and signal processing, highlighted by a mini-symposium on the subject, at the Mathematical Theory of Networks and Systems (MTNS 2002) biennial symposium held in August 2002 at the University of Notre Dame in South Bend, Indiana. The symposium provided a meeting ground of old as well as new members of the multidimensional systems and signal processing community. The attention of the readers is also directed to the June 2002 special issue on multidimensional signals and systems in the IEEE Transactions on Circuits and Systems I: Fundamental Theory and Applications. A future special issue on super-resolution image reconstruction is being planned for appearance in IEEE Signal Processing Magazine.

In the previous editorial I had mentioned that Marwan Simaan, one of our co-editors, was moving to Auburn University in Alabama. I am now told that for family reasons, he has decided to stay at the University of Pittsburgh. Finally, I wish to thank Dr. Chalie Charoenlarpnopparut from the Sirindhorn International Institute of Technology, Thammasart University, Thailand, for accepting my invitation to become an associate editor of our journal. He will be replacing Sanjit K. Mitra who served this journal creditably for many years and expressed a desire to be relieved of his duties at an opportune time. My thanks to Sanjit for his advice and help over the years.

N. K. Bose
Editor-in-Chief

Multidimensional Systems and Signal Processing, 14, 7–8, 2003

Guest Editorial

Radar has been around since World War II. As the technology develops, radar has been playing an increasingly more significant role in both military and civilian applications, especially due to its all-weather and day-and-night capabilities. As the role and goal of radar expand, many challenges emerge. The purpose of this special issue is to provide an overview of how signal processing methods can help tackle these challenges. It is evident from the papers that we have included into this special issue that radar is a fertile ground for signal processing applications. We will briefly discuss below, in a topic-wise manner, the papers included into this special issue.

As the radar bandwidth increases, the scattering centers of a target can be resolved to various degrees depending on the radar bandwidth. One may wonder how to use high range resolution (HRR) radar for automatic target recognition (ATR), especially for moving target identification. Wavelet transformation has been considered for selecting discriminant target features in (Nelson, Starzyk, and Ensley). For target scatterer extraction from HRR radar, data modelling and parameter estimation are investigated in (Ma, Du, and Ahalt).

Synthetic aperture radar (SAR) imaging is becoming increasingly more sophisticated. As we extend the length of the synthetic aperture for better cross-range resolution and expand the radar bandwidth for improved range resolution, aspect and frequency dependencies of the scattering objects must be addressed. The detection and analysis of anisotropic scattering are studied in (Kim, Fisher, and Willsky). Other interesting SAR related topics include two-dimensional adaptive extrapolation for improved resolution (Brito, Chan, and Cabrera), polarimetric calibration (Rajagopal and Potter), large bandwidth and long duration waveform SAR (Lin, Zeng, Bi, and Yeo), and ATR using SAR imagery in (DeVore and O'Sullivan).

Naturally we would like to use radar to collect as much information as possible, including the three-dimensional (3-D) information. Topics related to 3-D information retrieval and modelling include 3-D feature extraction of rigid moving body (Stuff, Sanchez, and Biancalana), absolute phase retrieval for SAR interferometry (Veneziani, Bovenga, and Refice), 3-D scattering model for HRR and SAR prediction (Denney and Figueiredo), and detection of 3-D target motion from ISAR data (Li, Ling, and Chen).

Finally, adaptive beamforming is known to have better resolution and much better interference rejection capability than the standard data-independent beamformers such as the delay-and-sum beamformer. Beamforming is needed for high-frequency over-the-horizon (OTH) radar systems, which suffer from multipath and scattering and propagation uncertainties. Various adaptive beamforming schemes are evaluated with respect to their interference cancellation performances when applied to OTH radar systems in (Fabrizio, Gray, and Turley).

Acknowledgments

We thank the authors who submitted papers to this special issue and the reviewers who spent their time to evaluate these papers. We also thank Ms. Jennifer Evans for initiating the development of this special issue and Dr. N. K. Bose for encouraging us along the way. Finally, we gratefully acknowledge the assistance of Ms. Michelle Misner and Ms. Melissa Sullivan of the Editorial Office.

Jian Li
Robert Hummel
Petre Stoica
Edmund G. Zelnio

Multidimensional Systems and Signal Processing, 14, 9–24, 2003

Wavelet Transformation and Signal Discrimination for HRR Radar Target Recognition

DALE E. NELSON Dale.Nelson@WPAFB.AF.MIL
Air Force Research Laboratory, 2241 Avionics Circle, AFRI/SNAT, Building 620 Rm C2S69, Wright Patterson AFB, OH 45433-7321

JANUSZ A. STARZYK starzyk@bobcat.ent.ohiou.edu
Department of Electrical Engineering and Computer Science, Ohio University, Stocker Center #347, Athens, OH 45701

D. DAVID ENSLEY David.Ensley@robins.af.mil
United States Air Force, WR-ALC/LUJE, 226 Cochran St, Robins AFB GA 31098-1622

Received November 7, 2000; Revised November 7, 2000; Accepted October 5, 2001

Abstract. This paper explores the use of wavelets to improve the selection of discriminant features in the target recognition problem using High Range Resolution (HRR) radar signals in an air to air scenario. We show that there is statistically no difference between four different wavelet families in extracting discriminatory features. Since similar results can be obtained from any of the four wavelet families and wavelets within the families, the simplest wavelet (Haar) should be used. We further show that a simple box classifier can be constructed from the extracted features and that any feature that classifies four or less training signals can be removed from the classifier without a statistically significant difference in the classifier performance. We use the box classifier to select the 128 most salient pseudo range bins and then apply the wavelet transform to this reduced set of bins. We show that by iteratively applying this approach, classifier performance is improved. The number of times the feature reduction and transformation can be performed while producing improved classifier performance is small and the transformed features are shown to quickly cause the performance to approach an asymptote.

Key Words: rough sets, wavelets, automatic target recognition, high range resolution radar, feature selection

1. Introduction

Most of the work in HRR target recognition has been done by or sponsored by the military. The approaches taken by various researchers as summarized by [8] appear to ignore the benefits that can be gained by proper transformations of the input signal. The wavelet transform is a new tool which has been used in image compression and more recently in target recognition. When wavelet transforms are used for image compression the most important goal is to minimize the loss of information. In ATR the most important objective is to be able to separate the various target classes [7]. Some researchers have explored the use of wavelets to provide a richer feature space [2], [3], [4], [7], [9], [12], [13]. However there is little evidence of widespread use of this technique. Mitchell himself explored transformations but he limited the analysis to an autoregressive approach to clean up (remove low information data) from the signature.

Famili states that preprocessing the data "... conditions the input data to allow easier subsequent feature extraction and increased resolution." [5]. In the past, engineers have used transforms such as the Fourier transform to move the signal from a time base to a frequency base [14]. Although this is useful for some applications, target recognition of HRR signals improved only a little under this transform. Wavelets bring a new tool to HRR signals classification. The benefits of using wavelets, according to Strang, are related to a local character of those transformations "...new transforms are much more local. An event stays connected to the time when it occurs. ...a time-frequency description" [11]. Researchers that have used wavelets for target recognition (especially for HRR) have found that the original feature space can be augmented by the wavelet coefficients and will yield a smaller set of more robust features in the final classifier [9], [13]. In addition to computational savings [4], investigators have also found that wavelet methods can improve radar performance (Pcc) [12], [13]. However, as pointed out by Stirman [13] even with improvement in Pcc there can be a bias of the wavelets toward one or two classes to the detriment of others.

In considering wavelets for ATR, serious consideration must be given to the selection of a wavelet family and a wavelet in the family. Lu explored this issue in the context of image coders [15]. In his paper, Lu compared two wavelets, one from the Biorthogonal family (B97) and the other from the Daubechies family (D8). Using two different metrics, Lu observed a slight advantage of the B97 versus the D8. In this paper, using the criterion of improving the probability of correct classification (Pcc), we show that there is no statistical advantage of one family (out of four) over any other family. Any difference in performance which can be observed in a particular application is due to the statistical nature of normal variations in the data. Stirman, using wavelets for ATR, explored the use of different wavelets from the Daubechies family, and found that results were similar among the three wavelets [13]. In this paper we show that here is no statistical advantage of one wavelet in a family over another in the same family, thus generalizing Stirman's observation.

Once the input data is transformed, the process of feature selection for the given type of classifier must begin. A very popular approach uses a quadratic classifier [8]. The quadratic classifier uses statistics of the signal to be classified and compares them to the statistics of a template for the various target classes. This method is fraught with problems since it uses the entire signal and thus tries to match noise to noise. In an effort to get around this problem, Mitchell [8] uses an autoregressive filter to remove noise and then uses the filter to help select important range bins for classification. It is not the purpose of this paper to explore the development of a classifier. However, in order to have a means to test the usefulness of the data transforms, we must have a classifier to test the performance. In order to reduce the problem of the quadratic classifier, we have chosen to use the simple generalized box classifier [1], [10] from which to evaluate results. Our main objective was to determine which, if any, family of wavelets provided the best feature set for a classifier. A secondary objective was to determine if further wavelet transformations would produce even better classification results. This required the use of a method for down selecting features from which to perform further wavelet analysis. In this paper, using wavelet transformations, we will show:

1) wavelets are useful for HRR ATR,
2) wavelet coefficients as features improve classifier performance,
3) what family of wavelets are best,
4) what wavelet in a family is best, and
5) how to mitigate or eliminate wavelet bias towards some target classes.

2. Signal Characterization

When constructing a classifier, the designer is often able to rely on intuition to select the best features to choose among the target classes. This works when the sensor used is a "literal" sensor. That is, the sensor provides an output similar to what the human senses are used to dealing with. When the sensor cannot do this, then automated means must be used to select the best features. This paper uses High Range Resolution (HRR) radar signals. A HRR signal is an n-dimensional vector $x = (a_1, a_2, \ldots, a_n)$, where $a_i \in \{0, 1, \ldots, 255\}$. The HRR radar provides a 1-D picture of what the sensor is looking at. HRR signals are particularly hard to use for target recognition, partly because the 3-D world is projected into just one dimension. When this is done, there are many ambiguities created which must be resolved. A further complication is that when HRR data is plotted as signal strength vs. range bin, the resulting graph is almost impossible for a human to use for target recognition, mostly because it is an image we are not used to interpreting. A better representation would be to present the HRR signal as an audio signal (similar to sonar) because humans recognize this kind of 1-D signal better. In fact, Szu points out that the human auditory system uses wavelets [7].

A further complication to target identification using HRR is that the signals change considerably with only a small change in azimuth and elevation. This is illustrated in Figure 1.

The signals in Figure 1 are from two different targets. The signals shown for each target were taken at 200 msec intervals. Their significant variations in a short time span illustrates how difficult it would be to construct a target identification system based on these signals.

3. Wavelet Transforms

Wavelet transforms have been found useful in a variety of applications. This is because they provide the analyst with an approximation of the signal and a detail of the signal as well. This helps to identify small anomalies which might be useful. For a complete description of wavelet analysis the reader should refer to [14] and [11]. A brief summary of how the wavelets were used is presented here.

The 1-D discrete wavelet transform (DWT) of a signal yields an approximation and a detail of the original signal. Passing the original signal through a low-pass filter then down

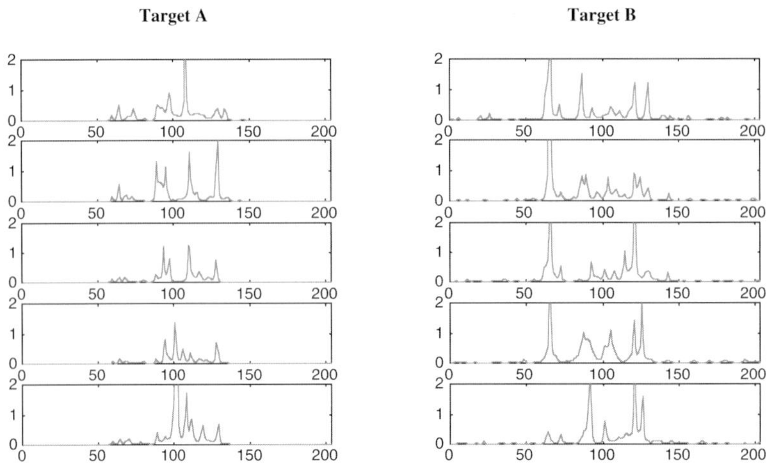

Figure 1. A comparison of two HRR target signals.

sampling produces the approximation. Passing the original signal through a high-pass filter then down sampling produces the detail. The corresponding wavelet functions may be obtained by an iterative process involving convolution of the filter coefficients.

Discrete wavelet packet (DWP) analysis begins with the DWT of the original signal. The next level of the DWP analysis calculates the DWT of the resulting approximation and the DWT of the resulting detail. Subsequent levels calculate the DWT of all the approximations and details of the previous level. Since the number of samples of each approximation and detail is approximately half of the number of samples of the input signal to the DWT, DWP analysis must cease when the approximations and details each contain a single sample.

The discrete wavelet packet (DWP) analysis is performed as follows. The length, L, of the normalized signal must be a power of 2. The number of levels of the DWP analysis is $N = \log_2(L)$, beginning with level N, and ending with level 1. At level j the 1-D discrete wavelet transform (DWT) of a signal of length 2^j is calculated using DWT.m from the MATLAB Wavelet Toolbox. The outputs of DWT were an approximation and a detail, each of length at least 2^{j-1}. The tails of the approximation and detail are all zero-valued, so zero values are trimmed from each tail to obtain lengths of 2^{j-1}. The trimmed approximations and details are used as inputs to the DWT at subsequent levels. The trimmed approximations and details at each level are appended to the normalized signal and placed in the training set.

The wavelet functions used in this paper for the DWT are shown in Table 1.

Prior to selecting features for the target classifier, it is useful to preprocess the original signal. Any operation which increases our ability to separate the classes is desirable. In this paper, we base feature selection on transformations derived from wavelets in Table 1. Training and test sets were constructed using each of the functions. The utility of each of these wavelets for enhancing performance of a classifier was then analyzed. An example

Table 1. Wavelet functions used in 1-D discrete wavelet transforms.

Db1 Haar Bior1.1	Bior3.5	Db5	Coif4
Bior1.3	Bior3.7	Db6	Coif5
Bior1.5	Bior3.9	Db7	Sym2
Bior2.2	Bior4.4	Db8	Sym3
Bior2.4	Bior5.5	Db9	Sym4
Bior2.6	Bior6.8	Db10	Sym5
Bior2.8	Db2	Coif1	Sym6
Bior3.1	Db3	Coif2	Sym7
Bior3.3	Db4	Coif3	Sym8

of the power of a wavelet transformation is illustrated in Figure 2 using the Haar wavelet transform on the original signal.

In Figure 2, the original signal is contained in the first 128 feature index points. The coefficients of the Haar transform are contained in the remaining feature index points. The original signal features show that the largest number of signals in the training set that can be classified by a single feature is 20 out of a maximum of 60. Selecting a single feature from the wavelet coefficients, it is possible to classify 50 out of 60 signals. This is a significant improvement!

4. Training and Test Data Sets

The data used for training and testing is critical to the classifier's performance. Obviously, the best data to be used are measured data. However, these data are in very short supply, very expensive to obtain, and are difficult to ground truth. Ground truth is the recording of

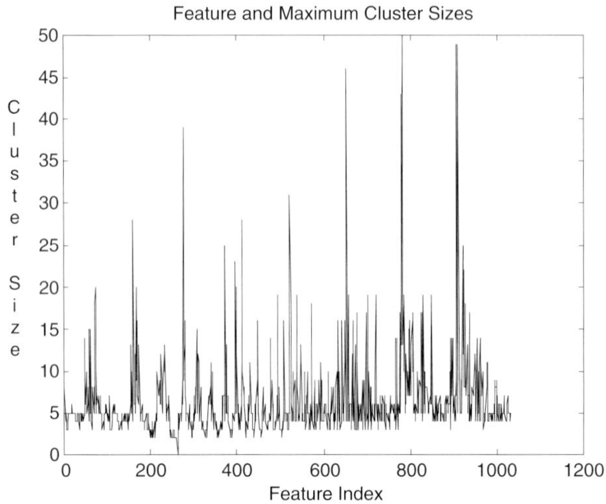

Figure 2. Maximum Cluster Sizes.

the radar parameters and the azimuth and elevation of the target aircraft. It is virtually impossible to gain measured data on enemy aircraft, since the enemy is unlikely to loan us an airplane for the measurements.

The data set used in this research consists of synthetic HRR returns on six targets. For each target there are 1071 range profiles consisting of 128 range bins. The value of each range bin is an integer between 0 and 255. The pose of the target is head-on with an azimuth range of $\pm 25^\circ$ and elevations of -20° to 0° in one degree increments as illustrated in Figure 3.

This data is divided into two sections, one for training and one for test. The training set consists of 25% of the data and the test set the 75% of the data (the remaining data), randomly selected. The small training set permits faster training, facilitating algorithm

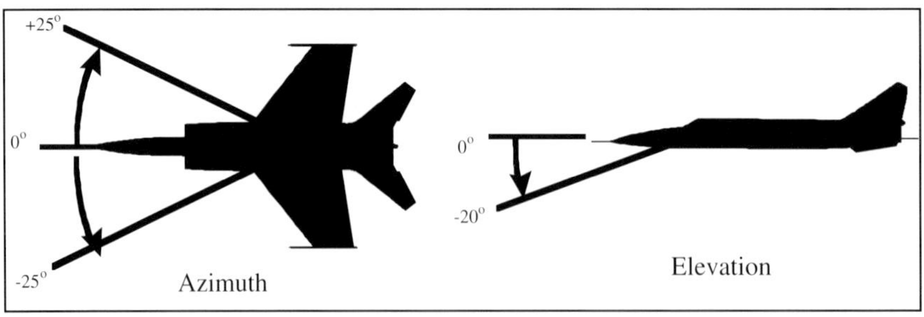

Figure 3. Azimuth and Elevation Ranges.

development and debugging. The training set was constructed by using a random number generator to select 25% of the azimuth and elevation angles and then by selecting signals from each target class with these angles. All remaining signals were placed into the test set.

We have seen that wavelets provide a powerful way of looking at the original signal so it makes sense to incorporate wavelet transforms and some statistical properties into the training and test set. The first step towards creating the training and test set is to normalize the original signal x_i using the l_2 norm yielding $\bar{x}_i$. We next calculate six statistical measures such as mean and standard deviation $Q(\bar{x}_i) = (q_{1i}(\bar{x}_i), q_{2i}(\bar{x}_i), \ldots, q_{6i}(\bar{x}_i))$ where Q represents the statistical vectors. Using similar notation, the wavelet transforms $W(\bar{x}_i)$ are constructed as described previously. The rows in the training set S are defined as the tuple $s_i = (Q(\bar{x}_i), \bar{x}_i, W(\bar{x}_i))$ where each $s_{ij} \in [0,1] \in \Re$. The training and test sets are conveniently represented as a matrix $S = \begin{bmatrix} \vdots \\ s_i \\ \vdots \end{bmatrix}$.

5. Classifier

The classifier used in this paper is a version of the generalized box classifier [1]. The training set produced as described above is used to construct the classifier. We refer to each row of the training and test sets as a signal. The training set S consists of signals having 1030 pseudo range bins.

The first step in constructing the classifier is to sort each column of S from the smallest value to largest value creating a new matrix $\bar{S}$. A matrix $\bar{M}$ is constructed with each element of $\bar{M}$ corresponding to the target type of each element of $\bar{S}$.

The algorithm determining a target classifier is as follows:

Let i denote the target class, and j the feature number. Set $i = j = 1$.

Step 1. Search all columns of $\bar{M}$ to find the column with the largest contiguous cluster of the selected target class i. Let $\sigma(j)$ denote the column determined by this procedure (σ is a permutation of the columns of $\bar{S}$). Let $\bar{s}_{n,\sigma(j)}$ denote the minimum value in the contiguous cluster and let $\bar{s}_{k,\sigma(j)}$ denote the maximum value in the contiguous cluster. The indices n and k correspond to the row indices of $\bar{S}$ with the minimum and maximum values. All signals contained in this cluster are removed from further consideration.

Step 2. Define the $\boldsymbol{j}^{th}$ feature of target class i as the set $f_{ij} = \left(\bar{s}_{n,\sigma(j)}, \bar{s}_{k,\sigma(j)}\right)$. Set $j = j + 1$ and repeat this process (go to step 1) until there are no more training signals from target class i.

Step 3. Increment target class $\boldsymbol{i}$ and set $j = 1$. Repeat this process (go to step 1) until all target classes are accounted for.

The elements of f_{ij} are called individual features. The feature set F is defined as the set of all f_{ij}. A transformed signal z is said to be classified as target class i when there exists a feature f_{ij} such that $z \in f_{ij}$.

Table 2. Performance of wavelets.

Wavelet name	P_{cc}	Wavelet name	P_{cc}	Wavelet name	P_{cc}	Wavelet name	P_{cc}
Bior1.3	0.72488	Haar	0.77130	Coif1	0.76115	Sym2	0.79153
Bior1.5	0.78045	Db2	0.79576	Coif2	0.78231	Sym3	0.75886
Bior2.2	0.75760	Db3	0.75886	Coif3	0.77133	Sym4	0.77458
Bior2.4	0.78150	Db4	0.79160	Coif4	0.78770	Sym5	0.75800
Bior2.6	0.77400	Db5	0.77567	Coif5	0.76943	Sym6	0.76345
Bior2.8	0.78600	Db6	0.78120			Sym7	0.76591
Bior3.1	0.70550	Db7	0.77460			Sym8	0.78275
Bior3.3	0.77030	Db8	0.76760				
Bior3.5	0.78020	Db9	0.79410				
Bior3.7	0.79410	Db10	0.77630				
Bior3.9	0.79290	Db11	0.75598				
Bior4.4	0.72990	Db12	0.76300				
Bior5.5	0.74150						
Bior6.8	0.73010						
Mean	0.76064		0.77550		0.77438		0.77073
Standard Deviation	0.02890		0.01329		0.01060		0.01272

A classifier is tested by classifying each of the transformed test signals, z. An $n \times n$ confusion matrix C is constructed to represent the results. To construct the confusion matrix, we first set $C = [0]$. Each test signal is classified and C is modified as follows. If the test signal i, known to be of class j is classified as target type j then $C_{ij} = C_{ij} + 1$. If the test signal i known to be of class j is classified as target type k, then $C_{jk} = C_{jk} + 1$. This process continues until all transformed test signals are classified. In this paper we used equal numbers of signals to represent each target class for both training and test. Therefore, to obtain the final confusion matrix, each element of C is divided by the number of signals for a target class. It should be noted that some test signals may not be classified as any target type. Therefore, it is possible the rows and columns of the confusion matrix will not sum to one. To evaluate the overall performance of the classifier the probability of correct classification, P_{cc}, is calculated. P_{cc} is defined for n target classes as $P_{cc} = \frac{1}{n}\sum_{i=1}^{n} C_{ii}$.

6. Wavelet Family Dependence

As observed in the prior discussion, a wavelet transform improves feature selection for target recognition. The natural question is to identify which wavelet improves target recognition the most. In this section we demonstrate that there is no single wavelet that outperforms all others in this task.

Proposition 1 No single wavelet transform has a statistically significant advantage over other wavelets in extracting features for target classification.

To verify proposition 1, classifiers were constructed using training sets from all the wavelet families. Table 2 shows the results obtained upon testing the classifier built from

Table 3. Wavelet family hypothesis test.

Wavelet name	Wavelet name	$\lvert Z \rvert$	Accept or reject H_0
Biorthogonal	Daubechies	1.723060	Accept
Biorthogonal	Coiflet	1.516130	Accept
Biorthogonal	Symlet	1.109050	Accept
Daubechies	Coiflet	0.183654	Accept
Daubechies	Symlet	0.775505	Accept
Coiflet	Symlet	0.540601	Accept

the original signal and the associated wavelet transform. In addition, the mean and standard deviation of P_{CC} for the wavelet family are presented in Table 2. To compare if there is any significant difference between the families we use hypothesis testing of the means [16]. The mean, μ, and the standard deviation, σ, of the population are calculated using:

$$\mu = \frac{1}{n} \sum_{i=1}^{n} P_{CC_i} \qquad \sigma = \sqrt{\frac{n \sum_{i=1}^{n} P_{CC_i}^2 - \left(\sum_{i=1}^{n} P_{CC_i} \right)^2}{n(n-1)}}.$$

When the mean and standard deviation are computed from samples, μ is replaced by $\overline{x}$ and σ is replaced by s respectively. We are testing the hypothesis; H_0: $\mu_1 = \mu_2$ against the alternative hypothesis; H_1: $\mu_1 \neq \mu_2$. We compute the test statistic as follows:

$$Z = \frac{\overline{x}_1 - \overline{x}_2}{\sqrt{\frac{\sigma_1^2}{n_1} + \frac{\sigma_2^2}{n_2}}}.$$

We will reject H_0 if $|Z| > 1.96$ (1.96 is for a two-tailed test where the results are significant at a level of .05). The results of this hypothesis testing are presented in Table 3.

From the analysis presented we must accept the null hypothesis, that there is no difference in the mean values. This means that there is no statistically significant difference in the performance of the classifiers when different families of wavelets are used to transform the input data. It would be best (from a computational standpoint) to use the simplest form of a wavelet possible. Since there is no difference between the families, the question arises is there significant difference within each family? By examining the size of the mean and the size of the standard deviation, we see that there is no significant difference between the wavelets within the families. It is safe to conclude that classifier performance would be the same no matter which wavelet we choose. Therefore, it benefits us to use the simplest form of wavelet possible, the Haar (Db1) wavelet.

Normally this type of analysis is limited to large samples where the standard deviations of the samples are known. A t-test was also performed which gave the same results. This indicates that the small number of samples did not give us a false acceptance of H_0.

7. Feature Size Dependence

When constructing the classifier, there are times when the classifier is selecting features to classify just a few training signals. However, each new feature increases the dimensionality of the statistical feature space in which the signal classification is performed. The increase in space dimensionality reduces the accuracy of the statistical representation of the training data. As a result, it is possible that when the classifier is choosing a feature to classify a few signals, the classifier performance may decrease on the test set. Based on our research we can make the following observation:

Proposition 2 Features which classify a small number of training signals do not significantly improve classifier performance.

An analysis was performed to determine what size feature (the number of signals that were classified from the training set) could be safely ignored. Elimination of features classifying a small number of signals allows for much faster training of the classifier and better generalization. Table 6 shows the results of the analysis. Each column in the table shows the probability of correct classification with the minimum feature size indicated in the first row.

Using the Z test, as before, we found that eliminating features that classify only one signal produces no statistically significant difference in the classifiers. However, it was suspected that because of the small sample size it might be necessary to perform a t-test. A t-test is used when either or both of the populations are small and the population variances are unknown. However, it must be assumed that the standard deviations of both populations are the same. The t statistic is defined as:

$$t = \frac{(\bar{x}_1 - \bar{x}_2)}{\sqrt{(n_1 - 1)s_1^2 + (n_2 - 1)s_2^2}} \sqrt{\frac{n_1 n_2 (n_1 + n_2 - 2)}{n_1 + n_2}}$$

and the null hypothesis is rejected if the value of t is greater than 2.074 or 2.85 depending on the level of significance chosen. As seen in Table 4, we can safely eliminate all features which classify less than five training signals.

The t-test depends on the standard deviations being equal so we need to test for this also. To accomplish this we use the statistic

$$F = \frac{s_1^2}{s_2^2}.$$

If this value is less than 2.85 we must accept the null hypothesis that the variances are equal. As seen from Table 3, we accept the null hypothesis for all the tests (the standard deviations are equal), i.e., we can accept the results of the t-test.

The results for the other families of wavelets are presented in the following Tables 5–7. The analysis was performed as before. These results show no matter what family of

Table 4. Significance of eliminating features for Daubechies wavelets.

Name	Test Pcc	>1 Pcc	>2 Pcc	>3 Pcc	>4 Pcc	>5 Pcc	>6 Pcc	>7 Pcc
Db1	0.77130	0.76903	0.75369	0.71595	0.67470	0.64463	0.61022	0.58327
Db2	0.79576	0.79474	0.77753	0.75514	0.71347	0.67884	0.65625	0.62203
Db3	0.75886	0.75762	0.73772	0.70496	0.67241	0.64131	0.61124	0.58242
Db4	0.79160	0.79060	0.77318	0.73483	0.70124	0.67408	0.65355	0.62640
Db5	0.77567	0.77443	0.75888	0.72965	0.69026	0.65667	0.63760	0.61002
Db6	0.78120	0.78065	0.76759	0.73462	0.69689	0.67035	0.62702	0.60214
Db7	0.77460	0.77381	0.75784	0.72695	0.69378	0.66351	0.63739	0.61624
Db8	0.76760	0.76655	0.74748	0.71327	0.68113	0.63822	0.61313	0.58058
Db9	0.79410	0.79247	0.77712	0.74913	0.70974	0.68175	0.65666	0.62411
Db10	0.77630	0.77526	0.75764	0.72073	0.68549	0.65522	0.62993	0.61210
Db11	0.75598	0.75452	0.73192	0.69771	0.65687	0.62743	0.58783	0.56087
Db12	0.76300	0.76157	0.74456	0.71409	0.66391	0.64318	0.61478	0.56875
Mean	0.77550	0.77427	0.75710	0.72475	0.68666	0.65627	0.62797	0.59908
Std Dev	0.01329	0.01340	0.01496	0.01708	0.01762	0.01762	0.02146	0.02285
Z		0.22515	3.18521	8.12297	13.94261	18.71136	20.24809	23.11965
Alpha = .05	1.96	NOT Sig.	Sig.	Sig.	Sig.	Sig.	Sig.	Sig.
Alpha = .01	2.58	NOT Sig.	Sig.	Sig.	Sig.	Sig.	Sig.	Sig.
t s1 = s2, normal		0.02601	0.37924	1.00870	1.75033	2.34906	2.74159	3.21468
t .05/22	2.074	NOT Sig.	NOT Sig.	NOT Sig.	NOT Sig.	Sig.	Sig.	Sig.
t .01/22	2.819	NOT Sig.	NOT Sig.	NOT Sig.	NOT Sig.	NOT Sig.	NOT Sig.	Sig.
F		1.00785	1.12539	1.28459	1.32554	1.32564	1.61402	1.71875
Alpha = .05 2.85 (11, 12)		Accept	Accept	Accept	Accept	Accept	Accept	Accept
Alpha = .01 4.54 (11, 10)		Accept	Accept	Accept	Accept	Accept	Accept	Accept

Table 5. Significance of eliminating features for Symlets wavelets.

Name	Test Pcc	>1 Pcc	>2 Pcc	>3 Pcc	>4 Pcc	>5 Pcc	>6 Pcc	>7 Pcc
Sym2	0.79153	0.79039	0.77297	0.74601	0.70704	0.68008	0.64691	0.61726
Sym3	0.75886	0.75762	0.73772	0.70496	0.67241	0.64131	0.61124	0.58242
Sym4	0.77458	0.77277	0.75763	0.72985	0.69129	0.65314	0.60380	0.58390
Sym5	0.75800	0.75701	0.73855	0.71056	0.67428	0.64194	0.62079	0.58409
Sym6	0.76345	0.76261	0.74665	0.71949	0.69461	0.65853	0.61811	0.59883
Sym7	0.76591	0.76509	0.74830	0.71927	0.67822	0.63571	0.60585	0.58118
Sym8	0.78275	0.78148	0.76801	0.73919	0.70975	0.67762	0.63906	0.61418
Mean	0.77073	0.76957	0.75283	0.72419	0.68966	0.65548	0.62082	0.59455
Std Dev	0.01272	0.01261	0.01384	0.01492	0.01526	0.01773	0.01646	0.01564
Z		0.22402	3.29655	8.22262	14.13325	18.29834	24.95777	30.26778
Alpha = .05	1.96	NOT Sig.	Sig.	Sig.	Sig.	Sig.	Sig.	Sig.
Alpha = .01	2.58	NOT Sig.	Sig.	Sig.	Sig.	Sig.	Sig.	Sig.
t s1 = s2, normal		0.02521	0.38028	0.96966	1.67871	2.28800	3.03959	3.62363
t .05/22	2.074	NOT Sig.	NOT Sig.	NOT Sig.	NOT Sig.	Sig.	Sig.	Sig.
t .01/22	2.819	NOT Sig.	NOT Sig.	NOT Sig.	NOT Sig.	NOT Sig.	Sig.	Sig.
F		1.00860	1.08822	1.17248	1.19974	1.39325	1.29408	1.22955
Alpha = .05 2.85 (11, 12)		Accept	Accept	Accept	Accept	Accept	Accept	Accept
Alpha = .01 4.54 (11, 10)		Accept	Accept	Accept	Accept	Accept	Accept	Accept

Table 6. Significance of eliminating features for Coiflets wavelets.

Name	Test Pcc	>1 Pcc	>2 Pcc	>3 Pcc	>4 Pcc	>5 Pcc	>6 Pcc	>7 Pcc
Coif1	0.76115	0.75991	0.74104	0.71139	0.67635	0.64028	0.60566	0.57041
Coif2	0.78231	0.78147	0.76551	0.72819	0.69896	0.67346	0.63241	0.60877
Coif3	0.77133	0.77008	0.75225	0.72675	0.69461	0.65937	0.62371	0.59240
Coif4	0.78770	0.78625	0.77008	0.73835	0.70456	0.68030	0.66371	0.62743
Coif5	0.76943	0.76862	0.74954	0.71699	0.68859	0.65541	0.63386	0.60753
Mean	0.77438	0.77327	0.75568	0.72433	0.69261	0.66176	0.63187	0.60131
Std Dev	0.01060	0.01056	0.01191	0.01047	0.01081	0.01572	0.02105	0.02128
Z		0.25889	4.06272	11.63783	18.70997	20.58158	20.95222	25.22373
Alpha = .05	1.96	NOT Sig.	Sig.	Sig.	Sig.	Sig.	Sig.	Sig.
Alpha = .01	2.58	NOT Sig.	Sig.	Sig.	Sig.	Sig.	Sig.	Sig.
t s1 = s2, normal		0.02663	0.43176	1.19448	1.93590	2.40503	2.77536	3.35826
t .05/22	2.074	NOT Sig.	NOT Sig.	NOT Sig.	NOT Sig.	Sig.	Sig.	Sig.
t .01/22	2.819	NOT Sig.	NOT Sig.	NOT Sig.	NOT Sig.	NOT Sig.	NOT Sig.	Sig.
F		1.00351	1.12434	1.01191	1.02042	1.48319	1.98604	2.00790
Alpha = .05 2.85 (11, 12)		Accept	Accept	Accept	Accept	Accept	Accept	Accept
Alpha = .01 4.54 (11, 10)		Accept	Accept	Accept	Accept	Accept	Accept	Accept

Table 7. Significance of eliminating features for Biorthogonal wavelets.

Name	Test Pcc	>1 Pcc	>2 Pcc	>3 Pcc	>4 Pcc	>5 Pcc	>6 Pcc	>7 Pcc
Bior1.3	0.72488	0.72466	0.71160	0.68548	0.64961	0.61934	0.58160	0.55175
Bior1.5	0.78045	0.77899	0.76178	0.72985	0.68486	0.65355	0.62349	0.60048
Bior2.2	0.75760	0.75639	0.73255	0.70332	0.65377	0.62744	0.59385	0.56337
Bior2.4	0.78150	0.77961	0.76365	0.73047	0.69253	0.66288	0.63697	0.59634
Bior2.6	0.77400	0.77194	0.75577	0.72591	0.69584	0.65874	0.64091	0.60566
Bior2.8	0.78600	0.78397	0.76551	0.73006	0.69482	0.66248	0.63345	0.59551
Bior3.1	0.70550	0.70455	0.68008	0.64484	0.60440	0.56750	0.53391	0.49556
Bior3.3	0.77030	0.76944	0.75182	0.71823	0.68195	0.64381	0.61188	0.58534
Bior3.5	0.78020	0.77899	0.76406	0.73462	0.69461	0.66807	0.63883	0.62515
Bior3.7	0.79410	0.79226	0.77567	0.75224	0.71783	0.69316	0.65936	0.62120
Bior3.9	0.79290	0.79247	0.77837	0.74789	0.70891	0.68361	0.65189	0.62349
Bior4.4	0.72990	0.72944	0.71783	0.69067	0.66848	0.62744	0.60753	0.58473
Bior5.5	0.74150	0.74146	0.73192	0.70869	0.68755	0.65167	0.61622	0.59736
Bior6.8	0.73010	0.73006	0.72032	0.69565	0.66538	0.63553	0.61936	0.58680
Mean	0.76064	0.75959	0.74364	0.71414	0.67861	0.64680	0.61780	0.58805
Std Dev	0.02890	0.02844	0.02857	0.02855	0.02887	0.03113	0.03232	0.03386
Z		0.08970	1.44924	3.96527	6.95551	9.28304	11.41263	13.42897
Alpha = .05	1.96	NOT Sig.	NOT Sig.	Sig.	Sig.	Sig.	Sig.	Sig.
Alpha = .01	2.58	NOT Sig.	NOT Sig.	Sig.	Sig.	Sig.	Sig.	Sig.
t s1 = s2, normal		0.01518	0.24566	0.67206	1.18217	1.60943	1.99979	2.38636
t .05/22	2.074	NOT Sig.	NOT Sig.	NOT Sig.	NOT Sig.	NOT Sig.	NOT Sig.	Sig.
t .01/22	2.819	NOT Sig.	NOT Sig.	NOT Sig.	NOT Sig.	NOT Sig.	NOT Sig.	NOT Sig.
F		1.01615	1.01172	1.01229	1.00088	1.07728	1.11825	1.171765
Alpha = .05 2.85 (11, 12)		Accept	Accept	Accept	Accept	Accept	Accept	Accept
Alpha = .01 4.54 (11, 10)		Accept	Accept	Accept	Accept	Accept	Accept	Accept

Table 8. Results of iterative application of Haar transform.

Iteration	P_{cc}	Target 1	Target 2	Target 3	Target 4	Target 5	Target 6
0	.7713	.9490	.6219	.8219	.6853	.8134	.7363
1	.81361	.9552	.6741	.8555	.6692	.8893	.8383
2	.84762	.9552	.7724	.9128	.7027	.9104	.8321
3	.86421	.9453	.7823	.9452	.7363	.9154	.8607
4	.87976	.9453	.7935	.9465	.7774	.9328	.8831
5	.88618	.9453	.8172	.9552	.7550	.9316	.9129
6	.88453	.9453	.8197	.9601	.7376	.9316	.9129
7	.89095	.9391	.8507	.9664	.7450	.9316	.9129
8	.89261	.9391	.8246	.9664	.7799	.9316	.9142
9	.88867	.9391	.8346	.9651	.7488	.9316	.9129
10	.89717	.9391	.8706	.9639	.7512	.9391	.9192
11	.89365	.9353	.8570	.9639	.7512	.9391	.9154
12	.90049	.9353	.8483	.9689	.7749	.9465	.9291

wavelet is used there is no statistical performance difference between the classifiers when features which classify 4 signals or less are removed from the classifier. Removing these features significantly reduces the time required to create the classifier.

If the original signals are transformed and then 128 of the most informative pseudo range bins selected for further transformation, a new linear transformation of the input data is created [6].

Proposition 3 By iteratively selecting the most informative pseudo range bins and transforming them, the informative value of the range bins increase and therefore yields a better classifier.

An experiment was performed to verify this conjecture.

The original 128 range bin signal was transformed (using the Haar wavelet) as previously discussed creating 1024 pseudo range bins. A box classifier was constructed. The range bins used as features were chosen for further transformation. If there are more than 128 pseudo range bin features, then the features which classify the most training signals are selected. If there are fewer than 128 features, then additional pseudo range bins are selected from the middle of the pseudo signal. These 128 range bins were wavelet transformed, a classifier was constructed and tested. This procedure was repeated a total of twelve times and the results are presented in Table 8 and Figure 4.

When using just one wavelet transform on the original signal Stirman showed an increase in Pcc of 6 percentage points [12] and 7.53 percentage points over the baseline classifier [13]. This difference, over the baseline classifier, may have resulted from changing the type of classifier or the use of wavelets. Stirman did not attribute the increase in performance to one or the other, neither did he analyze the significance of using the wavelet transform. In the results presented here, we find that using the same classifier, the improvement in Pcc after one application of the wavelet transform is

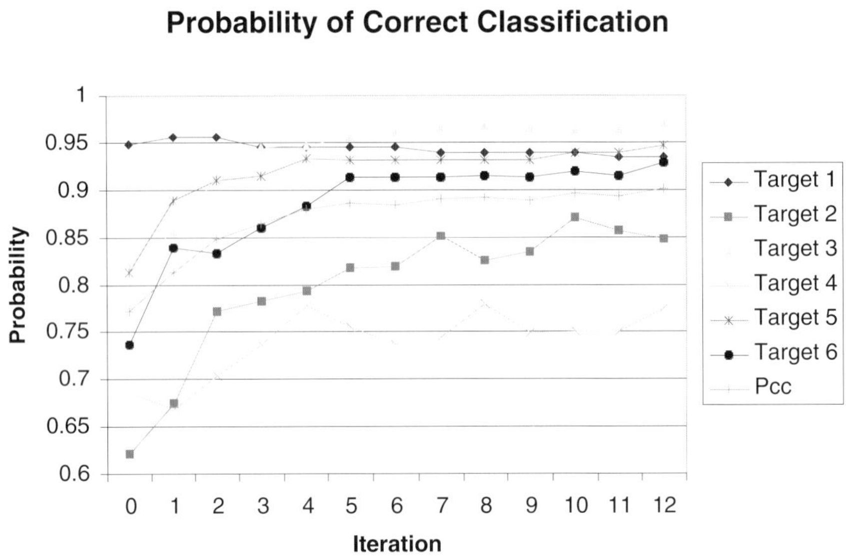

Figure 4. Classification Improvement with Iterated Wavelet Transforms.

4.2 percentage points. This improvement is smaller than the one observed by Stirman, but we can attribute this difference to our use of a different classifier and wavelet (Haar). The most important curve in Figure 4 is the one for Pcc. This curve shows an increase in overall classifier performance from .7713 to .89717 by iteration 10. This represents an improvement of 12 percentage points Furthermore, target 2 improved by 25 percentage points and target 6 by 18 percentage points. This is a significant improvement in performance over a single wavelet transform and confirms the benefit of using the iterated wavelet transform.

We questioned why there would be a decrease in performance on some of the target such as seen on target 2 between iterations 7 and 8. It is apparent that the iterated wavelet transforms yield an increasing performance in the entire classifier. Individual targets may sacrifice performance while overall performance increases. In general, the momentary decreases are recovered in later iterations. This may be a manifestation of the biasing problem reported by Stirman [13]. If so, by iterating the wavelet transform this problem appears to either be mitigated or eliminated. For this problem, the maximum advantage of iterating the wavelets happens at about ten iterations.

We suspected that performance would increase as fewer features are required and the features that are chosen classify more signals. This supposition was confirmed by the experiment and is graphically illustrated in Figure 5. It is easily seen that the targets with the fewest features have the highest performance (Figure 4). The targets with the lowest performance require the most features. This happens when two targets are very similar and the same range bins contain the information required to segment the targets.

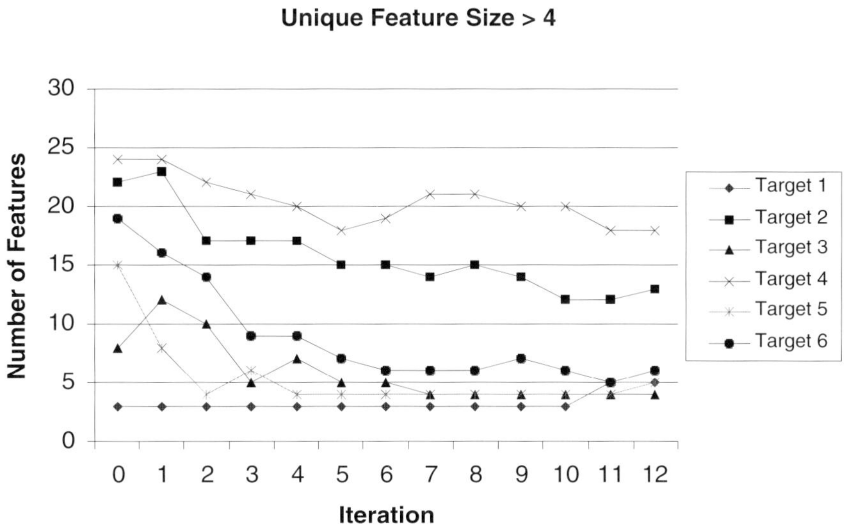

Figure 5. Feature Sizes.

In such case, an individual feature provides little differentiation between targets, and additional features are required. We also note that performance on a given target improves most significantly when the number of features required for classification decreases as in Target 2, Target 5, and Target 6.

8. Summary of Results

In this paper we have shown that for the HRR target recognition problem the use of wavelets to enrich the feature space improves classifier performance. We further show that there is no statistically significant difference in performance of the classifier when different wavelets are chosen. This means that the simplest wavelet to implement will do as good a job as any other wavelet, at least for the HRR target recognition problem. We presented results that show that (using a box classifier) features that classify fewer than five training signals can safely be ignored without producing a statistically significant change in the classifier's performance.

The most significant contribution of this paper is the idea of using iterated wavelet transforms to improve classifier performance. The authors have also used information content methods to down select the pseudo range bins, instead of using the box classifier features. This information based method has shown similar improvement in classification performance.

The application of the iterative wavelet method used here to improve performance could potentially be used in other problem domains.

References

1. M. Ichino, "Optimum Feature Selection for Decision Functions," *Systems and Computers in Japan*, vol. 21, no. 1, 1990.
2. M.W. Koch, M.M. Moya, L.D. Hostetler, and R.D. Fogler, "Cueing, Feature Discovery, and One-Class Learning for Synthetic Aperture Radar Automatic Target Recognition," *Neural Networks*, vol. 8, no. 7/8, 1995, pp. 1081–1102.
3. E.C. Botha, E. Barnard, and C.J. Barnard, "Feature-Based Classification of Aerospace Radar Targets Using Neural Networks," *Neural Networks*, vol. 9, no. 1, 1996, pp. 129–142.
4. A.J. Devaney, R. Raghavan, H. Lev-Ari, E. Manolakos, and M. Kokar, Automatic Target Detection and Recognition: A Wavelet Based Approach, Northeastern University Technical Report AD-A329696, Defense Technical Information Center, 1997.
5. A. Famili, S. Wei-Min, R. Weber, and E. Simoudis, "Data Preprocessing and Intelligent Data Analysis," *Intelligent Data Analysis*, vol. 1, no. 1, http://www.elsevier.com/locate/ida, 1997.
6. J.A. Starzyk, Evolutionary Feature Extraction for SAR Air to Ground Moving Target Recognition – Statistical Approach, Sverdrup Technology, Inc. Advanced Systems Group Report. August 1998.
7. H.H. Szu, "Review of Wavelet Transforms for Pattern Recognition," Wavelet Applications III, Proceedings of SPIE, vol. 2762, 1996, pp. 2–22.
8. R.A. Mitchell, "Hybrid Statistical Recognition Algorithm for Aircraft Identification," Dissertation Proposal, University of Dayton, February 1997.
9. D.E. Nelson and J.A. Starzyk, "Advanced Feature Selection Methodology for Automatic Target Recognition," 29th Southeastern Symposium on System Theory, Cookville, TN, March 1997.
10. J.A. Starzyk and D.E. Nelson, "Independent Classifiers in Ontogenic Neural Networks for ATR," Adaptive Distributive Parallel Computing Symposium, Dayton, OH, August 1996.
11. G. Strang and T. Nguyen, *Wavelets and Filter Banks*, Wellesley-Cambridge Press, 1996.
12. C. Stirman and A. Nachman, "Applications of Wavelets to Radar Data Processing," Technical Report AD-A239297, Defense Technical Information Center, 1991.
13. C. Stirman and A. Nachman, "Applications of Wavelets to Automatic Target Recognition," Technical Report AD-A294854, Defense Technical Information Center, 1995.
14. M. Misiti, Y. Nisiti, G. Oppenheim, and J. Poggi, *Wavelet Toolbox*, The MathWorks, Inc., March 1996.
15. J. Lu, V.R. Algazi, and R.R. Estes Jr., "Comparative Study of Wavelet Image Coders," *Optical Engineering*, vol. 35, no. 9, 1996, pp. 2605–2619.
16. I. Miller and J.E. Fruend, *Probability and Statistics for Engineers*, Prentice-Hall Inc, Englewood Cliffs, NJ, 1965.

Multidimensional Systems and Signal Processing, 14, 25–48, 2003

2D HRR Radar Data Modeling and Processing

JUNSHUI MA
Department of Electrical Engineering, the Ohio State University, 205 Dreese Laboratory, 2015 Neil Ave., Columbus, OH 43210 USA

XUN DU
Department of Electrical Engineering, the Ohio State University, 205 Dreese Laboratory, 2015 Neil Ave., Columbus, OH 43210 USA

STANLEY C. AHALT
Department of Electrical Engineering, the Ohio State University, 205 Dreese Laboratory, 2015 Neil Ave., Columbus, OH 43210 USA

Received November 29, 2000; Revised November 14, 2001; Accepted November 20, 2001

Abstract. High Range Resolution (HRR) -based Automatic Target Recognition (ATR) has attracted increasing attention due to a number of potential advantages over alternative radar techniques in moving target identification. Most current HRR-based ATR studies have been conducted using 1D HRR signatures. However, these 1D HRR signatures are generally plagued by scintillation effects, and thus demonstrate highly irregular behavior that dramatically degrades the performance and robustness of algorithms based on these signatures. In order to circumvent this difficulty, an alternative HRR radar data representation and processing technique is presented in this paper. This technique models and extracts the target characteristics directly, based on the 2D HRR raw data. In this paper, we first derive a general, but complex HRR radar model, and then simplify this model by instantiating a set of real-world radar and target parameters for the model. This simplification process produces two HRR radar models with different degrees of simplicity. After establishing this set of models, the typical HRR data processes, such as feature extraction and clutter suppression, are reduced to one problem, which is model-parameter estimation. Based upon the most simplified HRR model we proposed, we devise two model- parameter estimation algorithms. One is a scatterer extraction algorithm based on available 1D Parameter Estimation (1DPE), while the other is based on 2D discrete Fourier Transform (2DFT). In order to examine the performance of these two algorithms a set of simulations are conducted. The experimental results are presented, and the performance comparison between 1DPE and 2DFT is presented.

Key Words: HRR, ATR, modeling, clutter suppression, feature extraction, parameter estimation

1. Introduction

Techniques for automated moving-target recognition are needed in many military and civilian applications. Although Synthetic Aperture Radar (SAR) has been successfully applied to stationary target identification, its performance can be degraded considerably when applied to moving target scenarios. Fortunately, a relatively new radar technique named High Range Resolution (HRR) radar appears to be an attractive alternative because HRR radar requires less dwell time, and provides target information with high range resolution.

Currently, most of HRR-based ATR research has been conducted using 1D HRR signatures [1], [2], [3], [4], [5], [6], [7], [8], [9] as moving-target features, which are generated from 2D raw HRR data. This line of research is based on the belief that 1D signatures are capable of capturing the key characteristics of the underlying targets. Unfortunately, the resultant algorithms are commonly plagued by the highly variable behavior of 1D HRR signatures, which largely occur due to scintillation effects arising when multiple scatterers fallen in a single range bin and interact either constructively or destructively even over very small variations in aspect angle. One possible solution to this problem is to extract the target characteristics directly from 2D raw HRR data, and avoid using the (sensitive) 1D HRR signature as target features [10]. The method presented in this paper is one step in this direction. The "2D raw HRR data" we refer to is formed by lining-up all of the 1D reflected pulses received by the HRR radar during one dwell period. Thus, one axis of the 2D raw HRR data is the range, and the other axis is the position of the reflected pulse in the sequence of radar returns.

Presently, one widely adopted framework for typical HRR data processing tasks, including feature extraction and clutter suppression, is the model-based paradigm. That is, the HRR data is first assumed to abide by a physics-based HRR data model, and subsequent data processing algorithms are developed based on this model [10], [11], [12], [13], [14], [15]. A review of the relevant literature on this topic can be found in [35]. This methodology enables researchers to convert complex, non-parametric data into a simple parametric form, and to easily incorporate prior knowledge of the underlying targets. Obviously, a physics-based radar model is the prerequisite for this category of research. Devising a reliable 2D HRR model for moving targets has thus become an increasingly important topic.

A set of physics-based models for rigid body and non-rigid body moving targets has been previously presented in [10] and [11] respectively. A modified version of a moving target model was presented in [12], with an assumption that the chirp rate γ is reasonably small, which is, however, not the general case. Moreover, the non-rigid body moving target model presented in [11] is simplified by assuming that the radar and target parameters satisfy certain conditions. However, those assumptions seem to be violated in practical systems because unrealistic values may be obtained when applying a set of realistic radar parameters to the model. Furthermore, classifying targets as rigid body and non-rigid body makes application of the model difficult because there is no clear dividing line between rigid body and non-rigid body targets. Distinctions can become blurred in very common application scenarios, such as in the case where both rigid body and non-rigid body targets appear at the same time, or in the case where a rigid body target includes a rotating part such as a wheel or track.

In this paper we derive and simplify a new scattering-based HRR moving target model. The contributions of this new model are as follows:

- The proposed models are valid for both rigid and non-rigid body targets.
- These models make no assumptions regarding the distribution of the clutter.
- We provide two simplified HRR moving-target models with different degrees of simplification, and furnish some rudimentary checks on the validity of our simplifications.
- Performance bounds on the subsequent feature extraction algorithms are naturally obtained during the simplification procedure.

One point is worthy of clarification. Although our model derivation assumes that the HRR radar operates at X-band, exactly the same methodology employed here can be easily applied to other radar system and a different operational frequency band to obtain different forms of models.

After a physics-based model is made available, subsequent feature extraction and clutter suppression are reduced to a model parameter estimation problem. Therefore, based on our HRR model, we also propose two parameter estimation algorithms to achieve these goals.

In Section 2 we first derive a moving target model, and then simplify this model by instantiating a set of real-world radar and target parameters to the model in Sections 3 and 4. Two feasible model parameter estimation algorithms are derived and proposed in Sections 5 and 6 respectively, and experiments and performance comparison are presented in Section 7. A brief summary of this paper is provided in Section 8.

2. Derivation of a Physics-Based HRR Moving Target Model

Here we briefly restate portions of the derivations contained in [11] and [12] in order to facilitate the reading of the remainder of this paper. Our model derivation is based on four assumptions regarding HRR radar system and targets:

- The radar returns can be approximated by a sum of the scattered signals from a group of discrete scatterers if the radar operates at sufficiently high frequencies [16].
- The scatterers on targets and the scatterers from clutter are difficult to distinguish solely by their scattering properties, such as their respective scattering coefficients [17].
- Targets are moving and the clutter is stationary [10], [11], [12].
- The radar angle is fixed at 90 degree to the moving-platform trajectory. This assumption ensures that the Doppler frequencies of the stationary clutter cluster around 0 Hz [18].

Radars obtain target information by transmitting an electromagnetic (EM) field toward interesting targets and by acquiring the backscattered EM field, which encodes information about those targets. The transmission antenna of the radar is employed to transform the electronic radar signals, or radar pulses, into EM waves. It is the EM waves that propagate through free space and are reflected when hitting materials with different propagation properties. The receiving antenna of the radar is used to convert the received EM waves back into electronic signals. Because our research only focuses on the information stored in the electronic signal, we ignore the processes of converting the electronic signals to and from EM waves, and directly adopt the radar signal representation to describe the whole transmission-scattering-receiving procedure. This liberates our analysis from utilizing complex EM field descriptions, and greatly simplifies our subsequent analysis and processing. Of course, this simplification is achieved at the cost of losing some potentially useful physical properties manifested only in the EM waves, such as the polarization diversity. For those who have interest in these physical properties, please refer to [17], [19], and [20], [21], [22], [23], [24], [25].

The radar signal is a chirp pulse, which is denoted by $\tilde{s}(t)$

$$\tilde{s}(t) = \begin{cases} \exp\{j(\omega_0 t + \gamma t^2)\}, \ |t| \leq T_0/2 \\ 0, \ |t| > T_0/2 \end{cases}, \tag{1}$$

where T_0 is the pulse width, ω_0 is the center carrier frequency, and 2γ is the chirp rate of the linear FM pulse. During a dwell period a sequence of radar pulses is transmitted by an HRR radar. The sequence of L pulses can be expressed by $s_i(t)$:

$$s_i(t) = \sum_{l=0}^{L-1} \tilde{s}(t - lT), \tag{2}$$

where T is the time interval between pulses, and L is the number of pulses sent each time. The scattered signal from kth scatterer is denoted by $s_{bk}(t)$,

$$s_{bk}(t) = \sum_{l=0}^{L-1} a_k(t)\tilde{s}_{bk}(t - lT), \tag{3}$$

where $a_k(t)$ is the kth scatterer's amplitude determined by the frequency.

Because the radar signal is a linear FM waveform, the relationship between time and frequency can be represented as:

$$\omega = \omega_0 + 2\gamma t, \text{ where } |t| < T_0/2. \tag{4}$$

According to the GTD [16], [26], [27], [28], the frequency-dependence of the scatterer's amplitude has the form of:

$$\bar{a}_k(\omega) = A_k \left(\frac{\omega}{\omega_0}\right)^{\alpha_k}. \tag{5}$$

Therefore, $a_k(t)$ can be expressed in the form of:

$$a_k(t) = A_k \left[1 + \frac{2\gamma}{\omega_0}(t - lT)\right]^{\alpha_k}. \tag{6}$$

In order to facilitate the subsequent derivation, the complex scalar A_k is also represented in its amplitude-phase form as $\rho_k exp\{j\psi_k\}$ with ρ_k denoting the amplitude and ψ_k the phase part of the A_k. The parameter α_k is a "dimensionless, discrete type parameter characterizing the geometry" [28] of the kth scatterer. The correspondence between the value of α and the scatterer geometry is provided in [28].

Because the dwell time of HRR radar is usually very short, it is unlikely that the aspect angle of the target changes significantly during one dwell period. Thus, it is reasonable to ignore the effects of aspect angle on the scatterer amplitude $a_k(t)$.

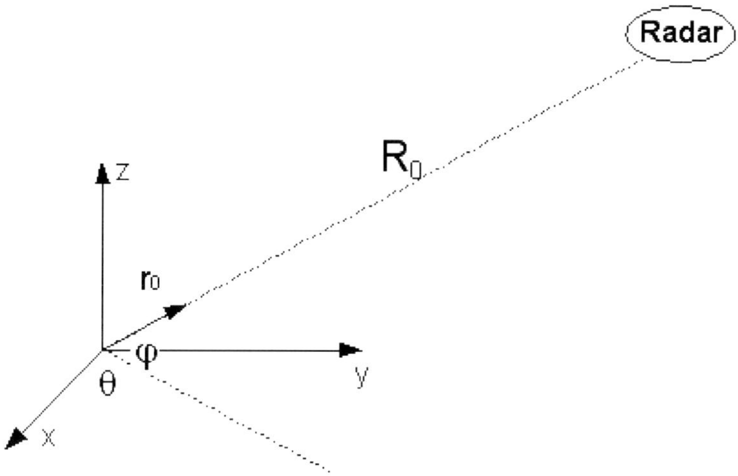

Figure 1. Coordinate System.

One of the scattered pulses from kth scatterer is denoted:

$$\tilde{s}_{bk}(t - lT) = \exp\left\{ j\omega_0\left(t - lT - \frac{2R_{rk}(t)}{c}\right) + j\gamma\left(t - lT - \frac{2R_{rk}(t)}{c}\right)^2 \right\}. \quad (7)$$

We denote the direction of radar line-of-sight (LOS) as $r_o = [\cos\varphi\cos\theta, \cos\varphi\sin\theta, \sin\varphi]^T$, where θ is the aspect angle, and φ is the depression angle of the radar to the targets, which is illustrated by Figure 1.

If we assume the kth scatterer is at ground location of L_k at t = 0, and it is moving at a speed of V_k, its relative location to the radar at time t, or $R_{rk}(t)$, is

$$R_{rk}(t) = R_0 - L_k^T r_0 - V_k^T r_0 t. \quad (8)$$

Substituting (8) into (7), we obtain

$$\begin{aligned} \tilde{s}_{bk}(t - lT) &= \exp\left\{ j\omega_0\left(t - lT - t_0 + t_k + \frac{2V_k^T r_0 t}{c}\right) \right. \\ &\quad \left. + j\gamma\left(t - lT - t_0 + t_k + \frac{2V_k^T r_0 t}{c}\right)^2 \right\} \\ &= \exp\left\{ j\omega_0(\hat{t} + t_k) + j\gamma(\hat{t} + t_k)^2 \right\} \exp\{ j\omega_{Dk} t\} \\ &\quad \times \exp\{ j2\gamma(\hat{t} + t_k) C_{vk} t\} \exp\left\{ j\gamma(C_{vk} t)^2 \right\} \end{aligned} \quad (9)$$

where $t_0 = \frac{2R_0}{c}$, $t_k = \frac{2L_k^T r_0}{c}$, $\omega_{Dk} = \omega_0 \frac{2V_k^T r_0}{c}$, $C_{vk} = \frac{2V_k^T r_0}{c}$, and, $\hat{t} = t - t_0 - lT$.

We mix the scattered pulse $\tilde{s}_{bk}(t - lT)$ with a delayed and conjugated transmitted pulse $\tilde{s}^*(t - t_0 - lT)$ to obtain the de-chirped scattered pulse $\tilde{y}_k(t - t_0 - lT)$ of the kth scatterer:

$$\begin{aligned}\tilde{y}_k(t - t_0 - lT) &= \tilde{s}_{bk}(t - lT)\tilde{s}^*(t - t_0 - lT)\\ &= \exp\{j\omega_0 t_k + j\gamma t_k^2\}\exp\{j2\gamma t_k\hat{t}\}\exp\{j\omega_{Dk}t\}\\ &\quad\times\exp\{j2\gamma(\hat{t} + t_k)C_{vk}t\}\exp\{j\gamma(C_{vk}t)^2\}.\end{aligned} \tag{10}$$

If we replace $t = \hat{t} + t_0 + lT$ into (10), we obtain

$$\begin{aligned}&\tilde{y}_k(t - t_0 - lT) = \exp\{j[\omega_0 t_k + \gamma t_k^2 + \omega_{Dk}t_0 + 2\gamma C_{vk}t_k t_0 + \gamma(C_{vk}t_0)^2]\}\\ &\exp\{j[2\gamma t_k + \omega_{Dk} + 2\gamma C_{vk}(t_0 + t_k) + 2\gamma C_{vk}^2 t_0]\hat{t}\}\\ &\exp\{j[\omega_{Dk} + 2\gamma C_{vk}t_k + 2\gamma C_{vk}^2 t_0]Tl\}\\ &\exp\{j[2\gamma C_{vk} + 2\gamma C_{vk}^2]Tl\hat{t}\}\\ &\exp\{j[2\gamma C_{vk} + \gamma C_{vk}^2]\hat{t}^2\}\\ &\exp\{j\gamma C_{vk}^2 T^2 l^2\}.\end{aligned} \tag{11}$$

We can discretize $\tilde{y}_k(t - t_0 - lT)$ by defining $\hat{t} = T_s n$, and rewrite $\tilde{y}_k(t - t_0 - lT)$ as a 2-D discrete function $\tilde{D}_k(n, l)$:

$$\tilde{D}_k(n, l) = \exp\{j\phi_k\}\exp\{j\bar{\omega}_k n\}\exp\{j\bar{\omega}_{Dk}l\}\exp\{j\bar{\omega}_{rk}nl\}\exp\{j\bar{\omega}_{2k}n^2\}\exp\{j\bar{\omega}_{2Dk}l^2\}$$

$$n = -\frac{N}{2}\ldots 0\ldots\frac{N}{2};\, l = 0\ldots L - 1 \tag{12a}$$

where

$$\text{phase } \phi_k = \omega_0 t_k + \gamma t_k^2 + \omega_{Dk}t_0 + 2\gamma C_{vk}t_k t_0 + \gamma(C_{vk}t_0)^2, \tag{12b}$$

$$\text{position frequency } \bar{\omega}_k = [2\gamma t_k + \omega_{Dk} + 2\gamma C_{vk}(t_0 + t_k) + 2\gamma C_{vk}^2 t_0]T_s, \tag{12c}$$

$$\text{speed frequency } \bar{\omega}_{Dk} = [\omega_{Dk} + 2\gamma C_{vk}t_k + 2\gamma C_{vk}^2 t_0]T, \tag{12d}$$

$$\text{mixed frequency } \bar{\omega}_{rk} = [2\gamma C_{vk} + 2\gamma C_{vk}^2]TT_s, \tag{12e}$$

$$2^{\text{nd}} \text{ order position frequency } \bar{\omega}_{2k} = [2\gamma C_{vk} + \gamma C_{vk}^2]T_s^2, \tag{12f}$$

$$2^{\text{nd}} \text{ order speed frequency } \bar{\omega}_{2Dk} = \gamma C_{vk}^2 T^2, \tag{12g}$$

and n denotes the range bin position, N the total number of range bins included, l the pulse number, and L the total number of pulses sent each time.

Accordingly, $a_k(t)$ can be discretized as:

$$a_k(n) = A_k\left(1 + \frac{2\gamma}{\omega_0} T_s n\right)^{\alpha_k}. \tag{13}$$

When the radar operates at a sufficiently high frequency, the radar returns can be approximated by a sum of scattered signals from a group of discrete scatterers [16]. Also, additive noise, $e(n, l)$, can be introduced to represent the totality of effects of many unpredictable nuisance factors, such as measurement errors and interference in the radar system, variance of the propagation media, etc. Therefore, the moving target model in 2D discrete form is obtained.

$$D(n,l) = \sum_{k=1}^{K} a_k(n)\tilde{D}_k(n,l) + e(n,l), n = -\frac{N}{2}\ldots 0 \ldots \frac{N}{2}; l = 0 \ldots L-1, \tag{14}$$

where K is the number of "visible" scatterers[1] within the illuminated radar swatch.

3. Parameters of the Radar System and the Target

We utilize a set of target and radar system parameters commonly adopted by HRR radar in our following model simplification process [5], [13], [14], [18], [36]. The parameters are as follows:

1) The radar is operating in the HRR mode.
2) The radar signal consists of linear FM chirp pulses at X-band.
3) The range resolution is 1 foot, or 0.3m.
4) There are 128 pulses in each burst transmission of the radar signal, or $L = 128$.
5) The propagation speed of the radar signal is the speed of light, or $c = 3 \times 10^8\ m/s$.
6) The radar platform is about 90km away from the targets, or $R_0 = 90 \times 10^3\ m$.
7) The depression angle of the radar system is 15° or 17°.
8) The radar platform is moving with a horizontal velocity of 180m/s, or $V_{ph} = 180 m/s$.
9) The sampling frequency of the radar A/D converter is 50MHz, or $f_s = 50 \times 10^6\ Hz$, which is assumed according to the specification of available commercial A/D converters [36]. That is, the sampling time interval is 0.02μs, or $T_s = 0.02 \times 10^{-6}\ s$.
10) We assume that minimum detectable target radial speed is 2.5 m/s, or $V_{r\min} = 2.5 m/s$, according to our common sense.
11) Given that the maximum detectable target ground speed is 198 km/hour, or about 124 mile/hour, the maximum target radial speed is about $V_{r\max} = 53 m/s$. This value also determines the selection of the interval time T of two pulses.
12) The total length of the range swatch is 500m, or $L_{\max} = 500 m$.

Table 1. Simplify $\overline{\omega}_k$.

Term	$2\gamma t_k T_s$	$\omega_{Dk} T_s$	$2\gamma C_{vk}(t_0 + t_k)T_s$	$2\gamma C_{vk}^2 t_0 T_s$
Max. Value (rad)	2.7143	4.1737×10^{-4}	3.4622×10^{-4}	1.2199×10^{-9}
Min. Value (rad)	3.2572×10^{-3}	1.8112×10^{-5}	1.4983×10^{-5}	2.2974×10^{-12}
Keep this term?	Yes	No	No	No

From the above basic parameter set, we can further obtain the following parameters.

1) Each chirp pulse has a bandwidth of around 600 MHz, or $B = 600 \times 10^6\ Hz$, which is obtained from $B = \frac{c}{2r_c}$, where the theoretical range resolution r_c is set to 0.25 m to guarantee that the resolution specification of 0.3m can be achieved.
2) The minimum Doppler frequency: ω_{Dmin} is 905.61 rad, from $\omega_{D\min} = \frac{4\pi V_{r\min} f_0}{c}$, while the maximum Doppler frequency: ω_{Dmax} is 20869 rad, from $\omega_{D\max} = \frac{4\pi V_{r\max} f_0}{c}$.
3) The time interval of radar pulses is 150μs, from $T = \frac{\pi}{\omega_{D\max}}$.
4) We assume that the width of each pulse is one third of the time interval of radar pulses. That is $T_0 = T/3 = 50\mu s = 50 \times 10^{-6}\ s$.
5) The chirp rate γ of the chirp signal is $4 \times 10^{13} rad/s$, from $\gamma = \frac{2\pi B}{2T_0}$.
6) The minimum time difference between two resolvable scatterers is 2ns, from $t_{k\min} = \frac{2r_c}{c}$, while the maximum time difference is 1.67μs, from $t_{k\max} = \frac{L_{\max}}{c}$.

4. Model Simplification

Note that model (14) consists of five frequency components, and each frequency component is further composed of several terms. Our basic idea is to instantiate the values of previously defined parameters to the model, and to observe that values of those frequency components are generally dominated by only a small number of terms. Therefore, the original model can be simplified by retaining only those terms with dominating values.

From (12c), the position frequency $\overline{\omega}_k$ is composed of the terms listed in Table 1.

Similarly, following (12d), the terms comprising the speed frequency $\overline{\omega}_{Dk}$ are listed in Table 2.

According to (12e), terms of the mixed frequency $\overline{\omega}_{rk}$ are listed in Table 3.

According to (12f), terms of the 2nd order position frequency $\overline{\omega}_{2k}$ are listed in Table 4.

Finally, according to (12g), terms of the 2nd order speed frequency $\overline{\omega}_{2Dk}$ are listed in Table 5.

Table 2. Simplify $\overline{\omega}_{Dk}$.

Term	$\omega_{Dk} T$	$2\gamma C_{vk} t_k T$	$2\gamma C_{vk}^2 t_0 T$
Max. Value (rad)	3.1303	7.1930×10^{-3}	9.1495×10^{-7}
Min. Value (rad)	0.1358	3.7458×10^{-7}	1.7231×10^{-9}
Keep this term?	Yes	No	No

Table 3. Simplify $\bar{\omega}_{rk}$.

Term	$2\gamma C_{vk}TT_s$	$2\gamma C_{vk}^2 TT_s$
Max. Value (rad)	8.6316×10^{-5}	3.0498×10^{-11}
Min. Value (rad)	3.7458×10^{-6}	5.7435×10^{-14}
Keep this term?	Depends...	No

Therefore, Eq. (12a) can be simplified by deleting all the negligible terms in the above five tables. A simplified $\tilde{D}_k(m, l)$can be thus represented as follows:

$$\tilde{D}_k(n,l) = \exp\{j\phi_k\}\exp\{j\bar{\omega}_k n\}\exp\{j\bar{\omega}_{Dk} l\}\exp\{j\bar{\omega}_{rk} nl\} \tag{15}$$

where $\bar{\omega}_k = 2\gamma t_k T_s$, $\bar{\omega}_{Dk} = \omega_{Dk} T$, and $\bar{\omega}_{rk} = 2\gamma C_{vk} TT_s$.

If we retain only those terms with dominating values, we can further simplify (15) as:

$$\tilde{D}_k(n,l) = \exp\{j\phi_k\}\exp\{j\bar{\omega}_k n\}\exp\{j\bar{\omega}_{Dk} l\}, \tag{16}$$

where $\bar{\omega}_k = 2\gamma t_k T_s$, and $\bar{\omega}_{Dk} = \omega_{Dk} T$.

Substituting (13) and (16) into (14), a simplified version of model (14) can be obtained as follows:

$$\begin{aligned} D(n,l) &= \sum_{k=1}^{K} a_k(n - N/2)\exp\{j\phi_k\}\exp\{j(\bar{\omega}_k n - N/2)\}\exp\{j\bar{\omega}_{Dk} l\} + e(n,l) \\ &= \sum_{k=1}^{K} \rho_k \exp\{j\bar{\phi}_k\}\left(1 - \frac{\gamma T_s N}{\omega_0} + \frac{2\gamma}{\omega_0} T_s n\right)^{\alpha_k} \exp\{j\bar{\omega}_k n\}\exp\{j\bar{\omega}_{Dk} l\} + e(n,l), \end{aligned} \tag{17}$$

where $n = 0 \ldots N$; $l = 0 \ldots L - 1$, and $\bar{\phi}_k = \psi_k + \phi_k - \bar{\omega}_k N/2$.

Furthermore, we observe that $\frac{2\gamma}{\omega_0} T_s$ is as small as 5.518×10^{-5} by instantiating the values of the radar and target parameters to it. Therefore, if the total number of range bins, N, is not too large, (13) can be simplified as:

$$a_k(n) = A_k. \tag{18}$$

That is, the effect of the type parameter, α_k, is negligible.

Table 4. Simplify $\bar{\omega}_{2k}$.

Term	$2\gamma C_{vk}T_s^2$	$\gamma C_{vk}^2 T_s^2$
Max. Value (rad)	1.1509×10^{-8}	2.0332×10^{-15}
Min. Value (rad)	4.9944×10^{-10}	3.8290×10^{-18}
Keep this term?	No	No

Table 5. Simplify $\overline{\omega}_{2Dk}$.

Term	$\gamma C_{vk}^2 T^2$
Max. Value (rad)	1.1437×10^{-7}
Min. Value (rad)	2.1538×10^{-10}
Keep this term?	No

Substituting (18) in (17), we obtain a very simplified model:

$$D(n, l) = \sum_{k=1}^{K} \rho_k \exp\{j\overline{\phi}_k\}\exp\{j\overline{\omega}_k n\}\exp\{j\overline{\omega}_{Dk} l\} + e(n, l), \tag{19}$$

where $n = 0 \ldots N$; $l = 0 \ldots L - 1$, and $\overline{\phi}_k = \psi_k + \phi_k - \overline{\omega}_k N/2$.

From (12b) and (17), we know that the initial phase term $\overline{\phi}_k$ is affected by many subtle factors, such as the surface material, structure, range position, radial speed, and propagation channel of the kth scatterer. Small change in any of these factors will probably lead to large deviations in $\overline{\phi}_k$. Therefore, it is reasonable for us to consider $\overline{\phi}_k$ as a random variable. Here, we assume that $\overline{\phi}_k$ is a uniform-distributed random variable over $[0, 2\pi]$.

In addition, Table 1 shows that that the minimum value of $\overline{\omega}_k$ is 0.003257 rad., which is determined by the resolution of the HRR radar system at X-band. This value can be considered as a criterion for the performance of subsequent feature extraction algorithms applied to this model. That is, the estimation errors of subsequent feature extraction algorithms based on this model should be less than this minimum value if we wish to extract all available target information encoded in the radar data.

5. Matrix Form of the HRR Data Model and Its 2D Auto-Correlation Matrix

According to (19), if we define

$$\Omega_k = [1 \quad \exp\{j\overline{\omega}_k\} \cdots \exp\{j\overline{\omega}_k N\}]^T$$

$$E = [\Omega_1 \; \Omega_2 \cdots \Omega_K],$$

$$\Omega_{Dk} = [1 \quad \exp\{j\overline{\omega}_{Dk}\} \cdots \exp\{j\overline{\omega}_{Dk}(L-1)\}]^T,$$

$$E_D = [\Omega_{D1} \quad \Omega_{D2} \cdots \Omega_{DK}],$$

$$\Lambda = \begin{bmatrix} \rho_1 \exp\{j\overline{\phi}_1\} & & & \\ & \rho_2 \exp\{j\overline{\phi}_2\} & & \\ & & \ddots & \\ & & & \rho_K \exp\{j\overline{\phi}_K\} \end{bmatrix}, \text{ and}$$

$$W = \begin{bmatrix} e(0,0) & \cdots & e(0,L-1) \\ \vdots & \ddots & \vdots \\ e(N,0) & \cdots & e(N,L-1) \end{bmatrix}.$$

The model (19) can be simply expressed in matrix form as:

$$D = E\Lambda E_D^T + W. \tag{20}$$

Let us define the 2D auto-correlation matrix R as

$$R = \begin{bmatrix} R(0,0) & \cdots & R(0,L-1) \\ \vdots & \ddots & \vdots \\ R(N-1,0) & \cdots & R(N-1,L-1) \end{bmatrix}, \tag{21a}$$

where

$$R(m_1, m_2) = E\{D(n+m_1, l+m_2)D^*(n,l)\},$$

$$m_1 = 0, 1, \cdots, (N-1); m_2 = 0, 1, \cdots, (L-1), \text{and} \tag{21b}$$

$E\{\bullet\}$ denotes the expectation of a random variable.

By substituting (19) into (21b), we obtain

$$\begin{aligned} R(m_1, m_2) &= E\left\{\sum_{b=1}^{K}\sum_{c=1}^{K} a_b a_c \exp\{j[\overline{\omega}_b(n+m_1) - \overline{\omega}_c n + \overline{\omega}_{Db}(l+m_2) - \overline{\omega}_{Dc} l \right. \\ &\qquad \left. + (\phi_b - \phi_c)]\}\right\} + E\{e(n+m_1, l+m_2)e^*(n,l)\} \\ &= \sum_{k=1}^{K} a_k^2 \exp\{j[\overline{\omega}_k m_1 + \overline{\omega}_{Dk} m_2]\} + \sigma^2 \delta_{m_1,m_2}, \end{aligned}$$

$$m_1 = 0, 1, \cdots, (N-1); m_2 = 0, 1, \cdots, (L-1), \tag{22}$$

where σ^2 is the variance of additive white noise $e(n, l)$.Therefore, the 2D auto-correlation matrix R is:

$$R = E\Sigma E_D^T + \Theta \tag{23}$$

where

$$\Sigma = \begin{bmatrix} a_1^2 & & & \\ & a_2^2 & & \\ & & \ddots & \\ & & & a_K^2 \end{bmatrix}, \text{ and } \Theta = \begin{bmatrix} \sigma^2 & 0 & \cdots & 0 \\ 0 & 0 & \cdots & 0 \\ \vdots & \vdots & \ddots & 0 \\ 0 & 0 & \cdots & 0 \end{bmatrix}.$$

In practice, the estimation of the 2D auto-correlation matrix, $\hat{R}$, is defined as:

$$\hat{R} = \begin{bmatrix} \hat{R}(0,0) & \cdots & \hat{R}(0,L-1) \\ \vdots & \ddots & \vdots \\ \hat{R}(N-1,0) & \cdots & \hat{R}(N-1,L-1) \end{bmatrix}, \tag{24a}$$

where

$$\hat{R}(m_1, m_2) = \frac{1}{NL} \sum_{n=0}^{N-m_1-1} \sum_{l=0}^{L-m_2-1} D(n+m_1, l+m_2) D^*(n,l). \tag{24b}$$

6. Parameter Estimation

From the model (19) we know there are three parameters, $\{\rho_k, \overline{\omega}_k, \overline{\omega}_{Dk}\}$, that need to be estimated for each scatterer. The whole set of estimable parameters of the model is thus denoted as:

$$P = \{\rho_k, \overline{\omega}_k, \overline{\omega}_{Dk}\}_{k=1}^{K}. \tag{25}$$

We refer to frequency $\overline{\omega}_{Dk}$ as the *speed frequency* of the kth scatterer. This is because $\overline{\omega}_{Dk}$ is proportional to the Doppler frequency of the kth scatterer and is therefore proportional to the radial speed of the kth scatterer. Here we assume that the targets are moving while the clutter is stationary [29]. Also, the radar angle is supposed to be perpendicular to the platform-flying trajectory. These assumptions ensure that the Doppler shift frequency of the clutter will be close to zero. As a result, we can discriminate whether a scatterer is a target-scatterer or a clutter-scatterer via the value of $\overline{\omega}_{Dk}$. That is, scatterers with sufficient small $\overline{\omega}_{Dk}$ are considered as the clutter-scatterers, while the others can be considered as target-scatterers. Meanwhile, the parameters defined in Eq. (25) capture the major information of the underlying targets, and can be a set of promising representative features. Consequently, the feature extraction and the clutter suppression process of HRR signals are reduced to a single fundamental problem — devising reliable parameter estimation algorithms based on the model.

In the subsequent parts of this section, we derive and present two feasible parameter estimation algorithms for this model. The first algorithm reduces the 2D-estimation problem to two 1D-estimation problems, and solves the problems by employing some mature 1D-estimation techniques. The second algorithm utilizes the 2D Discrete Fourier Transform (DFT) to estimate the model parameters.

In fact, the parameter estimation problem of this model is a well-studied topic. More than four different algorithms have been proposed to deal with this problem. A brief review of these algorithms was provided in Section 2 of Chapter 2 of [35].

6.1. Scatterer Extraction Using 1D Parameter Estimation (1DPE)

Note that the model represented in Eq. (19) is a 2D function in a very special form, which can be easily rewritten in a 1D functional form. This property naturally suggests the use of a straightforward parameter estimated algorithm, 1DPE. The basic idea of this 1DPE algorithm is (1) to reduce the original scatterer-detection and parameter-estimation problem to two 1D sub-problems, and then (2) to combine the two 1D estimation results to obtain the parameters of the scatterers encoded in the 2D data.

6.1.1. 1D Scatterer Detection and Parameter Estimation

The model can be easily re-written as a 1D function if we fix one dimension, for example, the dimension of variable l.

$$D_l(n) = D(n, l) = \sum_{k=1}^{K} \beta_k(l)\exp\{j\bar{\omega}_k n\} + e_l(n), \tag{26}$$

where $\beta_k(l) = \rho_k \exp\{j\bar{\omega}_{Dk} l + j\phi_k\}$, and $e_l(n) = e(l, n)$.

Although it is possible that more than one scatterer will have the same *position frequency* $\bar{\omega}_k$, we will delay consideration of this case until later. To simplify our current derivation, we first assume that all of the *position frequency* components, $\{\bar{\omega}_k\}_{k=1}^{K}$, are unique.

In fact Eq. (26) is a very typical and intensively studied 1D signal model if all the frequency components $\{\bar{\omega}_k\}_{k=1}^{K}$ are unique [30]. Many mature parameter estimation algorithms, such as Nonlinear Least Square (NLS), the High-order Yule-walker algorithm (HOYW) [31], MUSIC [32], and ESPRIT [33], which address this model, are readily available. Here we adopt the ESPRIT algorithm in our 1DPE for its robustness and accuracy with moderate SNR.

However, before we employ the ESPRIT algorithm, we must first estimate the model order K in (26), because the ESPRIT algorithm requires the model order K as an input parameter. Fortunately, there are also quite a few model order estimation algorithms applicable to this model (26). We select the Minimum Description Length (MDL) algorithm to implement our initial estimation of model order, K. Unfortunately, experimental results show that the output of MDL, $\hat{K}_1$, is not always reliable, and is bigger than the actual model order K in quite a few cases. In order to make our current derivation simpler we choose to initially tolerate this estimation error and solve it in a latter stage of the algorithm.

Thus we can directly apply the result of the MDL algorithm, $\hat{K}_1$, to the ESPRIT algorithm. The output of the ESPRIT algorithm is a set of estimated parameters of (8), which are the $\hat{K}_1$ *position frequency* components.

$$S = \{\hat{\bar{\omega}}_{l_k}, \hat{\beta}_{l_k}(l)\}_{k=1}^{\hat{K}_1}, \tag{27}$$

where $\{l_1 \quad l_2 \quad \cdots \quad l_{\hat{K}_1}\}$ is a permutation of the number set of $\{1 \quad 2 \quad \cdots \quad \hat{K}_1\}$.

Because $\hat{K}_1$ is not necessarily equal to K, some spurious frequency components will be introduced in the estimation results. We remove these spurious components according to their magnitudes. The magnitude of each frequency component is defined as $\hat{a}_k = |\hat{\beta}_k(l)|$. It is reasonable to consider the components with very small magnitudes as spurious frequency components. Therefore, we can use an adaptive threshold, *Th*, to remove the spurious frequency components. That is, we consider all the frequency components whose magnitudes are smaller than the threshold, *Th*, as spurious components, and remove them from the estimation set *S* to get a new estimation set, S',

$$S' = \{\hat{\bar{\omega}}_{l_k}, \hat{\beta}_{l_k}(l)\}_{k=1}^{\hat{K}'}, \tag{28}$$

where $\{l_1 \quad l_2 \quad \cdots \quad l_{\hat{K}_1}\}$ is a permutation of the number set of $\{1 \quad 2 \quad \cdots \quad \hat{K}'\}$. The threshold currently employed in our 1DPE algorithm is simply:

$$Th = \max\{0.1, \sigma/\sqrt{2}\}.$$

In fact, this 1D parameter estimation can be conducted on $D_l(n)$ with different fixed l. In order to get more accurate estimation, we recommend implementing this 1D algorithm along more than one l and then selecting the best estimation results as the final 1D estimation results. The best result can be determined by the error between input data and the data generated from the estimation results. The estimation result that produces minimal error is naturally regarded as the best one. Of course, this improvement of the estimation accuracy is at the cost of computation efficiency.

Similarly, the model (19) can also be re-written as a function of variable l if we fix the variable n in (19).

$$D_n(l) = D(n, l) = \sum_{k=1}^{K} \beta_{Dk}(n)\exp\{j\bar{\omega}_{Dk}l\} + e_n(l), \tag{29}$$

where $\beta_{Dk}(n) = a_k \exp\{j\bar{\omega}_k n + j\phi_k\}$ and $e_l(n) = e(l, n)$.

By following exactly the same procedure mentioned above, we can also obtain a set of estimated parameters of Eq. (29), which are $\hat{K}'_D$ *speed frequency* components.

$$S'_D = \{\hat{\bar{\omega}}_{Dl_k}, \hat{\beta}_{Dl_k}(n)\}_{k=1}^{\hat{K}'_D}, \tag{30}$$

where $\{l_1 \quad l_2 \quad \cdots \quad l_{\hat{K}'_D}\}$ is a permutation of the number set of $\{1 \quad 2 \quad \cdots \quad \hat{K}'_D\}$.

6.1.2. *Combination of the Two 1D Estimation Results*

The remaining problem is how to obtain the estimation of the parameter set P from the two 1D estimation results, S' and S'_D. The most straightforward method is to match the *position frequency* $\hat{\bar{\omega}}_{lp}$ from S' with the *speed frequency* $\hat{\bar{\omega}}_{Dlq}$, from S'_D by employing the fact of $a_k = |\beta_k| = |\beta_{Dk}|$, where $k = lp = lq$. Unfortunately, this is problematic in practice because:

- The ESPRIT algorithm can only estimate all the unique frequency components contained in the 1D data. If a group of scatterers have the same *position frequency*, $\bar{\omega}_k$, or *speed frequency*, $\bar{\omega}_{Dk}$, the ESPRIT algorithm can only output one frequency component for this group of scatterers in S$'$ or S'_D. This means that our combining algorithm must have the capacity of matching one *position frequency* component $\bar{\omega}_{lp}$ in S$'$ to multiple *speed frequency* components in S'_0, or vice versa, which is quite difficult.
- Although $|\beta_k|$ is theoretically equal to $|\beta_{Dk}|$, its estimate, $|\hat{\beta}_{lp}|$, does not necessarily match $|\hat{\beta}_{Dlq}|$ due to estimation error.

Therefore, alternative combination methods are required to combine the two 1D estimation results.

In order to facilitate the development of a new combination method, we first define two matrix operations.

- ***Column-exchanging operation X_l***: X_l is the matrix such that, if $AX_l = B$, B is the matrix obtained by exchanging the first and the lth column of A.
- ***Column-duplicating operation C_l***: C_l is the matrix such that, if $AC_l = B$, B is the matrix obtained by duplicating the lth column of A, and inserting the new column after the lth column in A.

From the estimated 1D frequency components in S' and S'_D, we can form the following two exponential matrices $\hat{E}$ and $\hat{E}'_D$:

$$\hat{E}' = \left[\hat{\Omega}_{l_1} \quad \hat{\Omega}_{l_2} \quad \cdots \quad \hat{\Omega}_{l_{\hat{K}'}}\right], \tag{31}$$

where $\hat{\Omega}_k = \left[1 \quad \exp\{j\hat{\bar{\omega}}_k\} \quad \cdots \quad \exp\{j\hat{\bar{\omega}}_k(N-1)\}\right]^T$, and

$$\hat{E}'_D = \left[\hat{\Omega}_{Dl_1} \quad \hat{\Omega}_{Dl_2} \quad \cdots \quad \hat{\Omega}_{Dl_{\hat{K}'_D}}\right], \tag{32}$$

where $\hat{\Omega}_{Dk} = \left[1 \quad \exp\{j\hat{\bar{\omega}}_{Dk}\} \quad \cdots \quad \exp\{j\hat{\bar{\omega}}_{Dk}(L-1)\}\right]^T$.

Also, we define the estimates of E and E_D as $\hat{E}$ and $\hat{E}_D$

$$\hat{E} = \left[\hat{\Omega}_1 \quad \hat{\Omega}_2 \quad \cdots \quad \hat{\Omega}_K\right], \text{and} \tag{33}$$

$$\hat{E}_D = \left[\hat{\Omega}_{D1} \quad \hat{\Omega}_{D2} \quad \cdots \quad \hat{\Omega}_{DK}\right]. \tag{34}$$

From (31) through (34), as well as the definition of X_l and C_l, we obtain:

$$\hat{E} = \hat{E}'P, \tag{35}$$

where $P = C_{n1}C_{n2} \cdots C_{nb}X_{l1}X_{l2} \cdots X_{la}$, and

$$\hat{E}_D = \hat{E}'_D P_D, \tag{36}$$

where $P_D = C_{nD1}C_{nD2} \cdots C_{nDb}X_{lD1}X_{lD2} \cdots X_{lDa}$.

By substituting (35) and (36) into (23), we obtain:

$$\hat{R} = \hat{E}' A \hat{E}'^T_D + \Theta, \tag{37a}$$

where

$$A = P\Sigma P_D^T. \tag{37b}$$

From (37b) we know that A can be obtained by duplicating some columns and rows of Σ, and then exchanging some columns and rows of Σ. Therefore, the presumed form of A is that some non-zero elements are embedded in many zero elements, and the non-zero elements have the same values as the non-zero elements in Σ. In practice, because of the estimation error and numerical error, the estimate of A, $\hat{A}$, is a matrix composed of some large-value elements, or "nontrivial elements", embedded in many small-value elements, or "trivial elements".

From (23) we know that each non-zero element in Σ corresponds to a frequency pair combined by a *position frequency* component from S' and a *speed frequency* component from S'_D. Therefore, from the relationship between A and Σ, i.e. (37b), a new method to match *position frequency* and *speed frequency* can be obtained.

- We can simply use a heuristic method to select those "nontrivial" elements in $\hat{A}$. To put it clearer, we first select a maximal element from each row and each column of $\hat{A}$. Then we choose the smallest value among those selected elements as the threshold to separate the "nontrivial" elements from those "trivial" ones. If $\hat{K}$ "nontrivial" elements are finally selected, $\hat{K}$ will be considered as the final estimation of the number of scatterers, or K, encoded in the 2D radar data.
- Each selected element corresponds to a matched frequency pair in the way that its row index indicates the index of the *position frequency* in S', and its column index shows the position of the *speed frequency* in S'_D. Specifically, if the element $A(p, q)$ is a chosen element, the *position frequency* component $\{\bar{\omega}_{lp}, \hat{\beta}_{lp}\}$ from S', and the *speed frequency* component $\{\bar{\omega}_{Dlq}, \hat{\beta}_{Dlq}\}$ from S'_D will form a matched frequency pair, $\{\hat{\beta}_{lp}, \bar{\omega}_{lp};\ \hat{\beta}_{Dlq}, \bar{\omega}_{Dlp}\}$.

Note that we have two complex scattering coefficients, β_{lp} and β_{lp}, in the above matched pair. This requires that we determine which complex scattering coefficient should be retained as the final estimation of the scattering coefficient of the corresponding scatterer. The strategy we employ to accomplish this correspondence is as follows:

- If the *position frequency*, $\hat{\bar{\omega}}_{lp}$, is unique in the matrix (33), we can simply adopt the magnitude of $\hat{\beta}_{lp}$ as the estimated scattering coefficient of the corresponding scatterer, i.e. $\hat{\alpha}_k = |\hat{\beta}_{lp}|$;
- If the *position frequency*, $\hat{\bar{\omega}}_{lp}$, is not unique in (33), we will adopt the magnitude of $\hat{\beta}_{Dlp}$ as the estimated scattering coefficient of the corresponding scatterer, i.e. $\hat{\alpha}_k = |\hat{\beta}_{Dlp}|$.

The rationale behind this treatment is that it can reduce the estimation error caused by *scintillation effects* which occur when multiple scatterers exist at the same range position [34]. As a matter of fact, this is a fairly critical point. We know that we almost have no way to resolve the scatterers at the same range position if we only utilize the 1D HRR range signatures [34]. However, we can distinguish between these scatterers as long as all these scatterers have different radial speeds, or *Doppler frequencies*. This improvement in the scatterer resolution is simply due to our introducing a new radar sensor dimension, the *Doppler frequency*, compared with the 1D HRR range signatures. Hopefully, this improvement will inspire us to find more new radar sensor dimensions to continuously refine the radar resolution.

The remaining problem is how we can obtain the matrix A. According to (37) we can calculate A as:

$$A = (\hat{E}'^H \hat{E}')^{-1} \hat{E}'^H (\hat{R} - \Theta) \hat{E}_D'^C (\hat{E}_D'^T \hat{E}_D'^C)^{-1}, \tag{38}$$

where $(\bullet)^H$ stands for the Hermitian transposition operation, and $(\bullet)^C$ stands for the complex conjugate operation.

The overall implementation of 1DPE is summarized in Figure 2.

6.2. *Scatterer Extraction Using 2D Discrete Fourier Transform (2DFT)*

The second parameter-estimation algorithm is presented in this section. If we directly apply 2D DFT to Eq. (22), we obtain:

$$\begin{aligned}\dddot{R}(\varpi_1, \varpi_2) &= \sum_{m_2=0}^{L_1-1} \sum_{m_1=0}^{N_1-1} R(m_1, m_2) \exp\{-j\varpi_1 m_1\} \exp\{-j\varpi_2 m_2\} \\ &= \sum_{k=1}^{K} \eta_k(\varpi_1, \varpi_2) \frac{\sin[N_1(\varpi_1 - \bar{\omega}_k)/2]}{\sin[(\varpi_1 - \bar{\omega}_k)/2]} \frac{\sin[L_1(\varpi_2 - \bar{\omega}_{Dk})/2]}{\sin[(\varpi_2 - \bar{\omega}_{Dk})/2]} + \sigma^2, \end{aligned} \tag{39}$$

where

$$\eta_k(\varpi_1, \varpi_2) = a_k^2 \exp\{j(\varpi_1 - \bar{\omega}_k)(N_1 - 1)\} \exp\{j(\varpi_2 - \bar{\omega}_{Dk})(L_1 - 1)\},$$

$$N_1 \geq N, \text{ and } L_1 \geq L.$$

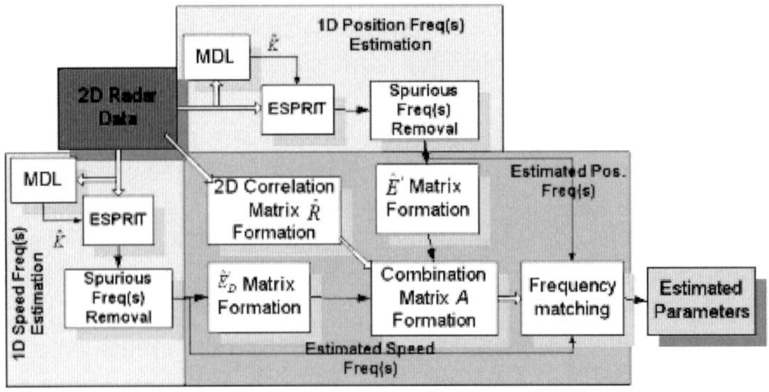

Figure 2. Implementation of 1DPE algorithm.

Eq. (39) suggests a very straightforward way to estimate the scatterer parameters:

- Apply 2D FFT to the correlation matrix, R, of the 2D radar data, D;
- Pick the K biggest peaks from the 2D FFT results.
- If we denote the locations of the K peaks as $\{\varpi_{1k}, \varpi_{2k}\}_{k=1}^{K}$, and denote the corresponding complex values at those peaks as $\{v_k\}_{k=1}^{K}$, the parameter set of the K scatterers, P, can be obtained as:

$$\bar{\omega}_k = \varpi_{1k}, \tag{40a}$$

$$\bar{\omega}_{Dk} = \varpi_{2k}, \text{ and} \tag{40b}$$

$$a_k = \sqrt{|v_k|}. \tag{40c}$$

Because we have not determined a reliable method to estimate the number of scatterers, K, in this algorithm, we have to specify K as a known parameter. Another primary problem with this algorithm is that it has limited frequency resolution due to the low-resolution property of the 2D DFT. If we employ Eq. (24) to compute the correlation matrix R of the 2D radar data, this is equivalent to adding a 2D triangular window on the 2D radar data. Therefore, the main lobe width for each scatterer will be $2\pi/N$ by $2\pi/L$, where N and L are the size of the 2D radar data. That is, any two scatterers whose relative distances are within an area of $[2\pi/N, 2\pi/L]$ are difficult to extract (differentiate) using this algorithm.

7. Simulation Experiment and Performance Comparison

In order to evaluate the performance of the above two parameter-estimation algorithms, we tested the algorithms on a set of synthetic 2D HRR radar data. We assume that there are

Table 6. Configuration of 6 experimental scatterers.

Scatterer Index (k)	**1**	**2**	**3**	**4**	**5**	**6**
Scat. Coef. (a_k)	1.45	0.83	1.11	0.99	1.39	1.0
Position Freq. ($\bar{\omega}_k$)	−1.62	−1.0	−0.5	0.06	0.6	−1.0
Speed Freq. ($\bar{\omega}_{Dk}$)	−0.8	0.36	0	0	−1.3	0.5

6 scatterers encoded in the radar returns. The parameters of these scatterers are listed in Table 6.

Note that the speed frequencies of both scatterer 3 and 4 are 0. That is, these two scatterers are supposed to be the clutter-scatterers. Also, note that the position frequencies of both scatterer 2 and 6 are the same, which implies that these two scatterers are at the same range position, and this will cause scintillation effect in 1D HRR signatures. Moreover, the distances between scatterer 2 and 6 are [0, 0.5−0.36], which are smaller than [2π/N, 2π/L]. Therefore, 2DFT algorithm is expected to have difficulty in properly extracting these two scatterers.

We employ Eq. (20) to generate a series of 2D radar data with N = L = 32, and with different SNRs. The SNR is controlled by changing the standard deviation (STD) of the CGAW noise from 0.15 to 0.51 with an incremental step of 0.02. The relationship between SNR and STD in our experiments is shown as follows:

$$SNR = 10\log_{10}(1.3226/STD^2)\ \text{dB},$$

where 1.3226 is the average power of the 6 scatterers employed in our experiments. Therefore, the range of the SNR is approximately from 7.1 dB to 17.7 dB.

To achieve consistent evaluation results, we generate 200 realizations for each STD value, and run each algorithm on each of these 200 data sets, and average all the 200 experimental results to serve as the final result of the algorithm at that STD, or SNR.

In Experiment 1, we illustrate 1DPE's capacity of estimating K. We ran the 1DPE algorithm a total of 3600 (200 × 19) times, and the distribution of the wrong estimations of K is illustrated in Figure 3. From the results we know that the probability of estimation error is only about 0.47% (18/3600) when the SNR is greater than 10 dB. Therefore, the 1DPE algorithm has relatively reliable performance in estimating K when the SNR is not inordinately challenging.

As we mentioned previously, 1DPE can estimate the scatterer number, K, while 2DFT cannot. Therefore, to keep the experimental results comparable, we only compare 1DPE's results in the case where K is correctly estimated with 2DFT's results in the following two experiments.

We define the estimation of the *position* and *speed frequency* of a scatterer as *frequency estimation*. If the *position frequency estimation* error of a scatterer is bigger than 0.05, or the *speed frequency estimation* error of a scatter is bigger than 0.065, this scatterer is considered as a *mis-estimated* scatterer. In order to illustrate and compare these two

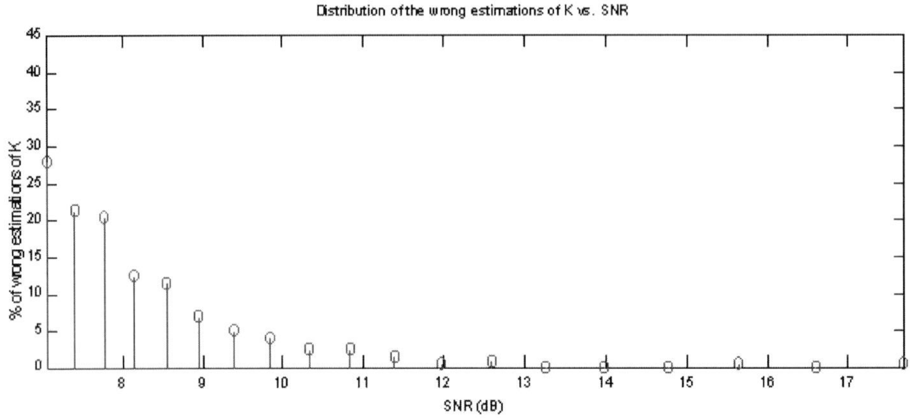

Figure 3. Distribution of the wrong estimations of K.

algorithms scatterer estimation capacities, we define the probability of scatterer estimation error, P_r, as:

$$P_r = M/T, \quad (41)$$

where M is the number of *mis-estimated scatterers*, and T is the total number of estimated scatterers.

The results of Experiment 2 are illustrated in Figure 4.

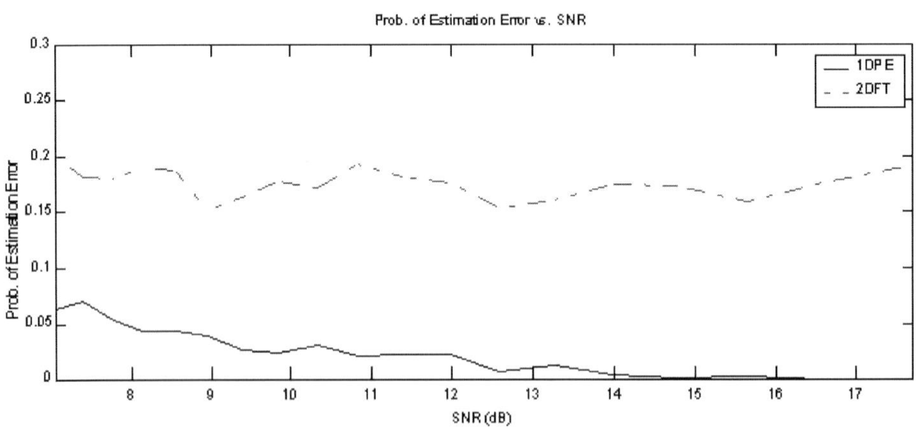

Figure 4. Probability of estimation error.

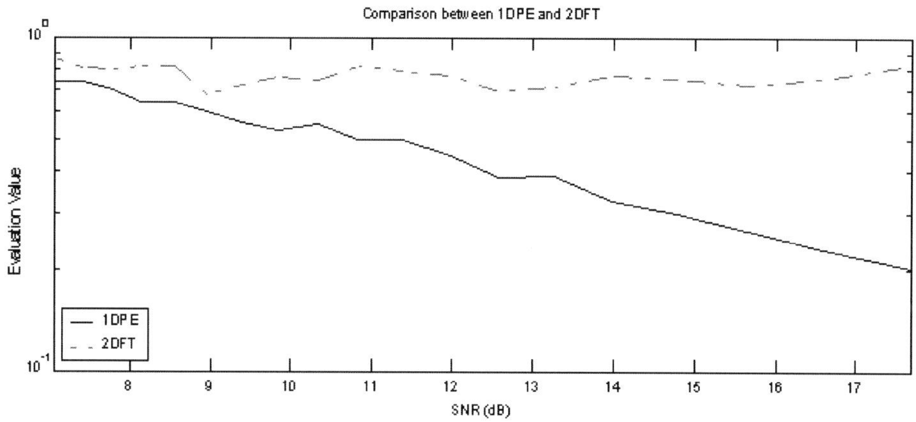

Figure 5. Comparison betwen 1DPE and 2DFT.

Figure 4 clearly suggests that 1DPE exhibits significantly better scatterer estimation accuracy.

Because P_r cannot illustrate the estimation accuracy when the scatterers are correctly estimated, we need another method of quantifying the algorithms' estimation accuracy for the properly estimated scatterers. Therefore, we define a new expression to meet this need:

$$V = MC + H, \tag{42}$$

where V is the evaluation value, and C is a constant, which quantifies the penalty for each mis-estimated scatterer. We let C equal to 0.7^2 in our experiments.

H is the Mean Square Error (MSE) of those properly estimated scatterers. The results of Experiment 3 are illustrated in Figure 5.

From the above experimental results we observe that 1DPE algorithm significantly outperforms the 2DFT algorithm under reasonably high SNR conditions. Moreover, 1DPE has reliable performance in automatically detecting the number of scatterers, K, when the SNR is not too low. However, 2DFT also has some attractive properties [18], which include:

- 2DFT assumes weaker requirements on the distribution of the additive noise, i.e. it is not necessary to assume the additive noise to be circularly Gaussian distributed.
- 2DFT can tolerate greater model mismatching.

8. Conclusions

In this paper, based on four basic assumptions regarding underlying radar systems and moving targets, we derived a general HRR radar model. In order to facilitate the application of this proposed model, we simplified the model by instantiating a set of real-world radar and target parameters to the model, and generated two HRR radar models with different degrees of simplicity. Because conventional HRR data processing, including feature extraction and clutter suppression, are naturally reduced to model parameter estimation after establishing this set of models, we also provided two model parameter estimation algorithms, 1DPE and 2DFT, based on the most simplified HRR model we proposed. Simulation results demonstrate that 1DPE can reliably estimate the number of scatterers embedded in the 2D raw HRR data, and dramatically outperforms 2DFT in estimation accuracy when the SNR is moderate. However, 2DFT also possesses some attractive features, such as insensitivity to the noise distribution and tolerance of model mismatching.

Acknowledgments

This work was supported by the Mission Research Cooperation under subcontract SC-1406-98-0001. Additional support was provided by the ARL PET program funded by the HPCMO.

Notes

1. Generally speaking, "visible" scatterers are those scatterers with relatively large magnitudes. A more precise definition depends heavily on the particular feature extraction algorithm employed.
2. Because (a) there are 3 parameters, *position frequency*, *speed frequency*, and *scattering coefficient*, for each scatterer, (b) if the *position frequency* estimation error of a scatterer is bigger than 0.05, and the *speed frequency* estimation error is bigger than 0.065, it is considered as a *mis-estimated scatterer*, and (c) the penalty to the scattering coefficient of a *mis-estimated scatterer* is defined as 0.585, the Penalty/scatterer $= 0.05 + 0.065 + 0.585 = 0.7$.

References

1. S.P. Jacobs and J.A. O'Sullivan, "High Resolution Radar Models for Joint Tracking and Recognition," *SPIE Proceeding*, vol. 3069, April 1997.
2. H.-J. Li and S.-H. Yang, "Using Range Profiles as Feature Vectors to Identify Aerospace Objects," *IEEE Transactions on Antennas and Propagation*, vol. 41, March 1993, pp. 261–268.
3. D. Chen and Z. Bao, "High Range Resolution Radar Target Identification Using Multilayer Feedforward Neural Network," *IEEE CIE International Conference of Radar Proceedings 1996*, Piscataway, NJ, pp. 215–218.
4. R. Williams, J. Westerkamp, D. Gross, A. Palomino, and T. Fister, "Automatic Target Recognition of Time Critical Moving Targets Using 1D High Range Resolution (HRR) Radar," *The Record of the 1999 IEEE Radar Conference*, 1999, pp. 54–59.
5. R. Williams, D. Gross, A. Palomino, J. Westerkamp, and D. Wardell, "1D HRR Data Analysis and ATR Assessment," *Air Force Research Laboratory Technical Report*, 1998.

6. C.J. Ying, H. Chiang, R.L. Moses, and L.C. Potter, "Complex SAR Phase History Modeling Using Two Dimensional Parametric Estimation Techniques," *SPIE Symposium on Algorithm for Synthetic Aperture Radar III*, June 1996, pp. 174–185.
7. S.P. Jacobs, J.A. O'Sullivan, M. Faisal, A. Srivastava, D.L. Snyder, and M.I. Miller, "Automatic Target Recognition Using Sequences of High-Resolution Radar Range-Profiles," *Electrical System and Signal Research Laboratory Monograph*, no. ESSRL-95-26.
8. S.P. Jacobs, J.A. O'Sullivan, M. Faisel, and D.L. Snyder, "Automatic Target Recognition System Using High Resolution Radar," *SPIE Proceeding*, vol. 2845, August 1995.
9. H.J. Li, Y.D. Wang, and L.H. Wang, "Matching Score Properties Between Range Profiles of High-Resolution Radar Targets," *IEEE Transactions on Antennas and Propagation*, vol. 44, April 1996.
10. Z. Bi, R. Wu, J. Li, and R. Williams, "High Range Resolution Feature Extraction of Moving Targets in the Presence of Stationary Clutter," Submitted to *IEE Proceeding on Radar, Sonar, and Navigation*, August 1998.
11. R. Wu and J. Li, "HRR Data Model for Non-Rigid Body Moving Targets in the Presence of Range Migration," *TRUMPETS Research Note*, Oct. 1998.
12. J. Ma, and S.C. Ahalt, "Moving Target Modeling, Clutter Suppression, and Feature Extraction," *TRUMPETS Research Note*, May 13, 1999.
13. D.R. Wehner, *High-Resolution Radar*, 2nd ed., Boston: Artech House Inc., 1995.
14. A.W. Rihaczek, *Principles of High-Resolution Radar*, Boston: Artech House Inc. 1996.
15. B. Borden, *Radar Imaging of Airborne Targets: A Primer for Applied Mathematicians and Physicists*, Inst. of Physics Pub., 1999.
16. J.B. Keller, "Geometrical Theory of Diffraction," *J. Opt. Soc. Amer.*, January 1962, pp. 116–130.
17. L.M. Novak, S.D. Halversen, G.J. Owirka, and M. Hiett, "Effects of Polarization and Resolution on the Performance of a SAR Automatic Target Recognition System," *The Lincoln Laboratory Journal*, vol. 8, no. 1, 1995, pp. 48–69.
18. M.B. Ressler, R. Williams, D. Gross, and A. Palomino, "Bayesian Multiple-Look Updating Applied to the SHARP ATR System," *SPIE Proceeding*, 1999. (http://www.toyon.com/papers/BayesSPIEpaperFinal.pdf)
19. W.M. Steedly and R.L. Moses, "High Resolution Exponential Modeling of Fully Polarized Radar Returns," *IEEE Transactions on Aerospace and Electronic Systems*, vol. AES-27, May 1991, pp. 459–569.
20. N.F. Chamberlain, E.K. Walton, and F.D. Garber, "Radar Target Identification of Aircraft Using Polarization-Diverse Features," *IEEE Transactions on Aerospace and Electronic System*, vol. 27, January 1991, pp. 58–67.
21. N.F. Chamberlain, "Recognition and Analysis of Aircraft Targets by Radar, Using Structural Pattern Representations Derived from Polarimetric Signatures," *Ph.D. Thesis*, the Ohio State University, June 1989. Also: *ElectroScience Laboratory Report* 719710-3.
22. W.M. Steeply and R.L. Moses, "High Resolution Exponential Modeling of Fully Polarized Radar Returns," *IEEE Transactions on Aerospace and Electronic Systems*, vol. 27, no. 3, May 1991, pp. 459–468.
23. S.K. Chaudhuri and W. Boerner, "A Polarimetric Model for the Recovery of the High-Frequency Scattering Centers from Bistatic-Monostatic Scattering Matrix Data," *IEEE Transactions on Antennas and Propagation*, vol. 35, no. 1, June 1987, pp. 87–93.
24. L.M. Novak, M.B. Sechtin, and M.J. Cardullo, "Studies of Target Detection Algorithms That Use Polarimetric Radar Data," *IEEE Transactions on Aerospace and Electronic Systems*, vol. 25, no. 2, July 1989, pp. 150–165.
25. D. Giuli, "Polarization Diversity in Radars," 1986, pp. 245–269.
26. L.C. Potter and R.L. Moses, " Attributed Scattering Centers for SAR ATR," *IEEE Transactions on Image Processing, Special Issue on Automatic Target Recognition*, vol. 6, no. 1, January 1997, pp. 79–91.
27. C.J. Ying, R. Carriere, L.C. Potter, and R.L. Moses, "Radar Target Modeling: A GTD-Based Approach", *Proceedings of SPIE International Symposium on Automatic Object Recognition IV*, Orlando, FL, April 4–8 1994, pp. 67–78.
28. L.C. Potter, D. Chiang, R. Carriere, and M.J. Gerry, "A GTD-Based Parametric Model for Radar Scattering," *IEEE Transactions on Antennas and Propagation*, vol. 43, no. 10, October 1995, pp. 1058–1088.
29. J. Ma and S.C. Ahalt, "Derivation of Physics-Based HRR Moving Target Model," *Proceedings of SPIE*

International Symposium on Signal Processing, Sensor Fusion, and Target Recognition IX, Orlando, FL, April 24–28, 2000.

30. P. Stoica and R. Moses, *Introduction to Spectral Analysis*, PH, 1997.
31. P. Stoica, R.L. Moses, T. Soderstrom, and J. Li, "Optimal High-Order Yule-Walker Estimation of Sinusoidal Frequencies," *IEEE Transactions on Signal Processing*, vol. 39, no. 6, 1991, pp. 1360–1368.
32. P. Stoica and A. Nehorai, "MUSIC, Maximum Likelihood, and Cramer-Rao Bound," *IEEE Transactions on Acoustics, Speech, and Signal Processing*, vol. 35, 1989, pp. 582–584.
33. R. Roy and T. Kailath, "ESPRIT—Estimation of Signal Parameters via Rotational Invariance Techniques," *IEEE Transactions on Acoust. Speech, Signal Processing*, vol. 37, no. 7, 1989, pp. 984–995.
34. J. Ma and S.C. Ahalt, "Analysis of Complex HRR Range Signatures," *Proceedings of SPIE International Symposium on Targets and Backgrounds VI*, Orlando, FL, April 24–28, 2000.
35. J. Ma, *Doctoral Research Proposal*, Dept. of Electrical Engineering, the Ohio State University, Aug. 2000.
36. Lxthos Products Data Sheet: IXI8250 Dual 8-bit 50 MHz A/D (http://www.ixthos.com/technical/products/ixianalog/ixi8250_ds.html).

Multidimensional Systems and Signal Processing, 14, 49–82, 2003

Detection and Analysis of Anisotropic Scattering in SAR Data☆

ANDREW J. KIM ajkim@alum.mit.edu
Stochastic Systems Group, Massachusetts Institute of Technology, Cambridge, MA 02139

JOHN W. FISHER III
Stochastic Systems Group, Massachusetts Institute of Technology, Cambridge, MA 02139

ALAN S. WILLSKY[1]
Stochastic Systems Group, Massachusetts Institute of Technology, Cambridge, MA 02139

Received December 1, 2000; Revised September 9, 2001; Accepted September 26, 2001

Abstract. Scattering from man-made objects in SAR imagery exhibits aspect and frequency dependencies which are not well modeled by standard SAR imaging techniques. If ignored, these deviations will reduce recognition performance due to the model mismatch, but when appropriately accounted for, these deviations from the ideal point scattering model can be exploited as attributes to better distinguish scatterers and their respective targets. With this premise in mind, we have developed an efficient modeling framework that incorporates scatterer anisotropy. One of the products of our analysis is the assignment of an anisotropy label to each scatterer conveying the degree of anisotropy. Anisotropic behavior is commonly predicted for geometric scatterers (scatterers with a simple geometric structure), but it may also arise from volumetric scatterers (random arrangements of interfering point scatterers). Analysis of anisotropy arising from these two modalities shows a clear source-dependent relationship between the anisotropy classification and parameters of the scatterer. In particular, the degree of anisotropy is closely related to the size of the scatterer, and increasing the aperture size reduces the incidence of volumetric anisotropy but preserves the detection rate for geometric anisotropy. This result helps to address the question in the SAR community regarding the utility of wide-aperture SAR data for ATR since wide-aperture data reveals geometric anisotropy while resolving volumetric anisotropy into individual isotropic scatterers.

Key Words: SAR, anisotropy, wide-aperture, multi-resolution, canonical scattering, volumetric scattering, ATR

1. Introduction

Scatterers composing a target in synthetic aperture radar (SAR) imagery often exhibit nonideal scattering behavior in the form of aspect and frequency dependence, particularly in wide-aperture or wide-band data. Standard SAR image formation ignores this model deviation resulting in suboptimal use of the data. The modeling error may furthermore

☆ This work is sponsored by Air Force Office of Scientific Research grant F49620-96-1-0028 (subcontract BU GC123919NGD), Office of Naval Research grant N00014-00-1-0089, and DARPA under Wright Patterson grant F33615-97-1014.

introduce artifacts, such as peak scatterer instability, which unnecessarily complicate the recognition problem. These deviations from the ideal point scattering model should not be viewed as a nuisance and approximated away, but they should instead be seen as a useful attribute which can be used to distinguish scatterers and thus their respective targets. In this paper, we present a multi-scale analysis for detecting and classifying scatterer anisotropy in SAR data. This analysis is used for (i) studying the dependence of the proposed anisotropy attribution on scatterer phenomenology and (ii) assessing the utility of wide-aperture data in the attribution process and hence for automatic target recognition (ATR).

There are numerous benefits to knowing the anisotropy of a scatterer which the standard image formation process ignores. From an imaging standpoint, a better scattering model allows for more accurate reflectivity estimates since the weighting in the coherent averaging can be adjusted according to the azimuthal scattering pattern (e.g. with a matched filter) such as in the work done by Allen et al. [1] and by Chaney et al. [2]. Knowledge of anisotropy may also be used to better suppress interfering scatterers in a super-resolution image formation algorithm such as Capon's method [3] or the high definition imaging proposed by Benitz et al. [4], [5]. In addition to improving reflectivity estimation, knowledge of a scatterer's anisotropy is useful as a feature in and of itself. For a canonical scatterer, one can infer properties about its geometric shape and size from the observed anisotropy. For instance, a flat plate produces a strong specular response where the degree of specularity is directly related to the size of the plate in cross-range. A sphere however produces an isotropic response regardless of the size of the sphere. Knowledge of the geometry of the individual scatterers on an object may then be considered as a group to aid the classification of the target under investigation as described in [9], [10], [11]. Anisotropy information may also be of use in characterizing the stability of a scatterer. In particular, one would expect the specular response from an anisotropic canonical scatterer to appear only over a small range in azimuth leading to an observability that is highly sensitive to azimuthal orientation. Knowledge of the anisotropy however allows such sensitivity to be incorporated into algorithms, such as peak matching, which would improve ATR performance.

The focus of this paper is to present a method for classifying anisotropy and to study behavior of the attribution under different conditions. The analysis presented here is based on a general characterization of azimuthal anisotropy which is not tuned to a particular canonical scatterer. The basis of our analysis is the sub-aperture pyramid which is a set of sub-apertures arranged in a pyramidal fashion. The collection of sub-apertures provides a multi-resolution representation of SAR data. These multiple apertures allow for the observation of varying degrees of azimuthal anisotropy which are unobservable under standard SAR image formation. With this pyramidal representation of the data, we propose a hypothesis test for anisotropy which measures the concentration of scattering energy across the aperture.

Analysis of this characterization reveals aspects of the phenomenology associated with anisotropy and how the attribution can be utilized in ATR. In studying the behavior of our anisotropy characterization, we consider two well known sources of anisotropy: geometric and volumetric scatterers. Scattering models such as physical

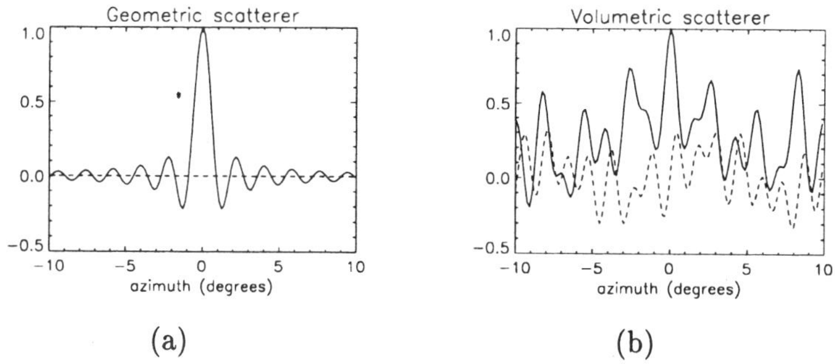

Figure 1. Instances of azimuthal scattering by a (a) geometric scatterer and (b) volumetric scatterer with the real and imaginary components of the response shown in the solid and dashed lines respectively.

optics and the Geometric Theory of Diffraction (GTD) give functional forms for radar scattering from canonical scatterers which clearly convey an azimuthal dependence on the shape, size, and orientation of a scatterer. These models are based on very specific assumptions about the configuration of scatterer, such as equal-sized plates in a dihedral or pairwise orthogonality of plates in a trihedral. As our analysis is concerned with producing a general characterization of anisotropy, we will not entangle ourselves with such fine structural details. This approach allows us to consider the broader class of what we call *geometric scatterers* which include not only canonical scatterers but also scatterers that deviate from canonicity. Besides geometric scatterers, anisotropy may also arise from a *volumetric scatterer* which is a collection of closely located point scatterers that interfere with each other. The anisotropy exhibited by volumetric scatterers, however, is not as stable as the anisotropy produced by geometric scatterers. As an example, Figure 1 shows the azimuthal response from a canonical flat plate measuring 1m in cross-range and a volumetric scatterer composed of 10 randomly located point scatterers distributed over 1m in cross-range. The flat plate response is clearly anisotropic. The volumetric scatterer is also anisotropic, but to a lesser degree as the scatterer's energy is more distributed across the aperture. As hinted to in Figure 1, the behavior between geometric and volumetric anisotropy[2] is different and ultimately leads to an ability to distinguish between the two sources of anisotropy to a degree. This distinction is highly desirable since it will then allow for tailored use of the anisotropy information based on the source. In particular, for geometric scatterers, we can infer physical properties of the scatterer and a high degree of stability, while for volumetric scatterers, we can infer the spatial support of the collection of scatterers and a low degree of stability.

The remainder of this paper elaborates on the ideas presented above. Section 2 presents the multi-resolution sub-aperture pyramid used to represent the SAR data. Section 3 then describes the set of hypotheses that we consider and develops the hypothesis tests on the sub-aperture pyramid. Section 4 presents experimental analysis demonstrating the utility of the anisotropy attribution and in particular the attribution's

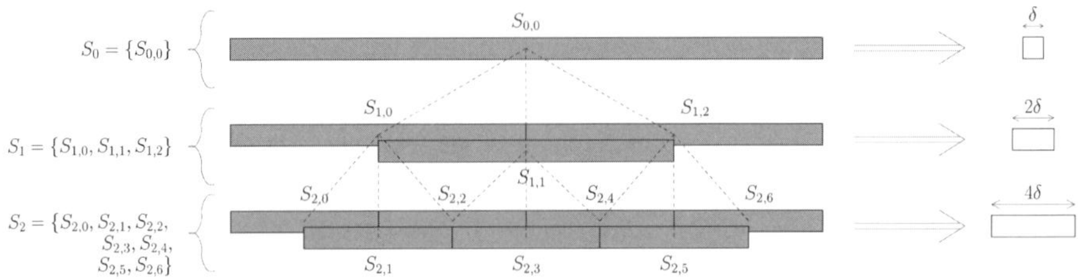

Figure 2. Illustration of the three level, half-overlapping, half-aperture pyramid. Left: Sub-apertures composing the pyramid. Right: Resolution cell size associated with reflectivity estimates formed from sub-apertures at the

relation to the scatterer phenomenology. The paper concludes with a summary and discussion in Section 5.

2. Sub-Aperture Analysis

The foundation of our analysis is the sub-aperture pyramid which we present in this section. This structure is motivated by the scattering physics involved in SAR and presents the data in a way that allows for simple and intuitively reasonable hypothesis tests. Because of the linear structure of the aperture, we associate it with an interval of the real line throughout this paper. In particular, the full-aperture is denoted by the interval $[0, 1) = \{t | 0 \leq t < 1\}$.

2.1. *Definition*

The intuitive idea of the sub-aperture pyramid is to generate an over-complete covering of the full-aperture with sub-apertures that can be arranged in a pyramidal structure. These sub-apertures will be used both to form our reflectivity estimates and to represent our set of candidate anisotropy hypotheses. The prototypical sub-aperture covering that we use throughout this paper is the half-overlapping, half-aperture pyramid shown in Figure 2.

Formally, we take a sub-aperture pyramid to be a set S of sub-apertures with the following structure. The set S is partitioned into smaller sets S_m all the elements of which have a particular sub-aperture length as illustrated in Figure 2. S_0 refers to the set consisting of the largest sub-apertures, and S_M refers to the set of the smallest sub-apertures. A second subscript on S denotes a specific sub-aperture at the given scale. Note that any sub-aperture can be used to form a SAR image. The images formed with a smaller value of m, i.e. with a larger aperture, have a finer cross-range imaging resolution because of the inverse relationship between resolution cell size and aperture length. This point is illustrated in Figure 2 where if the full-aperture produces a resolution cell of size δ, then the half and quarter-apertures produce resolution cells of sizes 2δ and 4δ, respectively. Note that this inverse relationship between sub-aperture size and resolution cell size may lead to ambiguity when terms such as "coarse" and "fine" are used. For instance, a "coarser" (i.e. larger) sub-aperture produces a "finer" resolution cell. To avoid any ambiguity, we will explicitly use the terms "smaller" and "larger" when referring to sub-aperture sizes and the terms "finer" and "coarser" when referring to resolution cell sizes.

The idea of the sub-aperture pyramid is to present SAR data in an over-complete representation at a variety of cross-range resolution versus azimuthal[3] resolution trade-offs. Recall that cross-range resolution is inversely proportional to aperture length. Thus, at lower levels of the sub-aperture pyramid, spatial resolution has been exchanged for azimuthal resolution, i.e. the ability to better observe anisotropic phenomena. This is the classic time–frequency resolution tradeoff in Fourier analysis, and each level of the pyramid represents the data under a particular cross-range–azimuth resolution. The presence of multiple resolutions is attractive because we expect the best representation to depend on the underlying anisotropy. In particular, it is the set which has the sub-aperture which best captures the scattering energy with minimal aperture length, i.e. the closest resemblance to a matched-filter using an indicator function over a sub-aperture. Since the scatterer anisotropy (and thus best sub-aperture representation) cannot be known apriori, an over-complete basis, such as that provided by the sub-aperture pyramid, is useful in finding the best representation.

Having conveyed the intuition behind the sub-aperture pyramid, we now establish the necessary conditions on the pyramid for what follows later. In particular, the following conditions are imposed on S:

(A) $\forall S_{m,i}$, $S_{m,i} \subset [0, 1)$,
(B) $\forall S_{m,i}$, $\exists$ a partition $\mathcal{P}(m, i) \subset S_M$ of $S_{m,i}$, and
(C) $\forall S_{m,i}$ with $m \geq 1$, $\exists S_{m-1,j}$ such that $S_{m,i} \subset S_{m-1,j}$.

The first condition simply restricts the sub-apertures to be a subset of the available aperture. Motivated by our search for concentrated unimodal scattering in azimuth, we only consider the special case where each sub-aperture is a connected interval. Moreover, we also assume that $S_{0,0}$ is the full-aperture to allow for consideration of the minimal degree of anisotropy and allow for the finest imaging resolution permissible. The second condition insures that each sub-aperture can be represented by a partition of smallest-sized sub-apertures. This condition will allow for the set of measurements obtained from S_M to form a sufficient statistic for all the measurements in S. Thus, although other sub-aperture measurements will be referred to in this paper for intuition and interpretation, all computation can be done using only measurements from the smallest sub-apertures. The third condition asserts that each sub-aperture, except those in S_0, has a parent sub-aperture which contains it. This last condition is not a fundamentally necessity for our anisotropy analysis, however, we use this condition because it permits us to construct an intuitive telescopic hypothesis test on a tree which will not only afford computational efficiency but also robustness. Herein, the term *sub-aperture pyramid* refers to one satisfying conditions (A)–(C).

Now, focusing our attention on a particular location in the scene, each sub-aperture $S_{m,i}$ can be used to generate an associated reflectivity measurement $q_{m,i}$. The collection of

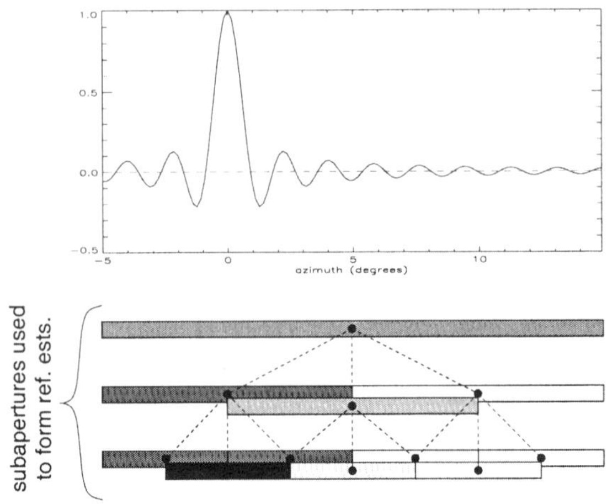

Figure 3. The response of a $1m \times 1m$ flat plate and a depiction of the reflectivity estimate for each of the sub-apertures. Darker shaded sub-apertures convey larger reflectivity estimates.

reflectivity measurements from the different offset sub-apertures in S_M are formed into a vector which is denoted by $\boldsymbol{q}_M$. The measured sub-aperture reflectivity $q_{m,i}$ is obtained in the same fashion as in standard SAR image formation, i.e.

$$q_{m,i} = \int_{S_{m,i}} a(t)dt \tag{1}$$

where $a(t)$ is the azimuthal response of the scatterer[4]. Although reflectivity measurements from smaller sub-apertures have an associated coarser resolution, they are less susceptible to azimuthal variations and provide a means for measuring scattering energy across the aperture. Note that the measured reflectivity is *not* normalized with respect to the sub-aperture length[5]. Thus, when interested in the actual reflectivity estimate, one should divide the measured sub-aperture reflectivity $q_{m,i}$ by the sub-aperture length $L_{m,i} = \lambda(S_{m,i})$, where λ denotes Lebesgue measure.

2.2. *Interpretation and Motivation*

Different types and sizes of geometric scatterers exhibit different aspect dependencies. The motivation for using the sub-aperture pyramid is that the pyramid is expected to reveal distinguishing aspect dependences in the scattering. For example, a metal sphere has a strong response in all directions and thus produces a strong consistent reflectivity estimate from each of the sub-apertures. However, as depicted in Figure 3, a flat plate produces a specular response which is significantly stronger when oriented broadside with respect to the radar. Thus, the reflectivity estimates vary across the sub-apertures with the largest estimate coming from the sub-aperture oriented broadside to the plate. Furthermore, because various sized sub-apertures are used, the duration of the broadside flash is also

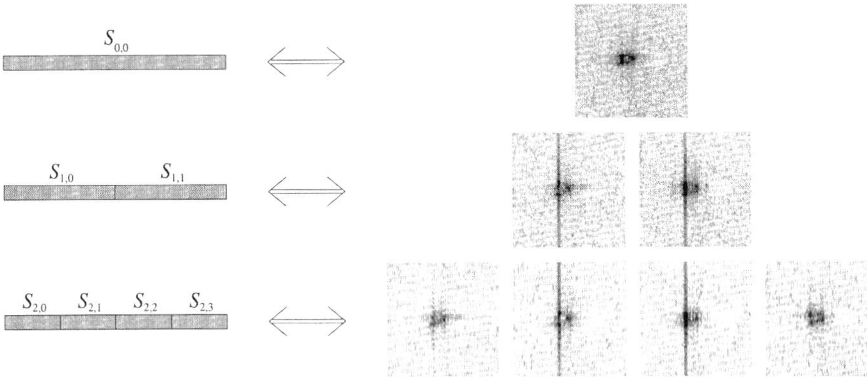

Figure 4. Left: Disjoint half-aperture pyramid. Right: Corresponding sub-aperture images of a BMP-2 at a 17° elevation and 0° azimuth. For each image, the front of the vehicle is the portion nearest the left edge of the image.

captured in this representation. To see this point, note that for all sub-apertures which reside within the main-lobe of the response, the reflectivity estimates are consistently large. As the sub-aperture is expanded, however, the additional signal energy received is relatively insignificant thereby lowering the reflectivity estimate which is normalized with respect to sub-aperture length. Not only does using an excessively large sub-aperture lower the normalized reflectivity estimate, but it also results in a noisier estimate because the additional portion of the aperture being incorporated is dominated by noise[6]. To illustrate that this anisotropic phenomena is clearly present in real data, we show in Figure 4 a three level, disjoint, half-aperture pyramid and the corresponding sub-aperture images for a BMP-2 from the MSTAR public release dataset [6] which uses a 2.8° aperture. Even in this narrow aperture setting, we clearly see that the response associated with the middle section of the aperture contains the majority of the energy from the large scatterer at the front of the vehicle.

3. Anisotropic Scattering Models

Having presented the sub-aperture pyramid, we now proceed to formulate our hypothesis testing problem for anisotropy where the hypotheses are formulated over the sub-aperture pyramid. Our goal is to use the sub-aperture pyramid to develop a general characterization of anisotropy that is not overly-sensitive to azimuthal dependencies such as those produced by geometric scatterers with non-canonical deviations. In particular, we only intend to measure the concentration of unimodal azimuthal scattering and are not concerned with the minor deviations in the azimuthal response that occur with slight changes in scatterer geometry. Two models are presented here that yield this characterization. The first is a simple isolated scatterer model with an intuitive sufficient statistic. This test, however, is susceptible to the interference from neighboring scatterers. This shortcoming motivates the second model which explicitly accounts for the effects of neighboring scatterers. The tests presented in this section are for a fixed scattering location which we assume to be specified. These locations could come from a peak extraction process (for peak attribution as is done in Section 4) or a pre-specified grid (for an image of anisotropy).

3.1. *Isolated Scatterer Model*

For each sub-aperture $S_{m,i}$, we define an associated scattering hypothesis $H_{m,i}$ over the aperture t via

$$H_{m,i} : a(t) = A 1_{S_{m,i}}(t) \tag{2}$$

where $1_{S_{m,i}}(\cdot)$ denotes the indicator function over $S_{m,i}$ and A is the unknown scattering amplitude of the signal. Thus, each hypothesis corresponds to a scattering response that is uniform over the sub-aperture in question and zero elsewhere. Clearly, this model is an idealization of anisotropic scattering, but because we are only interested in obtaining a general characterization of anisotropy, this model serves our purposes. The set of all possible hypotheses associated with the sub-aperture pyramid is denoted as $\mathcal{H}$.

A reasonable choice of features to test these hypotheses would be all the measured sub-aperture reflectivities $\{q_{m,i}\}$. From the definition of the $q_{m,i}$ in Eq. (1) and partition property (B), it is sufficient to consider the vector $\boldsymbol{q}_M$ of measurements from the smallest sub-apertures since any sub-aperture reflectivity $q_{m,i}$ can be computed from $\boldsymbol{q}_M$ by summing all the $q_{M,j}$ for which the associated $S_{M,j}$ form a partition of $S_{m,i}$, i.e.

$$q_{m,i} = \sum_{j | S_{M,j} \in \mathcal{P}(m,i)} q_{M,j}.$$

As an example, for the three level, half-overlapping, half-aperture pyramid in Figure 2, $q_{1,1} = q_{2,2} + q_{2,4}$ since $S_{2,2}$ and $S_{2,4}$ form a disjoint union of $S_{1,1}$, i.e. $S_{1,1} = S_{2,2} \cup S_{2,4}$ and $S_{2,2} \cap S_{2,4} = \emptyset$. Because $\boldsymbol{q}_M$ is a sufficient statistic for our sub-aperture reflectivities, we take it as our feature vector. The value of this feature vector under hypothesis $H_{m,i}$ and $A = 1$ is denoted as $\boldsymbol{b}(m, i)$ whose j^{th} element is given by

$$\begin{aligned} b(m,i)_j &= \int_{S_{M,j}} 1_{S_{m,i}}(t) dt \\ &= \lambda(S_{M,j} \cap S_{m,i}), \end{aligned} \tag{3}$$

i.e. $b(m, i)_j$ is the portion of the response $1_{S_{m,i}}(t)$ one expects to see over the jth sub-aperture at scale M. We now define our scattering model conditioned on anisotropy hypothesis $H_{m,i}$ as the signal plus noise model

$$q_{M,j} = \int_{S_{M,j}} A 1_{S_{m,i}}(t) + \eta(t)\, dt, \tag{4}$$

where $\eta(t)$ is circularly complex white Gaussian noise with spectral density $2\sigma^2$. This characterization leads to the model

$$H_{m,i} : \boldsymbol{q}_M = A\boldsymbol{b}(m,i) + \varepsilon, \quad \text{with } \varepsilon \sim \mathcal{N}(\mathbf{0}, 2\sigma^2\Lambda), \tag{5}$$

where Λ is the noise covariance structure inherited from the sub-aperture pyramid. The noise in the measured reflectivities in Eq. (5) are characterized as zero-mean circularly complex Gaussians with covariances dictated by the amount of sub-aperture overlap. The elements of the covariance matrix Λ are thus given by $[\Lambda]_{i,j} = \lambda(S_{M,i} \cap S_{M,j})$, which for the half-overlapping half-aperture pyramid in Figure 2 is

$$\Lambda = \frac{1}{2^M} \begin{bmatrix} 1 & .5 & & 0 \\ .5 & \ddots & \ddots & \\ & \ddots & \ddots & .5 \\ 0 & & .5 & 1 \end{bmatrix}.$$

To classify the anisotropy of a scatterer from our vector of sub-aperture measurements $\boldsymbol{q}_M$, we apply a log-likelihood ratio test to the model in Eq. (5) where each log-likelihood is compared to the full-aperture hypothesis $H_{0,0}$. Because there is the unknown reflectivity parameter A, we use a generalized log-likelihood ratio (GLLR) test where for each hypothesis, we take A to be the maximum likelihood (ML) estimate under that hypothesis. For $H_{m,i}$, the matched-filter gives the ML estimate as $\hat{A} = q_{m,i}/L_{m,i}$. Using this estimate results in the GLLR

$$\ell_{m,i} = \frac{1}{4\sigma^2}\left[\frac{1}{L_{m,i}}|q_{m,i}|^2 - |q_{0,0}|^2\right]. \tag{6}$$

Thus, the most likely sub-aperture in this case is the one whose average energy is largest. We note the similarity here to the approach taken by Chaney et al. [2] in which they replace, within the image, the standard full-aperture reflectivity estimate with the maximum sub-aperture reflectivity estimate $|q_{m,i}/L_{m,i}|$, thus using normalized reflectivity (instead of normalized energy) as their criterion for choosing anisotropy. Their approach however is based on intuitive arguments and not a derived statistic. Although this difference may be seemingly small, the effect is quite significant. In particular, using normalized reflectivity, instead of energy, over-compensates for decreasing aperture size. This over-compensation biases their test to choose smaller sub-aperture hypotheses and thus produces images with an excessively coarse imaging resolution. Furthermore, their technique does not address the very significant effect of neighboring scatterers that we discuss next. Thus, when used for image formation, our technique better preserves the fine

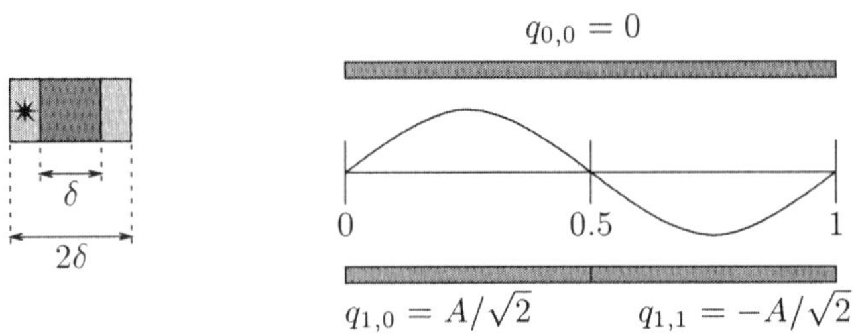

Figure 5. Illustration of how a scatterer that is not in the finest resolution cell can produce a false anisotropy classification from Eq. (6).

resolution of the imagery, where appropriate, since the technique produces more accurate anisotropy classifications.

Though simple and intuitive, the GLLR in Eq. (6) is susceptible to the effects of close proximity neighboring scatterers which are not accounted for in our model in Eq. (4). Recall that the images formed by the smaller sub-apertures have a coarser imaging resolution, and thus the hypotheses are not all consistent in terms of the resolution cell contents. In particular, if a scatterer were located outside the finest resolution cell but within a coarser resolution cell, then the finest resolution reflectivity $q_{0,0}$ would be 0, but $q_{m,i}$ would be large if the resolution cell associated with scale m included the scatterer. We illustrate this point with the example shown in Figure 5. Here a scatterer with amplitude A is located outside the finest resolution cell of size δ but is contained within the coarser resolution cell of size 2δ associated with half-aperture estimates. If this scatterer is isotropic, then the response is the complex exponential illustrated. Integrating over the full aperture gives a 0 reflectivity estimate as expected, but integrating over a half aperture produces a reflectivity with magnitude $A/\sqrt{2}$. Eq. (6) would then classify the center of the resolution cell as anisotropic, even though no scatterer is present in the finest resolution cell.

The problem above is a consequence of not modeling the influence of neighboring scatterers. One of the ways in which the neighboring scatterer manifests itself is through the corruption of the estimated reflectivity as the size of the resolution cell varies. Instead of choosing the ML estimate $\hat{A} = q_{m,i}/L_{m,i}$ as the unknown reflectivity, we can choose the best reflectivity estimate constrained to lie in the finest resolution cell associated with the full-aperture estimate $q_{0,0}$. Choosing $\hat{A} = q_{0,0}/L_{m,i}$ (where the normalization by $L_{m,i}$ accounts for the fact that the modeled azimuthal response is concentrated in a sub-aperture) for all hypotheses can be shown to produce the GLLR statistic

$$\ell_{m,i} = \frac{1}{4\sigma^2}\left[\frac{1}{L_{m,i}}|q_{m,i}|^2 - \frac{1}{L_{m,i}}|q_{0,0} - q_{m,i}|^2 - \frac{1}{L_{0,0}}|q_{0,0}|^2\right]. \quad (7)$$

This statistic is identical to that in Eq. (6) except for the extra term comparing the reflectivity estimates $q_{0,0}$ and $q_{m,i}$. Recall that Eq. (6) compared the average energy in a

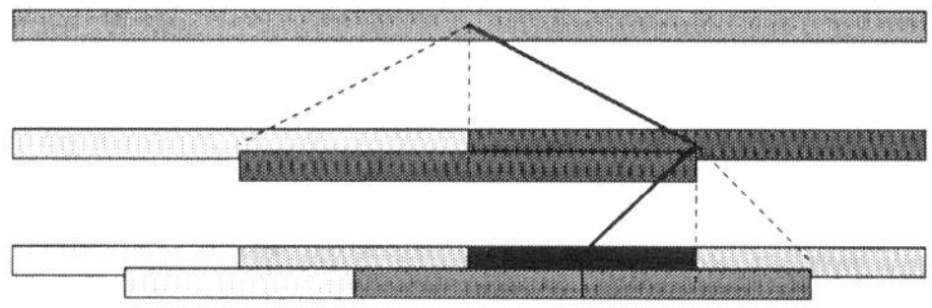

Figure 6. Illustration of how the anisotropy testing can be done in a decision-directed fashion by starting with the largest aperture and at each scale, inspecting only the children of the most likely sub-aperture. Darker shading of the sub-apertures indicate higher likelihoods. Dashed lines denote which hypotheses are tested, and solid lines denote the branch traversed.

sub-aperture to the full-aperture. This new GLLR accounts for the average energy outside the hypothesized sub-aperture as well. Viewed differently, under hypothesis $H_{m,i}$ and our scattering model in Eq. (4)–(5), the values of $q_{0,0}$ and $q_{m,i}$ should simply be noisy perturbations of each other. The new term penalizes when this is not the case thereby enforcing a consistency of the resolution cell contents. Under the example in Figure 5, since the the contribution of each of the half-apertures would be the same, the GLLR is equal to zero for all hypotheses, which is reasonable, since there is no underlying scattering at the focused location.

3.2. Multiple Scatterer Model

The modification in Eq. (7) addresses the problem when a neighboring scatterer is isotropic, however, when the neighboring scatterer is anisotropic, problems such as that illustrated by the example in Figure 5 can still arise as the contribution from the interfering scatterer will not integrate out over the full-aperture estimate. To account for this phenomenon, we generalize the model in Eq. (4) to explicitly include multiple scatterers. In particular, we consider the possibility of regularly spaced neighboring scatterers (each with its own anisotropy hypothesis) and incorporate their effects into the anisotropy test. Doing so, we obtain the GLL statistic[7]

$$\ell = \frac{1}{2\sigma^2} \boldsymbol{q}'_M \left[\Lambda^{-1} - 2\Lambda^{-1}BP + P'B'\Lambda^{-1}BP\right] \boldsymbol{q}_M \tag{8}$$

where B and P are defined in Eqs. (14) and (15) of Appendix A. We note that the hypothesis enters into Eq. (8) only via B and P, and thus, the weighting matrix in the brackets is data independent and can be precomputed allowing for a computationally simple GLL that corresponds to a norm on $\boldsymbol{q}_M$.

3.3. Telescopic Testing

The sub-aperture pyramid which we use to form our measurements and base our hypotheses is convenient not only for providing anisotropy information, but also for providing an efficient means of performing the hypothesis tests. We can obtain an efficient approximation to the test by evaluating only a small subset of the candidate hypotheses. Due to the nested structure of the sub-apertures (condition (C) in Section 2), we can perform the tests in a telescopic fashion by traversing down a branch in the tree of sub-apertures as depicted in Figure 6. On the pyramid, we expect the likelihoods to increase as the hypothesized sub-apertures "shrink down to" the correct sub-aperture, and then to decrease as the hypothesized sub-apertures "shrink beyond" the correct sub-aperture. This intuition motivates performing the hypothesis test in the following manner:

Step 1. Start with the set of largest sub-aperture(s) at scale $m = 0$. Find the most likely hypothesis at that scale and denote this hypothesis as H_{0,i_0^*}.
Step 2. Consider those hypotheses at scale $m+1$ for which $S_{m+1,i_{m+1}} \subset S_{m,i_m^*}$. Find the one which has the highest likelihood and denote this hypothesis as H_{m+1,i_{m+1}^*}.
Step 3. If the parent is more likely (i.e. $\ell_{m,i_m^*} > \ell_{m+1,i_{m+1}^*}$), then stop and return H_{m,i_m^*} as the estimated hypothesis.
Step 4. If $m+1 = M$, we are at the bottom of the tree, so stop and return H_{m+1,i_{m+1}^*} as the estimated hypothesis. Otherwise, increment m and goto step 2.

This procedure provides an efficient approximation to finding the ML hypothesis without evaluating all the likelihoods. If $H_{m,i}$ is the hypothesis with the highest likelihood, then all that is required for it to be chosen by this scheme is that:

(D) At each scale $n < m$, the sub-aperture at scale n with the highest likelihood is an ancestor of $S_{m,i}$.
(E) The sequence of likelihoods increases as the branch is traversed until $S_{m,i}$ is reached.

Intuitively, these are reasonable conditions that one would usually expect to hold given the interpretation that our tests measure average energy over a sub-aperture. This intuition can be justified under the isolated scattering models. In particular, consider the expected value of $\ell_{n,j}$ when the true hypothesis is $H_{m,i}$ and the proportion of overlap between $S_{n,j}$ and $S_{m,i}$ is given by $\alpha = \frac{\lambda(S_{n,j} \cap S_{m,i})}{\lambda(S_{m,i})}$. For Eq. (6), in which the estimate $\hat{A} = q_{m,j}$ is used, the expected value of the GLLR is

$$
\begin{aligned}
\mathbb{E}\left[\ell_{n,j} \mid H_{m,i}\right] &= \mathbb{E}\left[\frac{1}{L_{n,j}} |q_{n,j}|^2 - |q_{0,0}|^2 \middle| H_{m,i}\right] \\
&= \left(\frac{\alpha^2}{L_{n,j}} - 1\right) |A|^2 + (2L_{n,j} + 1)\sigma^2 \\
&\approx \left(\frac{\alpha^2}{L_{n,j}} - 1\right) |A|^2 \qquad (9)
\end{aligned}
$$

where the approximation is for high SNR scattering. From this equation, we see the intuitive behavior described above. In particular, to maximize the GLLR at each scale $n < m$, we would choose a sub-aperture that maximizes α (producing $\alpha = 1$) and thus completely overlaps the true sub-aperture, i.e. an ancestor of the true sub-aperture $H_{m,i}$.

Furthermore, for all ancestors of $H_{m,i}$, the expected value of the GLLR increases with decreasing sub-aperture length, i.e. the sequence of expected log-likelihoods increases as the branch is traversed. When the hypothesized sub-aperture becomes too small, the overlap is $\alpha = L_{n,j}/L_{m,i}$, and the expected GLLR decreases. So, as intuition would suggest, the expected value of the GLLR (i) is maximized for the true hypothesis and (ii) obeys conditions (D) and (E) for Eq. (6) under high SNR.

Similarly, for Eq. (7), in which the estimate $\hat{A} = q_{0,0}/L_{m,i}$ is used, the expected value of the GLLR is

$$\begin{aligned}\mathbb{E}\left[\ell_{n,j} \mid H_{m,i}\right] &= \mathbb{E}\left[\frac{1}{L_{n,j}}|q_{n,j}|^2 - \frac{1}{L_{n,j}}|q_{0,0} - q_{n,j}|^2 - |q_{0,0}|^2 \,\middle|\, H_{m,i}\right] \\ &= \left(\frac{2\alpha - 1}{L_{n,j}} - 1\right)|A|^2 - \left(\frac{1}{L_{n,j}} - 1\right)2\sigma^2 \\ &\approx \left(\frac{2\alpha - 1}{L_{n,j}} - 1\right)|A|^2 \end{aligned} \tag{10}$$

where the approximation is for high SNR scattering. The behavior with respect to α and $L_{n,j}$ follow the same trend as for Eq. (9) and thus again supports our intuition for using the telescopic testing. Because of the difficult form of our regularized multiple scatterer log-likelihood given in Eq. (16), we have not shown that the same pattern holds for this extended test, however, intuition and the success demonstrated in the empirical results of Section 4 lead us to believe this behavior holds here as well.

Besides computational efficiency, performing the hypothesis tests in a telescopic fashion also enforces a form of consistency in the sub-apertures tested. In particular, in order for a specific sub-aperture to be chosen, its ancestor hypotheses have to be chosen in the testing at the previous scales. Enforcing such a progression of likelihoods protects against anomalies where a small sub-aperture could possess a high GLLR while none of its ancestors do, such as in the example associated with Figure 5.

3.4. Boxcar Model Deviations

The boxcar hypotheses defined in Eq. (2) are simplified models to which real scatterers do not exactly correspond. One may question whether deviations from this model may drastically effect our hypothesis tests. For example, if the scattering has a sinc(·) like dependence in azimuth, then the sidelobes[8] will have a large response for a strong scatter and may not be well modeled as additive in the measurement noise ε. To address this issue, we incorporate deviations from the boxcar response into our model. In particular, we start by assuming the underlying *normalized* scattering pattern has been perturbed by white Gaussian noise. For the isolated scatterer model, this modification changes the model response in Eq. (3) to

$$\tilde{b}(m,i)_j = \int_{S_{M,j}} 1_{S_{m,i}}(t) + \nu(t)\, dt$$

where $\nu(t)$ is a circularly complex white Gaussian process with spectral density $2\rho^2$ and is independent of the measurement noise $\eta(t)$. Thus, our modeled measurement vector is now a random vector characterized as

$$\tilde{\boldsymbol{b}}(m,i) \sim \mathcal{N}(\boldsymbol{b}(m,i), 2\rho^2\Lambda).$$

The resulting measurement model is then

$$H_{m,i} : \boldsymbol{q}_M = A\boldsymbol{b}(m,i) + w, \text{ with } w \sim \mathcal{N}(\mathbf{0}, 2(|A|^2\rho^2 + \sigma^2)\Lambda).$$

Thus, we have essentially the same model as in Eq. (5) except that the variance of the noise now depends affinely on the square-magnitude of the underlying scatterer. For consistency, the full-aperture reflectivity estimate is used as the estimate of A in the overall noise variance under each hypothesis. Thus, the statistics for all the hypotheses are scaled by the same factor, and the hypothesis that maximizes the likelihood is not changed by the incorporation of this boxcar deviation.

4. Experimental Analysis

In this section, we examine the behavior of the anisotropy characterization developed in the previous section by studying the dependence on anisotropic phenomenology. In particular, the extended test based on the multiple scatterer model (assuming full-aperture interfering scatterers) is used within the telescopic testing procedure[9]. We start our analysis using narrow-aperture data because of its prevalence in the SAR community. Specifically, we use the MSTAR public release data [6] and data synthetically generated according to MSTAR operational parameters, i.e. a 0.32m cross-range resolution (2.8° aperture), 9.6GHz center frequency, and 0.25m range resolution (590MHz bandwidth). We emphasize that our use of MSTAR parameters is to help promote intuitive understanding, and the results can always be scaled with respect to the aperture size for other settings. For our experiments, we use the three level, half-overlapping, half-aperture pyramid[10] depicted in Figure 2. The number of neighboring scatterers considered is set by $K = 6$ and their spacing is set by the MSTAR oversampling rate of $\Delta_r/\Delta_p = 1.25$. The value of the regularization parameter on neighboring reflectivities is set by $\gamma = 0.5$. Deviations from the boxcar model are accounted for by setting $\rho = 0.1$.

To first demonstrate that our test reveals anisotropic behavior in real data, we show in Figure 7 the anisotropy attributions for peaks extracted from a BMP-2 tank at 0°, 30°, 60°, and 90° azimuths and a 17° depression. Each image displays the log-magnitude reflectivities using darker shades to indicate stronger reflectivities. Peak locations and anisotropy characterizations are given by the symbol overlays in each image. A circle denotes a full-aperture classification; a square denotes a half-aperture classification; a triangle denotes a quarter-aperture classification. Even though the aperture associated with

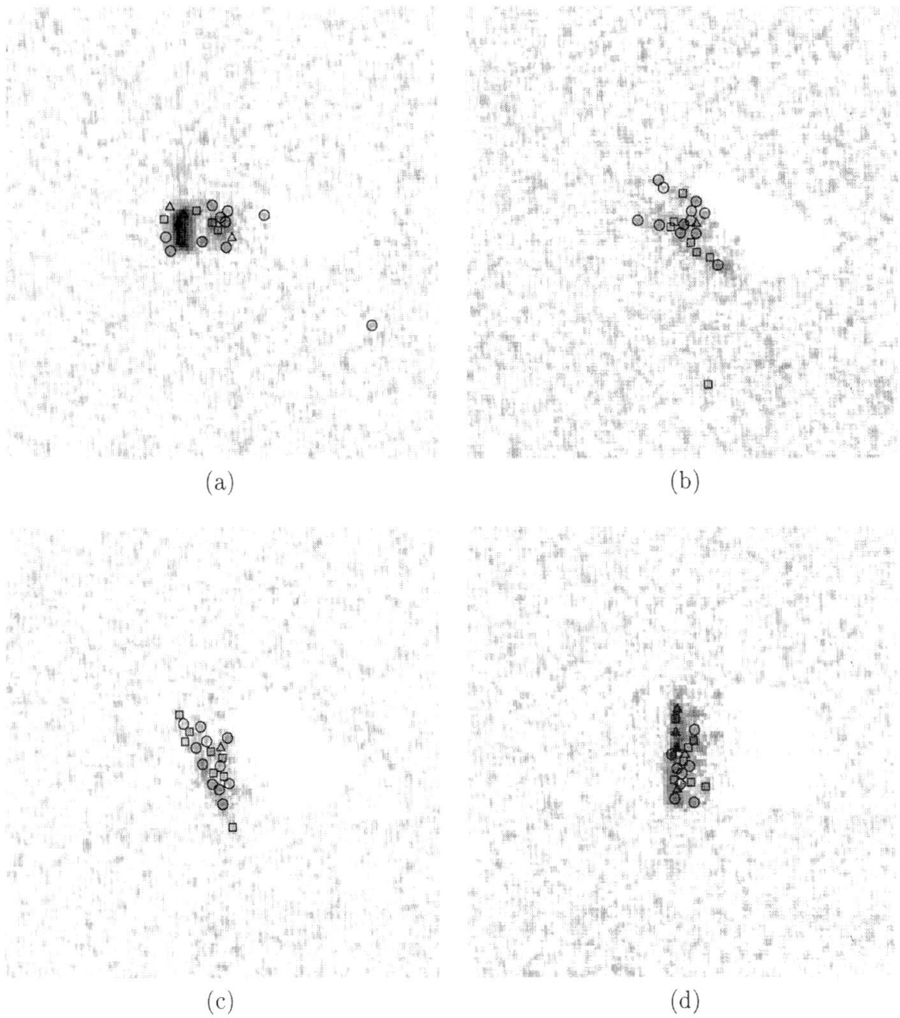

Figure 7. Anisotropy characterization of peaks extractions on a BMP-2 tank at (a) 0°, (b) 30°, (c) 60°, and (d) 90° azimuths.

this data set is relatively small at 2.8°, we note that we are still able to detect anisotropic scattering particularly at cardinal azimuthal orientations.

The images in Figure 7 show that scatterers are being classified as anisotropic; however, the images do not convey how useful that information is in characterizing targets. This topic is addressed by Kim, et al. [7] where anisotropy is used as an attribution in a peak-based Bayes classifier. One of the more interesting results in that work is that scatterer anisotropy is significantly more stable at near-cardinal target orientations than at off-

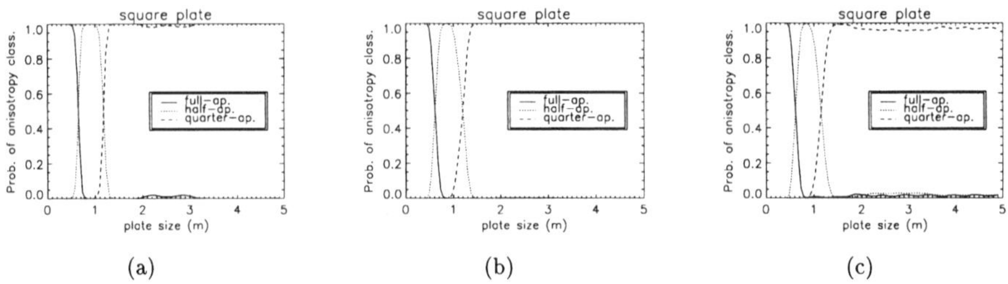

Figure 8. Anisotropy plots for a square plate in Gaussian noise at 20dB PSNR. (a) no azimuthal and location uncertainty; (b) maximal azimuthal uncertainty and no location uncertainty; (c) maximal azimuthal uncertainty and location uncertainty given by peak extraction.

cardinal orientations which argues for at least two distinct sources of anisotropy. In particular, we expect geometric scattering to dominate the near-cardinal orientations and produce consistent anisotropy attributions but volumetric scattering to dominate at off-cardinal orientations and produce erratic anisotropy attributions. To verify this intuition, we separately study our proposed anisotropy characterization for the canonical flat plate and volumetric scatterers using synthesized data. In both cases, we use what we call an anisotropy plot to convey the dependence of our anisotropy classification on an underlying parameter. In the case of the canonical flat plate, this parameter is the plate size, and in the case of the volumetric scatterer, this parameter is the distribution of constituent scatterers.

4.1. *Geometric Anisotropy*

To examine geometric anisotropy, we synthesize azimuthal data for the flat plate according to the physical optics model using MSTAR parameters. To generate the associated anisotropy plots, we use probability estimates based on Monte Carlo runs involving 8192 trials for each plate size where the scatterer is taken at broadside and white Gaussian noise is added across the aperture. The PSNR of the signal is kept constant at PSNR = 20dB, thus noise variance scales with the maximum peak reflectivity.

The anisotropy plot under these settings is shown in Figure 8(a). Contained in this figure are three curves each of which represents the probability of a particular anisotropy classification conditioned on the size of the scatterer. Thus, the three values for a given plate size represent the PMF of these anisotropy classifications (i.e. full, half, and quarter-aperture) conditioned on that plate size. As expected, the full-aperture classification dominates for the smaller plate sizes. As plate size is increased, the half-aperture classification starts to dominate. Between 0.8m and 1m, the half-aperture classification probability is approximately one. The noiseless azimuthal response for a plate size in the middle of this region (0.9m) is shown in Figure 9(a) which is quite reasonably classified as half-aperture. The solid curve in the figure represents the real component of the response, and the dashed curve represents the imaginary component. The vertical lines are a visual aid dividing the aperture into eighths. As plate size is increased beyond 1m, the classification transitions from half-aperture to quarter-aperture in the anisotropy plot of

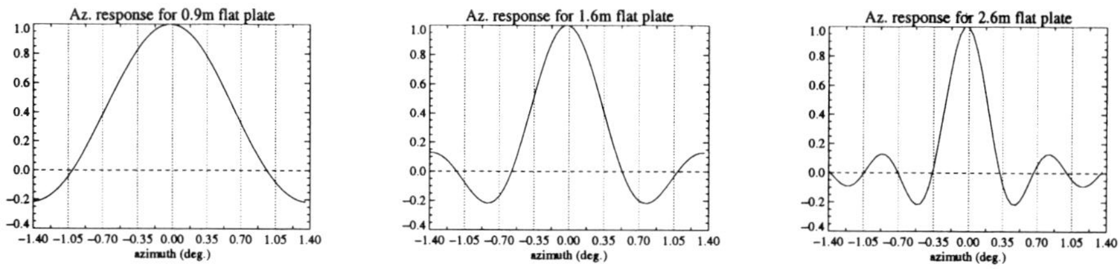

(a) (b) (c)

Figure 9. Azimuthal response for different plate sizes: (a) 0.9m, (b) 1.6m, and (c) 2.6m.

Figure 8(a). Between 1.3m and 1.9m, the quarter-aperture classification probability is approximately one. The azimuthal response for a plate size in the middle of this region (1.6m) is shown in Figure 9(b) which also agrees with what one would expect. Intuition would suggest that the probability of anisotropy classification would be unimodal and in particular that the probability of the quarter-aperture classification would be monotonically increasing with plate size (since quarter-aperture anisotropy is the most anisotropic classification available), however, this is not the case. In particular, starting at about 1.9m, the quarter-aperture classification rate begins to decrease. The middle of the range where the rate drops is about at 2.6m. The azimuthal response for this plate size is shown in Figure 9(c). Based on this response, we would expect the scatterer to be classified as quarter-aperture anisotropy. In fact, the null-to-null bandwidth is a quarter of an aperture. So, the question is why is this scatterer occasionally classified as full-aperture scattering. This result is an artifact from the scale-telescopic hypothesis testing. To see this point, observe the GLLR's for the noiseless signal given in the first column of Table 1. From these values, we see that the quarter-aperture hypothesis does in fact produce the highest GLLR, but because the full-aperture hypothesis is nearly as likely as the half-aperture hypothesis, the quarter-aperture hypothesis is not always not tested. To gain a better understanding of why the full-aperture hypothesis is nearly as likely as the half-aperture hypothesis, consider the GLLR for our simple test that does not consider neighboring scatterers, i.e. the one which corresponds to Eq. (7), that compares normalized sub-aperture energies. For the 2.6m plate, the second zero-crossings of the sinc-like azimuthal response are located such that they minimize the reflectivity in the center half-aperture, i.e. the first sidelobes maximally negate the influence of the mainlobe. Thus, the middle half-aperture reflectivity is reduced and produces a correspondingly lower GLLR for this hypothesis. Similarly, the quarter-aperture classification rate "dips" at plate sizes near multiples of 2.6m as the other "negative" sidelobes have a locally maximal negative effect on the middle half-aperture[11]. This problem could easily be alleviated by relaxing the hard decision made at each scale in the telescopic hypothesis test. However, we do not use such a test here because as is discussed next, this problem is not as significant in a more realistic setting that incorporates the presence of azimuthal uncertainty.

The anisotropy plot in Figure 8(a) is for when the plate is oriented exactly broadside to the radar. To incorporate the effects of azimuthal uncertainty, we modify the setup as follows. We generate SAR data with the same parameters as before except that the aperture

Table 1. GLLR's for noiseless 2.6m plate response (assuming an PSNR of 20dB).

	Without azimuthal uncertainty	With azimuthal uncertainty
full-aperture	0	0
half-aperture	0.60	1.0–2.3
quarter-aperture	8.4	4.2–7.5

size is increased by a factor of three, i.e. data is generated over a 8.4° aperture. From this extended aperture, we randomly take a section measuring 2.8° from a uniform distribution whose support is given by the following cases to maximize the azimuthal uncertainty subject to observing the mainlobe response.

- If the plate size is less than two wavelengths (i.e. the plate size is less than 6.25 cm), then the scatterer is considered isotropic and any 2.8° section of the extended aperture can be selected,
- If the plate size is larger than two wavelengths but still small enough such that the half-power bandwidth of the azimuthal response is larger than 2.8°, then any section of the extended aperture within the half-power band can be chosen, and
- If the plate size is sufficiently large such that the half-power bandwidth of the azimuthal response is smaller than 2.8°, then any section of the extended aperture containing the half-power band can be chosen.

The resulting anisotropy plot is shown in Figure 8(b). Interestingly, performance is improved for larger plate sizes by the addition of azimuthal uncertainty. Looking at the GLLR's in the second column of Table 1 for the 2.6m plate (assuming 20dB PSNR), we see that it is *not* that the quarter-aperture likelihood increases (in fact it decreases slightly to typically vary between 4.2 and 7.5) but that the half-aperture likelihood increases to typically vary between 1.0 and 2.3. This GLLR increases because the exact broadside orientation represents a worst-case scenario for the half-aperture hypothesis (relative to the full-aperture hypothesis) due to the maximal negative effect that the first (and other odd number) sidelobes have on the middle half-aperture reflectivity as previously described. Azimuthal uncertainty, however, perturbs this degenerate alignment.

In addition to azimuthal uncertainty, we can also observe the effect of location uncertainty which we simulate by using a peak extraction step[12] to determine the scatterer location. The resulting anisotropy plot is shown in Figure 8(c). Not surprisingly, there is slight degradation in performance, particularly for larger plate sizes. This degradation is due to the fact that large plates span several resolution cells over which their reflectivity is approximately the same. Thus, the peak extraction is susceptible to noise over these resolution cells and frequently reports a location that, while on the scatterer, does not correspond to the scattering center. The resulting peak location error introduces a modulation on the observed azimuthal response which distorts the anisotropy classification. In particular, for larger apertures, the linear phase on the azimuthal response prevents the sub-aperture reflectivity estimates from reaching their true value due to the destructive interference that occurs when integrating over the sub-apertures. Thus, the energy from larger sub-apertures is more heavily biased towards smaller values. Smaller sub-apertures, on the other hand, are less susceptible to this phase variation due to the shorter integration interval and thus give stronger reflectivity estimates. This interference is the reason that errors in scatterer location tend to bias the anisotropy classification towards higher degrees of anisotropy.

The anisotropy plots in Figure 8 show that even in the presence of azimuthal and location uncertainty, our anisotropy measure exhibits the strong dependence of anisotropy on plate size that is predicted by canonical scattering models. Furthermore, we see that the anisotropy classifications are consistent with these models as illustrated in Figure 9. One would naturally expect the dependencies in the anisotropy plot to be preserved as the aperture size is increased. As we discuss in Section 4.3, this is in fact true, but this seemingly trivial property turns out to be quite significant when contrasted to the behavior of volumetric anisotropy which we discuss next.

4.2. Volumetric Anisotropy

Similar to the anisotropy plots for the flat plate scatterer, we generate anisotropy plots for volumetric scatterers where the dependent variable corresponds to the "size" of the scatterer which in this case is the spatial support of the pdf generating the individual isotropic scatterers. The radar and sub-aperture parameters are the same as those used for the geometric anisotropy plots. For each Monte Carlo sample, a volumetric scatterer is taken as a collection of random point scatterers generated according to the following parameters:

- The density of the scatterers in terms of the average number of scatterers per resolution cell.
- The mean and variance of the normal distribution producing the real-valued reflectivity of the scatterers. Both of these values have been set to 1, and the distribution is truncated to prevent negative values.
- The support for the uniform distribution producing the down-range and cross-range location of the scatterers. The support of the pdf producing the down-range location of the scatterers is taken as one resolution cell. The support of the pdf producing the cross-range location is the dependent variable in the anisotropy plot.
- The azimuthal uncertainty which is produced using a uniform pdf with a support of $[-0.7^\circ, 0.7^\circ]$, i.e. shifts up to a quarter of the aperture are allowed.
- The PSNR of the collective scattering which is taken to be 20dB.

In Figure 10, we show the anisotropy plots for scattering densities of 1, 10, and 100 scatterers per resolution cell or equivalently 0.32, 3.2, and 32 scatterers per meter in cross-range under a 2.8° aperture. From these plots, it is immediately clear that the behavior of volumetric anisotropy is heavily dependent on the spatial density of the constituent scatterers. In particular, the plots go from saying that anisotropy is independent of size for low scattering densities to saying that anisotropy is highly dependent on

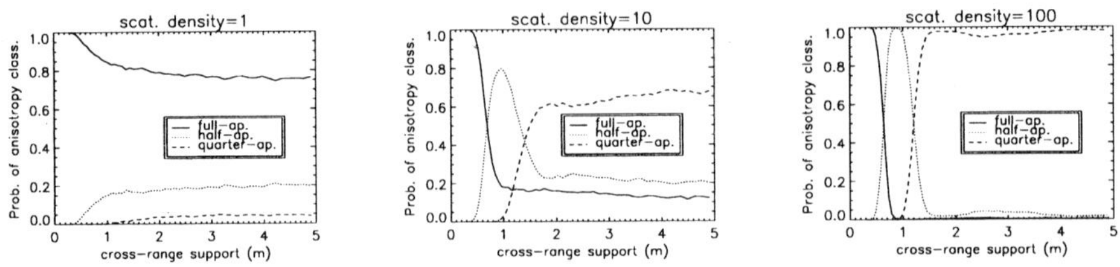

(a) (b) (c)

Figure 10. Anisotropy plots for volumetric scatterers with scattering density of (a) 1, (b) 10, and (c) 100 scatterers per resolution cell.

the size at densities sufficiently high that the aggregation resembles a flat plate. The behavior in these plots is quite reasonable. For an average of one scatterer per resolution cell, we would expect the model to usually be able to separate the individual isotropic scatterers, thus producing the full-aperture classification. However, this is not always the case because scatterers may occasionally be considerably closer than one resolution cell apart or the peak extraction may cause the algorithm to focus to the wrong location. For a scattering density of 100 scatterers per resolution cell, there are so many scatterers drawn from the uniform distribution that the aggregation very closely resembles a flat plate of the same size as the support. Deviations from the anisotropy plot for the flat plate can be attributed to

1. the scatterers are not regularly spaced since their locations are random and
2. the amplitudes on the scatterers are random.

The anisotropy plot for a density of 10 scatterers per resolution cell is a transitional phase between the two extreme cases of one and 100 scatterers per resolution cell.

From the anisotropy plots in Figure 10, we conclude that low density volumetric scattering is preferable to high density scattering because the former has a lower incidence of volumetric anisotropy attributions. This effect allows for anisotropic classifications to be more confidently associated with geometric scatterers from which reliable information can be inferred. One may at first be inclined to believe that the density of volumetric scattering is a physical parameter over which we have no control, but this is not the case. The scattering density in our analysis is with respect to the number of scatterers per resolution cell, and the size of the resolution cell is inversely related to the size of the aperture. Thus, by using wide-aperture data, we should be able to reduce volumetric scattering density and hence the detection rate for volumetric anisotropy as discussed in the next section.

4.3. *Extension to Wide-Aperture Data*

The results in Figures 8 and 10 are promising in that they imply that by increasing the size of the aperture, one can preserve the detection rate for geometric anisotropy classifications while reducing the rate of volumetric anisotropy classifications. This effect allows for a devoted analysis of predictable anisotropic geometric scatterers in an ATR algorithm.

For low density volumetric scattering (per resolution cell), Figure 10 shows that the anisotropy classification is usually full-aperture regardless of the cross-range support. However, at medium to high scattering densities, the number of anisotropy classifications increase. Wide-aperture data should thus reduce the number of anisotropy classifications because increasing the length of the aperture reduces the scattering density in terms of the number of scatterers per resolution cell. For example, increasing the aperture size by a factor of ten decreases the resolution cell size and hence scattering density by the same factor. From the anisotropy plots, we see that such a factor should have a significant impact on volumetric anisotropy classification. Geometric scatterers, however, can be considered to be composed of a continuum of infinitely small scatterers (with respect to practical imaging resolutions), so increasing the resolution by an order of magnitude does not detract from the high scatterer density thus preserving the structure of the anisotropy plot.

The above is an image (or spatial) domain motivation for why wide-aperture data would reduce the rate of volumetric anisotropy while preserving the rate of geometric anisotropy, but there is also an azimuthal domain interpretation. Consider a scatterer which is declared anisotropic based on narrow-aperture data. This classification is made because most of the energy is concentrated in one region of the aperture. When the size of the available aperture is increased for volumetric scatterers, there is likely to be a considerable amount of energy in the newly appended sub-apertures because (as depicted in Figure 1(b)) the underlying response is not truly unimodal. However, geometric scatterers with a unimodal azimuthal response should have little energy in the appended sub-apertures. Furthermore, many anisotropic geometric scatterers exhibit sinc-like oscillations in the appended sub-apertures (which nearly integrate to zero) while many volumetric scatterers do not. Thus, for volumetric scatterers there is likely to be considerable energy in newly appended sub-apertures which may change the anisotropy classification, but this is not the case for geometric scatterers.

To verify this conjecture empirically, we generate anisotropy plots for the geometric and volumetric scatterers in the same fashion as before except that the aperture size is increased by a factor of eight. This modification naturally changes the sub-aperture pyramid we use. To be consistent with the narrow-aperture results, we use a three level, half-overlapping, half-aperture pyramid[13] which consists of 1/8 apertures at scale 0, 1/16 apertures at scale 1, and 1/32 apertures at scale 2. These hypotheses correspond to the full, half, and quarter-apertures in the narrow-aperture case and are thus consistent in terms of the physical size of the underlying scatterer. To prevent any confusion between the wide and narrow-aperture cases, we will refer to these hypotheses with their measurements in degrees, i.e. 2.8°, 1.4°, and 0.7°. To be consistent in the geometric and volumetric settings, azimuthal uncertainty in both cases is modeled with a uniform distribution over 1/4 of the narrow-aperture, i.e. 0.7°. For consistency in the noise level between the narrow and wide-aperture settings, we use the same noise spectral density (in the azimuthal domain) as in the narrow-aperture case. Thus, for isotropic scatterers, the wide-aperture PSNR is actually higher than in the narrow-aperture case, and for anisotropic scatterers, the PSNR is lower due to the coherent averaging involved.

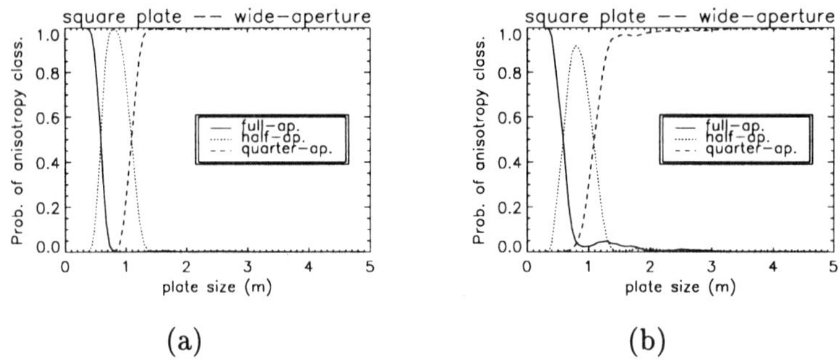

Figure 11. Anisotropy plots for the canonical flat plate scatterer using (a) coarse-resolution peak extraction and (b) high-resolution peak extraction.

The wide-aperture geometric anisotropy plot is displayed in Figure 11. The two plots in Figure 11 correspond to two different peak extraction methods. Figure 11(a) is for coarse-resolution peak extraction which is done using the center 2.8° narrow-aperture. Figure 11(b) is for high-resolution peak extraction which is done using the full-aperture. As predicted, Figure 11(a) looks quite similar to the narrow-aperture anisotropy plot in Figure 8(c) showing a strong dependence on the underlying plate size. Figure 11(b) also looks similar to Figure 8(c), but the classification rates for 1.4° and 0.7° anisotropy have unexpectedly dropped for larger plate sizes. One can also see that the "transition band" between anisotropy classifications is wider for the high-resolution peak extraction. This effect is due to the fact that in the high-resolution setting, the plate sizes corresponding to the degrees of anisotropy that we are testing span several resolution cells. Over these resolution cells, the reflectivity is nearly constant resulting in a peak location which comes from a uniform distribution over the size of the scatterer. The resulting location error for the scattering center induces a modulation over the azimuthal response which distorts the anisotropy classification as discussed in Section 4.1. This problem is not significant for the coarse-resolution peak extraction because the coarse-resolution cell size is the same order of magnitude as the plate sizes. Thus, the spatial averaging usually covers a significant portion of the scatterer and is more likely to return a peak location near the center of the flat plate.

We now consider the behavior of volumetric scatterers in the wide-aperture setting. The nominal scattering density that we examine is 10 scatterers per coarse-resolution cell or equivalently 1.25 scatterers per resolution cell in the high-resolution setting. Figure 12 displays the wide-aperture anisotropy plots. Figure 12(a) is for the coarse-resolution peak extraction using the center 2.8° narrow-aperture, and Figure 12(b) is for the high-resolution peak extraction using the full-aperture. The volumetric anisotropy plot in Figure 12(b) behaves as conjectured, bearing a close resemblance to the anisotropy plot for a scattering density of one scatterer per resolution cell in the narrow-aperture setting. However, Figure 12(a) shows virtually no reduction in the rate of anisotropy classifications. Recalling the image domain rationale that wide-aperture data reduces volumetric anisotropy because we are better able to resolve constituent scatterers, this result is not

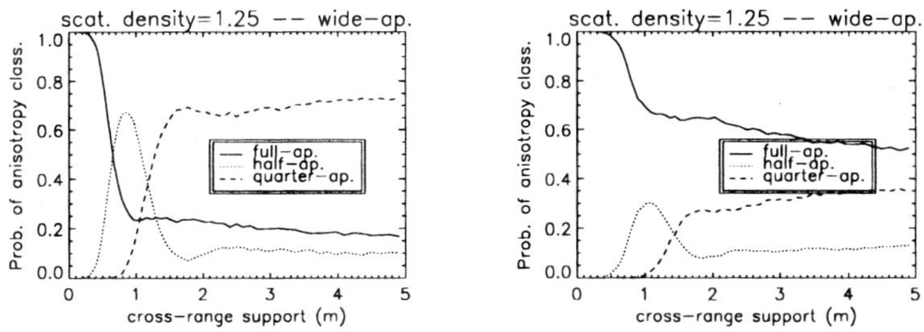

(a) (b)

Figure 12. Anisotropy plots for volumetric scatterers with scattering density of 10 scatterers per coarse-resolution cell using (a) coarse-resolution peak extraction and (b) high-resolution peak extraction.

surprising. When we use the coarse-resolution peak extraction, we are not focusing on peaks that could be resolved in the high-resolution regime thus disregarding that information which would benefit us.

The dependence on the peak extraction process can be summarized by saying that the coarse-resolution peak extraction preserves the anisotropy classification rates for both geometric and volumetric scatterers, while high-resolution peak extraction preserves the classification rate for geometric scatterers to a lesser degree but significantly reduces the rate for volumetric scatterers. However, by combining the coarse and high-resolution peak extractions, we should be able to obtain the desired behavior for both geometric and volumetric anisotropy. In particular, we can select which peak extraction method to use based on scatterer size inferred from the anisotropy attribution. Recall that the high-resolution peak extraction degrades the geometric anisotropy plot because the extraction provides a worse peak location estimate for larger plate sizes which causes the degree of anisotropy to be over-estimated. Because this effect only occurs for larger plate sizes and higher degrees of anisotropy are associated with larger scatterers, we know that in this situation, it can only be beneficial to repeat the anisotropy attribution of the scatterer using the coarse-resolution peak location which gives a more accurate estimate of the scattering center. Thus, we first use the peak location from the high-resolution peak extraction, and if the extraction is declared anisotropic (conveying that there is likely to be a large geometric scatterer there), then we defer to the peak location from the coarse-resolution peak extraction and use its associated anisotropy classification[14]. Thus, using this approach, the high rate of minimally anisotropic 2.8° classifications is preserved for volumetric scatterers since they do not defer to the coarse-resolution peak extraction. The 1.4° and 0.7° classifications for geometric scatterers, however, are also largely preserved since their initial anisotropy classification is usually over-estimated and thus defers to the location estimate given by the coarse-resolution peak extraction.

The anisotropy plots using this combination of coarse and high-resolution peak extractions are given in Figure 13. Not surprisingly, this approach is not better than using coarse-resolution for the geometric scatterers nor better than only using the high-resolution for the volumetric scatterers, but this technique offers a good trade-off resulting in an

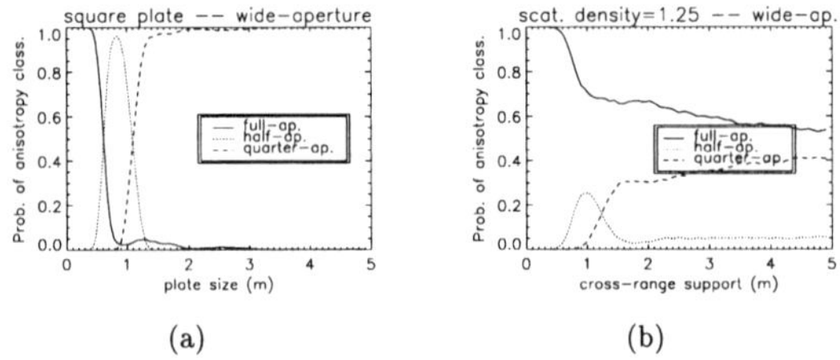

Figure 13. Anisotropy plots using both the high and coarse-resolution peak extractions for the (a) geometric flat plate scatterer and (b) volumetric scatterer with 10 scatterers per coarse-resolution cell.

algorithm which works well on both geometric and volumetric scatterers. Thus, as the reasoning at the beginning of this section implied, by using wide-aperture data, we can significantly reduce the rate of volumetric anisotropy while preserving the rate of geometric anisotropy.

Up to now, for clarity in the comparison of narrow and wide-aperture anisotropy attributions, we have use the same hypothesis set of $\{2.8°, 1.4°, 0.7°\}$ sub-apertures (which we call the *restricted hypothesis set*) for both cases. However, in addition to reducing the rate of volumetric anisotropy (while preserving that of geometric anisotropy), wide-aperture data allows for additional anisotropy hypotheses up to the size of the larger aperture. These additional anisotropy hypotheses do not effect the argument for reducing the rate of volumetric anisotropy, however, they does allow for a more refined anisotropy characterization of smaller scatterers by the corresponding additional degrees of anisotropy. To demonstrate this point, we show in Figure 14 the geometric anisotropy plots for the flat plate scatterer with the *augmented hypothesis set* of $\{22.5°, 11.3°, 5.6°, 2.8°, 1.4°, 0.7°\}$ sub-apertures from the half-overlapping, half-aperture pyramid. The domain of plate sizes here is reduced to 0m–2.5m to better illustrate the behavior for small scatterers. The behavior exhibited is a natural extension of the geometric anisotropy plots in Figures 11 and 13(a). In particular, the only significant effect is for the 2.8° classification which is the

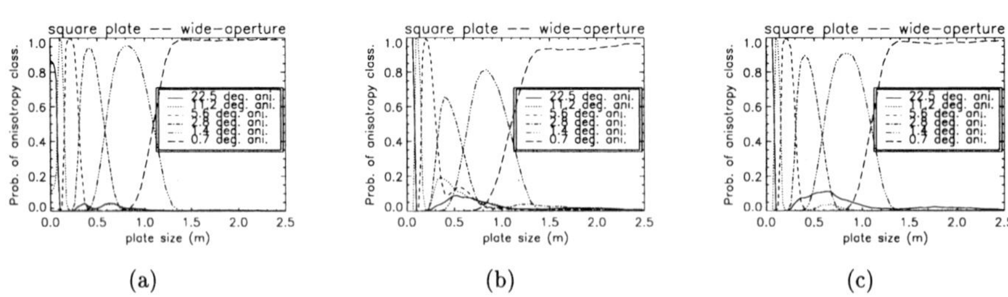

Figure 14. Anisotropy plots for the canonical flat plate scatterer using the augmented hypothesis set and (a) coarse-resolution peak extraction, (b) high-resolution peak extraction, (c) combination of coarse and high-resolution peak extractions.

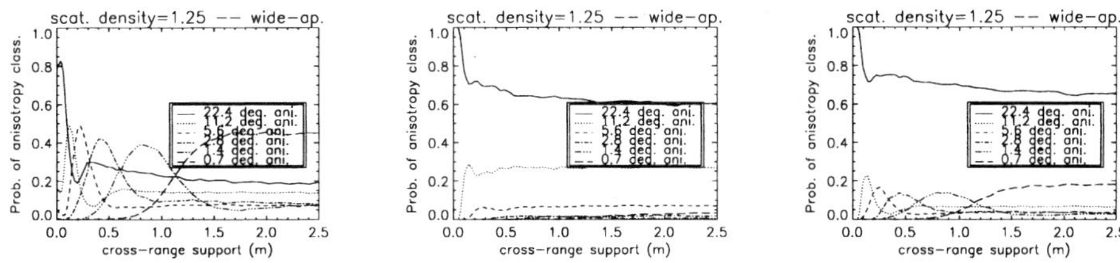

(a) (b) (c)

Figure 15. Anisotropy plots for volumetric scatterers with scattering density of 10 scatterers per coarse-resolution cell using the augmented hypothesis set and (a) coarse-resolution peak extraction (b) high-resolution peak extraction (c) combination of coarse and high-resolution peak extractions.

least anisotropic classification in the restricted hypothesis set. Under the new augmented hypothesis set, scatterers which were previously characterized as 2.8° are now classified as 22.5°, 11.3°, 5.6°, or 2.8°. The 22.5° classification dominates the smallest plate sizes, and the dominant attribution transitions to 11.3°, 5.6°, and 2.8° as plate size is increased. Interestingly the 22.5° classification probability starts out noticeably lower for the coarse-resolution peak extraction than for the other two methods. The reason for this error is that for the smaller plate sizes, the coarse-resolution peak extraction introduces location errors on the order of size of the coarse-resolution cell. As previously noted, location error introduces a modulation on the observed azimuthal response which biases the classification result to higher degrees of anisotropy. However, since this location error is small, the only effect is to misclassify some of the small plates as 11.5° instead of 22.5°. Note that in the combined coarse and high-resolution approach, the vast majority of these small plates are characterized as 22.5°. This correct attribution is declared because the anisotropy attribution from the high-resolution peak extraction is analyzed before the coarse-resolution extraction, and thus, the decision is not deferred to the coarse-resolution extraction.

In Figure 15, we show the volumetric anisotropy plots using the augmented hypothesis set. Again, the behavior of each plot is a natural extension of Figures 12 and 13(b). In particular, the coarse-resolution peak extraction shows a high rate of anisotropic classifications, but the high-resolution extraction allows for the scatterers to be frequently resolved into individual isotropic scatterers. The combination of the two methods preserves the separation of interfering scatterers to suppress the rate of volumetric anisotropy.

4.4. Collected Wide-Aperture Data

In this section, we apply our anisotropy attribution to peaks extracted from measured wide-aperture data. The data has a resolution of 1.9 inches in down and cross-range achieved using a 4GHz bandwidth, a 10GHz center frequency, and a 26° aperture. For narrow-aperture comparisons, we take a 3.3° section of this aperture (i.e. 1/8 of the wide-aperture) as our narrow-aperture data. For each narrow-aperture peak extraction, we apply our anisotropy test with the hypothesis set of {3.3°, 1.6°, 0.8°} sub-apertures whose elements correspond to the full, half, and quarter narrow-apertures. For the wide-aperture data, we use the combined coarse and high-resolution peak extraction method described in the previous section. Because the wider aperture allows for more anisotropy hypotheses to be defined, we have several choices for the hypothesis set in this regime. We examine two particular sets. The first is the restricted hypothesis set of {3.3°, 1.6°, 0.8°} sub-apertures used in the narrow-aperture case. The other set is the augmented hypothesis set obtained by taking all the sub-apertures in the half-overlapping, half-aperture pyramid for the wide-aperture. The sub-aperture sizes in the augmented hypothesis set are thus 26°, 13°, 6.5°, 3.3°, 1.6°, and 0.8°. In Figures 16 and 17, we show the log-magnitude imagery at MSTAR resolution[15] with extracted peaks and their associated anisotropy attribution for two different vehicles. The symbol overlays in Figures 16(a) and 17(a) specify the location and anisotropy attribution of each peak extraction based on narrow-aperture data. In these figures, a black circle is used to denote 3.3° anisotropy, a square is used to denote 1.6° anisotropy, and a triangle is used to denote 0.8° anisotropy. The symbol overlays in Figures 16(b) and 17(b) again specify peak location and anisotropy attribution, except they are now based on wide-aperture data. To denote the additional wide-aperture 26°, 13°, and 6.5° hypotheses, we again use a circle, square, and triangle, respectively, but display them in white instead of black.

For each of the vehicles, we see that peak extractions that are labeled as the same (or nearly the same) degree of anisotropy in the narrow and wide-aperture settings can frequently be associated with an apparent geometric scatterer. In particular, we note in Figure 16(b) the collections labeled "A" and "B" appear to be large geometric scatterers

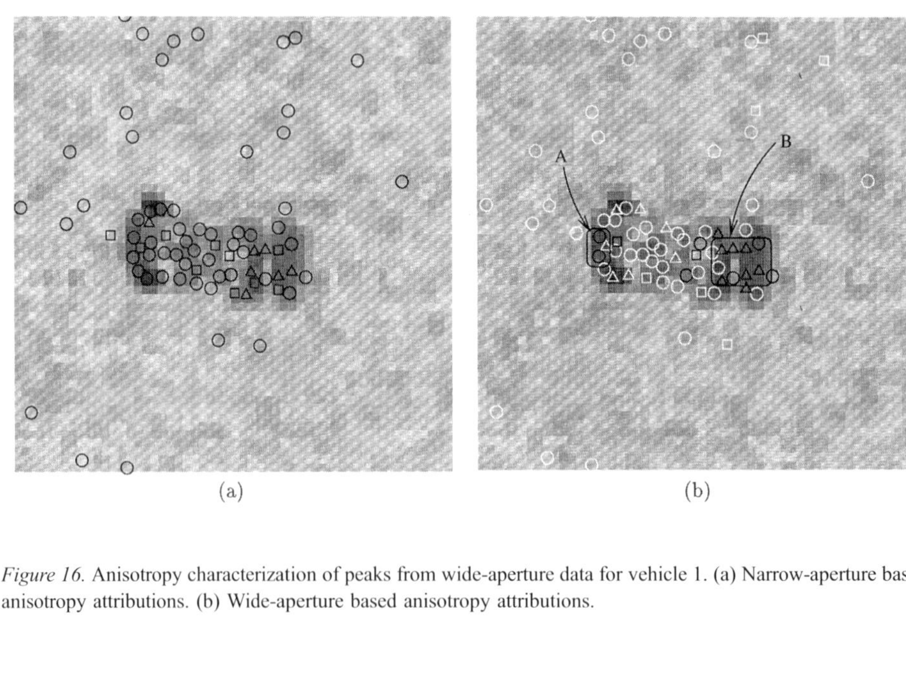

(a) (b)

Figure 16. Anisotropy characterization of peaks from wide-aperture data for vehicle 1. (a) Narrow-aperture based anisotropy attributions. (b) Wide-aperture based anisotropy attributions.

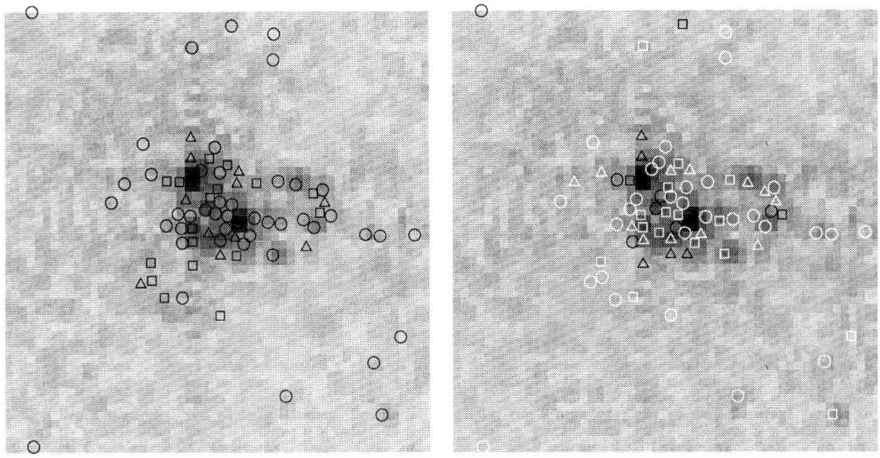

(a) (b)

Figure 17. Anisotropy characterization of peaks from wide-aperture data for vehicle 2. (a) Narrow-aperture based anisotropy attributions. (b) Wide-aperture based anisotropy attributions.

at the front and rear of the vehicle and are appropriately characterized as anisotropic. Note that based on the narrow-aperture data, the two scatterers labeled "A" are attributed as 3.3° in Figure 16(a), but there is no means of telling if the underlying scatterers are actually 3.3° or perhaps less anisotropic since 3.3° is the lowest degree of anisotropy available in the narrow-aperture setting. However, with the augmented hypothesis set for wide-aperture data, we see that in fact they are both approximately 3.3°, and not much more. This result demonstrates how the additional hypotheses afforded by the larger aperture allow more anisotropy information to be conveyed for smaller scatterers. We also point out that in Figures 16 and 17, those peak extractions which do not have the same anisotropy attribution in the narrow and wide-aperture regime are frequently not identifiable with a geometric scatterer and have a less anisotropic classification (usually the minimally anisotropic 26° classification) in the wide-aperture setting.

So far, we have observed the conjectured benefit of wide-aperture data over narrow-aperture data on the anisotropy characterization of geometric scatterers, i.e. the preservation of higher degrees of anisotropy in the narrow-aperture data and elaboration of the least anisotropic narrow-aperture hypothesis by augmenting the hypothesis set with larger sub-aperture hypotheses. This result is predicted by the analysis presented earlier. That analysis also predicted a reduction in the rate of volumetric anisotropy as aperture size is increased. To test this behavior with real data, we use the wide-aperture data focused on clutter using 26° and 3.3° apertures as before. In particular, we examine the three clutter scenes composed of trees and field shown in Figure 18 where the narrow-aperture anisotropy characterization is displayed in the left-hand side and the wide-aperture characterization is displayed in the right-hand side. The set of hypotheses and respective graphical labels are the same as in Figures 16 and 17. It is immediately clear that there is a significant

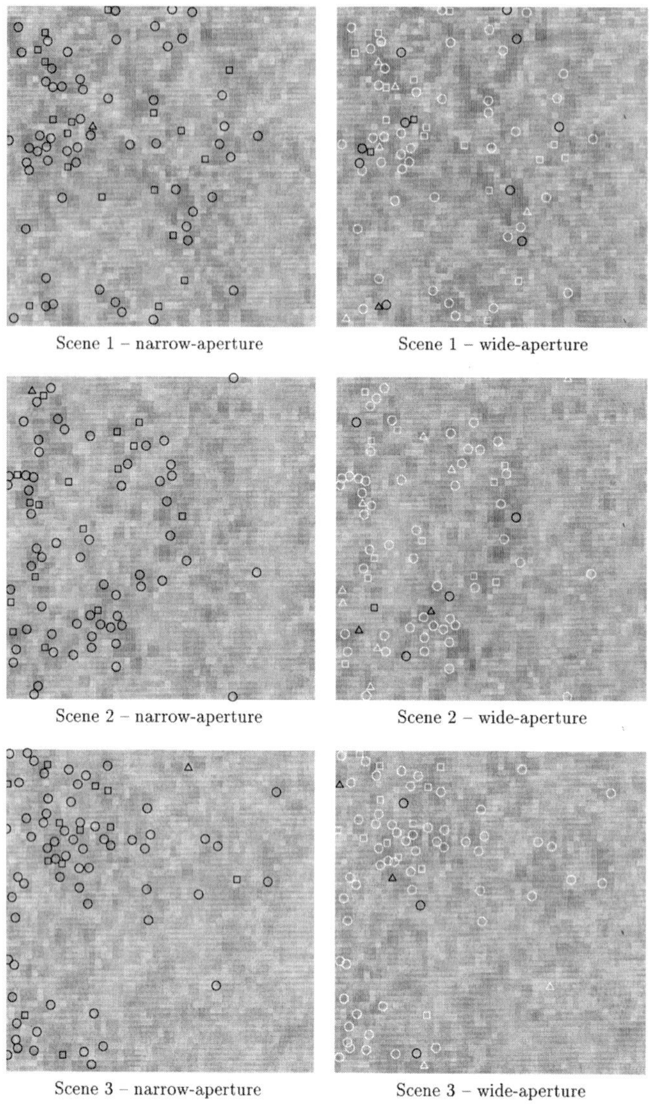

Figure 18. Anisotropy characterization of peaks from 3 instances of clutter in the wide-aperture data. Left: Narrow-aperture anisotropy attributions. Right: Wide-aperture anisotropy attributions.

reduction in the rate of higher degrees of anisotropy using wide-aperture data. To explicitly see that this reduction is due to the volumetric scatterers being resolved into individual isotropic scatterers (in contrast to their simply being independently distributed over the larger hypothesis set), we show in Tables 2 and 3 the tabulated anisotropy correspondences

Table 2. Anisotropy classification correspondences for clutter with narrow and wide-aperture data using the same anisotropy hypotheses.

Narrow-ap.\Wide-ap.	3.3°	1.6°	0.8°
3.3°	176	0	0
1.6°	45	0	1
0.8°	3	0	0

Table 3. Anisotropy classification correspondences for clutter with narrow and wide-aperture data using the augmented hypothesis set for wide-aperture data.

Narrow-ap.\Wide-ap.	26°	13°	6.5°	3.3°	1.6°	0.8°
3.3°	117	28	10	15	3	3
1.6°	29	8	5	2	0	2
0.8°	1	2	0	0	0	0

between the two cases. Table 2 shows the correspondences between the narrow and wide-aperture anisotropy classifications when both are restricted to use the same set of narrow-aperture hypotheses. The empirical results here strongly argue that the wider aperture has resolved the scatterers which were attributed as 1.6° and 0.8° under the narrow-aperture into individual minimally anisotropic 3.3° scatterers. Table 3 shows the same as Table 2 except that the correspondences here are for when the augmented hypothesis set is used for the wide-aperture data. Here we see that many of the 3.3° anisotropy classifications are further resolved into the less anisotropic classifications and the majority of them are assigned to the 26° minimal degree of anisotropy. Thus, we see the predicted reduction in the rate of volumetric anisotropy in real wide-aperture data.

5. Conclusions

We have proposed a general characterization of anisotropy based on concentrated scattering in azimuth. A sub-aperture pyramid is used to generate a tree of multi-resolution images at a variety of cross-range versus azimuthal resolutions allowing for the detection of anisotropic phenomena. With each sub-aperture in the pyramid, we associate a hypothesis that the azimuthal scattering is confined to and uniform over that sub-aperture. This model then leads to a sequence of hypothesis tests to characterize anisotropy which can be approximated with an efficient pruning algorithm due to the tree-structure over the sub-apertures.

This characterization of anisotropy allows us to explore some of the underlying phenomenology behind anisotropic scattering. In particular, we study the canonical flat plate and volumetric scatterer. Anisotropy plots clearly show a strong observability of the inverse relation between plate size and degree of anisotropy that is predicted by physics-based models. Anisotropy is also strongly dependent on the spatial distribution of a dense cluster of isotropic scatterers composing a volumetric scatterer, however, as the density of scatterers is reduced, this dependence decreases as the scatterers are more frequently resolved into constituent isotropic scatterers. The dependence of volumetric anisotropy on the density of the scatterers leads to the attribute that the probability of volumetric anisotropy classifications can be reduced by increasing the size of the aperture. The only significant effect on geometric anisotropy of extending the aperture is related to additional hypotheses (corresponding to larger sub-apertures) in the augmented hypothesis set. In particular, the larger sub-aperture hypotheses allow for a more accurate characterization of smaller scatterers which exhibit broader (but still anisotropic over the wide-aperture) azimuthal energy concentrations.

Besides demonstrating results on synthesized data, our anisotropy analysis is applied to real data. Peaks are extracted from data and attributed with a degree of anisotropy. The behavior of the peak attribution is consistent with the results of the synthetic data. In particular, the anisotropy characterization of geometric scatterers is consistent between narrow and wide-aperture settings when the underlying anisotropy is included in both hypothesis sets. Furthermore, in the wide-aperture setting with the augmented hypothesis set, we obtain a more detailed anisotropy characterization for smaller scatterers. A reduction in the rate of volumetric anisotropy with larger apertures is also observed using the real data.

The results presented here show a great deal of promise for the use of wide-aperture data. This benefit does not necessarily arise from simply using higher resolution imagery in existing ATR algorithms but comes from exploiting previously ignored anisotropy information. Not only can this information aid existing algorithms by providing better reflectivity estimates but also by characterizing scatterer stability as one can tell with more confidence which peak extractions are due to stable geometric scattering. Furthermore, the anisotropy information itself is a useful feature in ATR as this information allows one to infer properties of the underlying scatterer geometry.

Appendix A. Anisotropy Statistic Accounting for Neighboring Scatterers

In this Appendix, we present the details in extending the isolated scatterer anisotropy test to account for the presence of interfering scatterers by generalizing the model in Eq. (4) to explicitly include multiple scatterers. First, however, we must generalize the sub-aperture scattering model to account for the modulations produced by neighboring scatterers. Again, we assume that the sub-aperture measurements have been formed while being focused on a particular cross-range location y_0. The possibility of other scatterers are considered at discrete locations $y_0 + k\Delta_p$, where Δ_p specifies a cross-range sampling resolution for interfering scatterers. To model the effect of a scatterer with anisotropy H_{m_k,i_k} at location $y_0 + k\Delta_p$ on the measurements at location y_0, we simply modulate the azimuthal response $1_{H_{m_k,i_k}}(\cdot)$ to account for the shift in the image domain. The observed effect over the smallest sub-apertures is then given by

$$b^k(m_k, i_k)_j = \int_{S_{M,j}} e^{j2\pi\frac{k\Delta_p}{\Delta_r}t} 1_{H_{m_k,i_k}}(t)\, dt \tag{11}$$

where Δ_r is the null-to-null cross-range resolution associated with the full-aperture. Incorporation of the neighboring scatterers is now modeled via superposition, i.e.

$$\boldsymbol{q}_M = \sum_k A_k \boldsymbol{b}^k(m_k, i_k) + \varepsilon \tag{12}$$

where the noise model is the same as in Eq. (5). The summation over k in Eq. (12) should be over a range large enough to include all scatterers contained in the coarsest resolution cell, i.e. the resolution cell associated with the smallest sub-apertures. Thus, if L_* is the smallest aperture length, then we need to consider $k \in \{-K, \ldots, K\}$ where

$$\begin{aligned} 2K\Delta_p &\geq \frac{\Delta_r}{L_*} = \text{size of coarest resolution cell} \\ K &\geq \left\lceil \frac{\Delta_r}{2L_*\Delta_p} \right\rceil \end{aligned} \tag{13}$$

and $k = 0$ corresponds to the resolution cell under investigation. Rewriting Eq. (12) as a matrix equation we get

$$\begin{aligned} \boldsymbol{q}_M &= \left[\boldsymbol{b}^{-K}(m_{-K}, i_{-K}) \quad |\cdots| \quad \boldsymbol{b}^K(m_K, i_K)\right] \begin{bmatrix} A_{-K} \\ \vdots \\ A_K \end{bmatrix} + \varepsilon \\ &= B\boldsymbol{A} + \varepsilon \end{aligned} \tag{14}$$

where B and A are appropriately defined[16]. Thus, we can use weighted least squares (WLS) to implicitly estimate the values of the interfering A_k and account for their contribution to $\boldsymbol{q}_M$. In order to have our least squared error minimization correspond to ML, we weight our inner product by the inverse of the noise covariance to whiten the measurements, i.e. $\langle \boldsymbol{u}, \boldsymbol{v} \rangle = \frac{1}{2\sigma^2} \boldsymbol{u}^T \Lambda^{-1} \boldsymbol{v}$. The ML estimate for $\boldsymbol{A}$ is obtained from this model as

$$\hat{\boldsymbol{A}} = \arg \min_{\boldsymbol{A}} \{ \|\varepsilon\|^2_{\Lambda^{-1}} \} = (B'\Lambda^{-1}B)^{-1} B'\Lambda^{-1}\boldsymbol{q}_M$$

which simplifies to Eq. (6) in the case of limiting the model order to $K = 0$. One would like to then use ML to classify the anisotropy of the $k = 0$ scatterer, i.e. choose the sub-aperture hypothesis which minimizes the weighted norm of the estimated error $\varepsilon = \boldsymbol{q}_{Mz} - B\hat{\boldsymbol{A}}$. The model in Eq. (12), however, has too many reflectivity parameters and thus results in an under-constrained minimization problem. In particular, if one chooses K sufficiently large to account for all scatterers in the coarsest resolution cell, then the model order is greater than the number of sub-aperture measurements and the estimated $\hat{\varepsilon}$ can be made zero for all hypotheses. Thus, we need to regularize the model. In order for the estimated error $\hat{\varepsilon}$ to be made small under the incorrect model, many of the values in $\boldsymbol{A}$ generally have to be made unreasonably large and result in an unrealistic scenario. Thus, we impose a 2-norm regularization penalty in the estimation of $\boldsymbol{A}$. In particular, instead of minimizing the weighted squared error to estimate $\boldsymbol{A}$, we take as our estimate

$$\begin{aligned} \hat{\boldsymbol{A}} &= \arg \min_{\boldsymbol{A}} \{ \|\varepsilon\|^2_{\Lambda^{-1}} + \gamma \boldsymbol{A}' R \boldsymbol{A} \} \\ &= (B'\Lambda^{-1}B + \gamma R)^{-1} B'\Lambda^{-1}\boldsymbol{q}_M \\ &= P\boldsymbol{q}_M \end{aligned} \tag{15}$$

where P is defined accordingly, γ is the regularization parameter, and R is the regularization matrix that penalizes the energy in all A_k other than $k = 0$, i.e.

$$R = \boldsymbol{I} - \boldsymbol{e}_K \boldsymbol{e}'_K = \text{diag}(1, \ldots, 1, 0, 1, \ldots, 1).$$

This regularized estimate for A produces the following value for the weighted norm on the error estimate

$$\begin{aligned} \|\hat{\varepsilon}\|^2_{\Lambda^{-1}} &= \|\boldsymbol{q}_M - B\hat{\boldsymbol{A}}\|^2_{\Lambda^{-1}} \\ &= \frac{1}{2\sigma^2} \boldsymbol{q}'_M [\Lambda^{-1} - 2\Lambda^{-1}BP + P'B'\Lambda^{-1}BP] \boldsymbol{q}_M \end{aligned} \tag{16}$$

which can be used as the sufficient statistic $\ell_{m,i}$ for our hypothesis test as this statistic corresponds to a generalized log-likelihood with a Gaussian prior on the neighboring reflectivities.

Although the norm in Eq. (16) is a statistic which can be easily calculated, there are an excessively large number of extended-hypotheses for which the test statistic needs to be computed. In particular, if we allow $2K$ neighboring scatterers, the number of candidate extended-hypotheses is now $\mathcal{H}^{2K+1}$ since each location has an associated scatterer and corresponding anisotropy hypothesis. For the initial work presented in this paper, we restrict our attention to hypotheses where only the anisotropy for the focused resolution cell under investigation is allowed to vary. In particular, the hypothesis for each of the $2K$ neighboring pixels is fixed to be full-aperture anisotropy while all hypotheses for the focused location are tested. Thus, we escape the exponential growth in the number of hypotheses since our hypothesis space is effectively still $\mathcal{H}$. Certainly, this constraint can be relaxed in order to appropriately take into account the anisotropy of neighboring scatterers, but we do not do so here[17]. We make special note that although the model for neighboring scatterers is restricted to full-aperture scattering, this multiple scatterer model is more powerful than the isolated scatterer model presented in Section 3.1 as the latter does not explicitly consider *any* interference from neighboring scatterers.

Notes

1. Correspondence should be sent to: Andrew Kim.
2. We will refer to anisotropy due to geometric scatterers as *geometric anisotropy* and anisotropy due to volumetric scatterers as *volumetric anisotropy*.
3. For clarity, we will exclusively use the term "cross-range" when referring to cross-range in the spatial (image) domain, and we will exclusively use the term "azimuthal" when referring to the corresponding dimension in the sensor domain.
4. By azimuthal response, we mean the 1-D cross-range uncompressed signal for a given down-range location. The signal $a(t)$ is assumed to have already been appropriately demodulated to have zero phase modulation for the inspected cross-range location.
5. The lack of normalization simplifies the notation in what follows later.
6. To be precise, it is actually the mean-squared reflectivity estimate normalized by the variance of the estimate that decreases as too much of the aperture is used. The estimator variance itself actually decreases because the coherent averaging is done over a larger aperture, but the squared-mean of the estimate decreases even more so since no additional signal energy is being incorporated.
7. The details of the multiple scatterer model, and the derivation of the resulting statistic can be found in Appendix A.
8. The term "sidelobes" here refers to those on a sinc-like function and *not* those associated a neighboring scatterer.
9. A comparison between the test based on the multiple scatterer model and the test from the isolated scattering model can be found in Chapter 5 of Kim [8].
10. The hypotheses thus correspond to isotropic (i.e. constant over the full 2.8° aperture), half-aperture (1.4°), and quarter-aperture (0.7°) anisotropy.
11. These other ripples are too small to be seen on this plot but are apparent if the noise level is increased or the scaling of the vertical axis is changed.
12. For each Monte Carlo trial, the peak extraction step detects local maximum in the log-magnitude reflectivity image and refines the scattering center location using parabolic interpolation.
13. This sub-aperture pyramid does not have the full-aperture at the root, however, all the previous analysis in Section 3 still applies. We note that the use of this "cropped" pyramid is solely to aid our comparison of narrow and wide-aperture data. In applied settings, we recommend that the full-aperture always reside at the root node to make maximum use of the aperture.
14. Note that there is also the issue of consistency of the peak locations. In order to overcome this problem, we first estimate the coarse-resolution peak location and then limit the high-resolution peak location to reside within the resolution cell corresponding to the coarse-resolution extraction.
15. The coarse resolution is used for the wide-aperture case (even though a finer imaging resolution can be achieved) due to publishing restrictions on this data.
16. Although the matrix B clearly depends on the extended-hypothesis $\{(m_{-K}, i_{-K}), \ldots, (m_K, i_K)\}$, this dependence is omitted for notational simplicity.
17. Research along this direction can be found in Chapter 7 of the thesis by Kim [8].

References

1. M. Allen, D. Sofianos, and L. Hoff, "Wide-Angle Wideband SAR Matched Filter Image Formation for Enhanced Detection Performance," in *Proc. of the SPIE, Algorithms for SAR Imagery*, vol. 2230, Apr. 1994, pp. 302–314.

2. R. Chaney, A. Willsky, and L. Novak, "Coherent Aspect-Dependent SAR Image Formation," in *Proc. of the SPIE, Algorithms for SAR Imagery*, vol. 2230, Apr. 1994, pp. 256–274.
3. J. Capon, "High-Resolution Frequency-Wavenumber Spectrum Analysis," in *Proc. IEEE*, vol. 57, no. 8, Aug. 1969, pp. 1408–1418.
4. G. Benitz, "Adaptive High-Definition Imaging," in *Proc. of the SPIE, Algorithms for SAR Imagery*, vol. 2230, Apr. 1994, pp. 106–119.
5. G. Benitz and D. DeLong, "Extensions of High Definition Imaging," in *Proc. of the SPIE, Algorithms for SAR Imagery*, vol. 2487, Apr. 1995, pp. 165–180.
6. "MSTAR (Public) Targets: T-72, BMP-2, BTR-70, SLICY", see http://www.mbvlab.wpafb.af.mil/public/MBVDATA.
7. A. Kim, *Exploring Scatterer Anisotropy in Synthetic Aperture Radar via Sub-Aperture Analysis*, PhD thesis, Massachusetts Institute of Technology, Sept. 2001.
8. A. Kim, J. Fisher, A. Willsky, and P. Viola, "Nonparametric Estimation of Aspect Dependence for ATR," in *Proc. of the SPIE, Algorithms for SAR Imagery*, vol. 3721, Apr. 1999, pp. 332–342.
9. H. Chiang and R. Moses, "ATR Performance Prediction Using Attributed Scattering Features," in *Proc. of the SPIE, Algorithms for SAR Imagery*, vol. 3721, Apr. 1999, pp. 785–796.
10. A. Kim, S. Dogan, J. Fisher, R. Moses, and A. Willsky, "Attributing Scatterer Anisotropy for Model Based ATR," in *Proc. of the SPIE, Algorithms for SAR Imagery*, vol. 4053, Apr. 2000.
11. L. Potter and R. Moses, "Attributed Scattering Centers for SAR ATR," *IEEE Trans. on Image Proc.*, vol. 5, Jan. 1997, pp.79–91.

Multidimensional Systems and Signal Processing, 14, 83–104, 2003

SAR Image Superresolution via 2-D Adaptive Extrapolation☆

ALEJANDRO E. BRITO abrito@dsp.ece.utep.edu
Department of Electrical and Computer Engineering, The University of Texas at El Paso, El Paso, TX 79968 USA

SHIU H. CHAN
Department of Electrical and Computer Engineering, The University of Texas at El Paso, El Paso, TX 79968 USA

SERGIO D. CABRERA cabrera@ece.utep.edu
Department of Electrical and Computer Engineering, The University of Texas at El Paso, El Paso, TX 79968 USA

Received December 1, 2000; Revised March 27, 2002; Accepted April 9, 2002

Abstract. In this paper, we present a description of a nonparametric two dimensional (2-D) procedure to extrapolate a signal, an extension of the Adaptive Weighted Norm Extrapolation (AWNE) method, and illustrate its application to SAR image formation. The benefits of the AWNE procedure are shown for synthetic data and for MSTAR data. Once the phase history is recovered, the AWNE method is applied to a subaperture or to the full set of frequency samples to extrapolate them to a larger aperture from which a superresolved complex SAR image is obtained. Use of the 2-D AWNE procedure proves to be superior to its one-dimensional separable version by reducing undesirable effects such as sidelobe interference, and variability in energy of the extrapolated data from row to row and column to column. To assess the performance of AWNE in enhancing prominent scatterers, reducing speckle, and suppressing clutter, we compare the superresolved images to the images formed with the traditional Fourier technique starting from the same phase history data. Fourier images are also compared with superresolved images formed using less data in order to assess the quality of the extrapolation and to quantify the method's ability to recover lost resolution. We illustrate performance by visual comparison and by the use of a geometric constellation of prominent point scatterers of the targets extracted from the images. A brief comparison with the 2-D versions of Capon and the linear prediction methods is illustrated and a hybrid AWNE/Capon approach is proposed.

Key Words: image formation, superresolution, MSTAR, synthetic aperture radar, extrapolation

1. Introduction

There is an increasing interest in the use of superresolution methods to form Synthetic Aperture Radar (SAR) images [1], [2], [3], [4], [5], [6], [7], [8]. The motivation for applying these methods to SAR, as discussed by DeGraaf [1], is to improve resolution,

☆ This work was supported by ARO Grants DAAH04-95-1-0494 and DAAD19-99-1-0012, and by a grant from the Texas Instruments Foundation.

to suppress sidelobe artifacts, and to reduce speckle in comparison with traditional Fourier imaging techniques. Improvements in resolution of prominent scatterers, speckle reduction, enhancement of contrast, and CFAR detectability of targets in clutter are among the reported results [1]. Superresolution methods have also proven to increase significantly the performance of Automatic Target Recognition (ATR) systems [2]. A one-dimensional (1-D) nonparametric procedure to extrapolate a narrowband signal, denoted Adaptive Weighted Norm Extrapolation (AWNE) [9], has been used to enhance resolution in SAR imaging by row-column processing [7]. In [8], we reported the use of a two-dimensional (2-D) version of the algorithm where the emphasis was on objectively quantifying the quality of the superresolved images using a CFAR detector to segment target and clutter in order to compute metrics, such as peak location and matching, when comparing to a reference image. The AWNE procedure is based on the iterative use of re-weighting in minimum weighted norm extrapolation to produce a spectrally concentrated stationary extension of the given data [9]. Recently, a similar procedure has been extensively analyzed and promoted for use in geophysical data processing problems [10]. It has also been recently re-cast and related to a general procedure to solve a linear inverse problem with sparseness constraints [11].

In this paper, we present a description of the two-dimensional (2-D) version of AWNE and propose its application in SAR imaging using data obtained from the MSTAR [12] targets database of images. Once the phase history is obtained by reversing the windowing and Discrete Fourier Transform (DFT) steps of the Fourier image formation, the AWNE method is applied to a subaperture or to the full set of frequency samples to extrapolate them to a larger aperture. Then, the Inverse DFT is applied to obtain the new complex SAR image. The use of the 2-D AWNE procedure minimizes potential artifacts that can be introduced when the 1-D version [7] is used.

To experimentally assess the performance of the 1-D and 2-D AWNE in enhancing prominent scatterers, reducing speckle, and suppressing clutter, we compare the super-resolved (SR) images to images formed with the traditional Fourier techniques using the same amount of data. We also evaluate images formed with less data to assess the quality of the extrapolation and to quantify the benefits of using superresolution if less than the desired amount of phase history data is available. Thus, in our first set of experiments, we use a subaperture of the frequency samples to evaluate the performance of the 2-D AWNE in this *handicapped* situation in comparison to the 1-D AWNE algorithm and to the traditional Fourier approach. The other focus of our experiments is the illustration of the potential advantages of going beyond the traditional limits of resolution by extrapolating the full aperture of phase history to a larger size. We evaluate the results by visual comparison and the geometric constellation of prominent point scatterers of the targets extracted from the images. Finally, an experimental comparison of superresolved images generated by the 2-D AWNE, the Capon's minimum variance, and the 2-D linear prediction (AR) methods is presented. Particular features of the SR images are discussed and a hybrid scheme combining the 2-D AWNE and MVM methods is proposed to take advantage of the best features of each method.

2. The Adaptive Extrapolation Method

2.1. *Minimum Weighted Norm Extrapolation*

2.1.1. Previous Related Work

The nonparametric extrapolation problem has been the subject of extensive research and a diverse array of iterative and noniterative methods have been devised [9], [13], [14], [15], [16], [17], [18], [19]. Recently, generalizations of some of these methods to multidimensional signals with discontiguous passbands and discontiguous time-concentration regions have been developed [19]. Since the nonparametric extrapolation problem always admits an infinite number of solutions due to its underdetermined nature, several schemes to choose a solution have been proposed and good reviews of the most popular methodologies for this selection can be found in [17], [18]. Of particular interest are the minimum norm (energy) extrapolation algorithms [9], [13], [14], [15], [16] where the signal with the least energy is chosen from the set of admissible extrapolations. Typically, conventional least squares stabilization theory for dealing with ill-conditioned problems is used. Also, some type of constraint or assumed *a priori* information is used in order to reach a desirable type of solution. Frequency weighting based extrapolation algorithms [9], [15], [16], [17] are particular cases devised in order to obtain a minimum weighted norm (energy) solution that matches the given data (data consistent) and reflects the available frequency profile information, which describes the desirable energy distribution. The frequency weighting can be a given profile [15], [16] or it can be directly inferred from the data [9] via a spectrum estimator. These algorithms can be considered more general than the traditional band-limited extrapolation algorithms [13], [14] in the sense that ideal spectral support information is not necessary. However, the main data consistency constraint is always present in both types of algorithms.

2.1.2. Mathematical Formulation

We denote a general set of L given discrete-time data samples as $\{x(m_1), x(m_2), \ldots, x(m_L)\}$, where all m_k values are integers. A frequency weighting function $Q(f)$ can be used to define an inner product in the linear space V of signals with spectral support B, where $B \subset [-0.5, 0.5]$. Assuming that $Q(f) > 0$ for $f \in B$, and that $Q(f) = 0$ otherwise, this inner product is

$$\langle x(n), y(n) \rangle_Q = \int_{f \in B} \frac{X(f) Y^*(f)}{Q(f)} df.$$

The infinite set D of *valid extrapolations* is the subset of signals in V which go through the given time samples. An optimal unique choice $\hat{x}(n)$ for the associated linearly constrained optimization problem [11] is the *Minimum Weighted Norm Extrapolation*

(MWNE), which solves

$$\min_{x(n)\in D}\left\{\int_{f\in B}\frac{|X(f)|^2}{Q(f)}df\right\}=||\hat{x}(n)||_Q^2=\int_{f\in B}\frac{|\hat{X}(f)|^2}{Q(f)}df.$$

The extrapolation $\hat{x}(n)$ is thus the result of an optimal spectral shape match of $|\hat{X}(f)|^2$ to $Q(f)$. That is, $Q(f)$ can be used as a tool to influence the choice of $\hat{x}(n)$ from D and it provides a means for obtaining high resolution. The weighting function $Q(f)$ may not be available as *a priori* information but may be a spectrum estimate obtained using a simple technique that only requires the available time samples at hand, such as the periodogram. It can be shown [9] that defining the autocorrelation sequence $q(n)$ as the inverse-Fourier transform of $Q(f)$, a closed form solution for $\hat{x}(n)$ exists. This is the MWNE extrapolation which is a linear expansion of the form

$$\hat{x}(n)=\sum_{i=1}^{L} b_i q(n-m_i), \tag{1}$$

where the b_i's are the *extrapolation coefficients* that can be obtained directly from the L-constraints that make $\hat{x}(n)$ a data consistent extrapolation. To solve for the b_i 's, a $L \times L$ Gram matrix G is defined having entries $(G)_{r,c}=q(m_r-m_c)$ for row r and column c. In the case of uniformly-spaced, contiguous samples, G is a Hermitian Toeplitz matrix since $(G)_{r,c}=q(r-c)$. From Eq. (1), we note that the resulting extrapolation $\hat{x}(n)$ has a length given by $L+N_q-1$, for an autocorrelation $q(n)$ of length $N_q=2J-1$ and a given parameter $J\geq L$ that controls the desired extrapolation length. From Eq. (1), the frequency domain form of MWNE is given by

$$\hat{X}(f)=Q(f)\sum_{i=1}^{L} b_i e^{-j2\pi f m_i}. \tag{2}$$

The frequency weighting $Q(f)$ summarizes the prior or inferred knowledge and is a smooth function if $q(n)$ has finite duration as we assume. This spectrum heavily influences the solution, see Eq. (2).

2.2. *Adaptive Weighted Norm Extrapolation (AWNE)*

In general, the minimum norm extrapolation solution concentrates the energy around the observation indices, suppressing energy outside this region [19]. A good extrapolation results from this method only if most of the energy of the signal to reconstruct (extrapolate/interpolate) is actually concentrated in the region of the observations. This assumption is not always appropriate, specially in the recovery of sinusoidal signals. Therefore, schemes can be devised to improve the quality of the reconstructed signal. One algorithm that incorporates the benefits of the minimum energy solution and tries to overcome the concentration of energy around the observation indices by generating longer extrapolation sequences through an iterative procedure is the AWNE method [9]. AWNE, in each

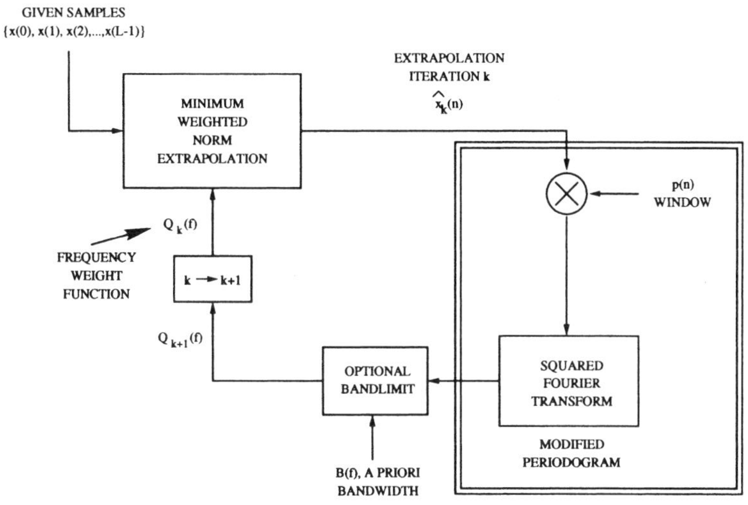

Figure 1. 1-D AWNE block diagram.

iteration, finds a minimum weighted norm extrapolation solution and uses it to compute a new, more concentrated weighting function. The fundamental idea is to infer the frequency weighting information from the extrapolation solution of the previous iteration in order to refine the current spectrum profile by making it more concentrated. Consequently, this increases the energy in the overall time (space) domain. This method provides a better solution for sinusoidal signals [11] than any of the *two-step* minimum energy extrapolation methods [13], [14], [15], [16].

AWNE is an iterative process that uses the original data, as well as some of the newly computed (extrapolated) data samples, to obtain a new $Q(f)$ at each iteration for use in the next one. To reduce sidelobe effects and have control of the resultant extrapolation length, the spectrum in the $(k+1)$-th estimate is computed via the modified periodogram as $Q_{k+1}(f) = |\mathcal{F}\{\hat{x}_k(n) \cdot p(n)\}|^2$, where $p(n)$ can be any standard non-negative symmetric window. There is a direct relationship between the resolution that the extrapolation can achieve and the type of window $p(n)$ used as well as a trade off between the stability of the extrapolation and the sidelobe attenuation implied by the window type. Analysis of these issues has been reported in [11], [20], [21] where the extrapolation resolution is related to an imposed window which is characterized as a Tukey-like window and a new algorithm has been devised to undo this effect achieving the highest resolution possible [11], [22].

The AWNE procedure is illustrated in Figure 1 for the case of contiguous data samples in the range $0 \leq n \leq L-1$, indicating that *a priori* spectral support (bandwidth) information can also be incorporated, if available. Using a fixed window $p(n)$ of length $J \geq L$, the iterative procedure produces increasing resolution until little change occurs between one iteration and the next. The final result of the AWNE procedure is the extrapolated signal at the last iteration. The maximum possible length N_E of this signal, and that of all intermediate results, is determined by the window length parameter J,

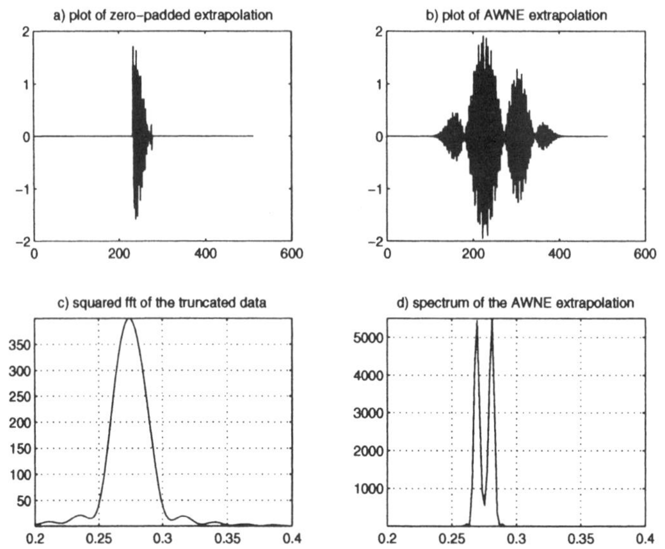

Figure 2. 1-D extrapolation example.

which produces an autocorrelation $q(n)$ of length $N_q = 2J - 1$. Since the number of given data samples is L, this leads to (see Eq.(1))

$$N_E = N_q + L - 1 = 2J + L - 2. \tag{3}$$

To save computations, the process can also be performed for a fixed number of iterations; typically 3–5 iterations, stopping before the convergence is reached. Such an approach has been used in some previous experiments with SAR imaging [7]. Because the length of all signals involved is known and finite, a DFT of length N_D can be used, with $N_D \geq N_E$. The zero-th extrapolation resulting from $Q_0(f) = 1$ is the zero-padded data $\hat{x}_0(n)$. The first iteration (non-trivial) of the procedure uses a weight $Q_1(f)$ which is the modified periodogram computed using only the zero-padded given data. As a consequence, the first extrapolation $\hat{x}_1(n)$ will be a linear combination of shifted versions of the autocorrelation $q_1(n)$ of the windowed zero-padded data. This result is a significantly different extrapolation than the zero-padded data. The length of $\hat{x}_1(n)$ is $N_1 = 3L - 2$ which corresponds to N_E when $J = L$ in Eq. (3). Although we are concentrating on contiguous uniformly spaced data samples, it is important to mention that AWNE can be applied to interpolate missing gaps of samples [9]. Figure 2 shows the results of applying the AWNE method to extrapolate 45 samples of the signal

$$x(n) = e^{j2\pi(0.27n)} + e^{j2\pi\left(0.28n + \frac{\pi}{4}\right)}.$$

In Figure 2, the real part of the 45 samples of the original signal are shown in a); the result of the AWNE method is shown in b); the spectrum of the truncated data is shown in c); and the spectrum of the extrapolated data is shown in d). This example shows the frequency domain high resolution capability of AWNE.

2.3. *Extension to a Two-Dimensional Version*

2.3.1. *Motivation*

In this paper, we provide an extension to 2-D of the 1-D AWNE method focusing on the case of uniformly spaced samples and on its use for SAR superresolution. We recently reported [7] the use of 1-D AWNE to enhance the resolution of MSTAR SAR images for further processing. We performed experiments to evaluate the performance of AWNE on SAR images taken from the MSTAR database focusing on the effects of using a reduced set of phase history samples to attempt to re-generate the same images, under this handicapped situation, with the help of AWNE. We used several criteria [7] to compare the super-resolved images to the original MSTAR images and showed how the method behaves as the phase history subaperture block size changes. Despite the handicapped situation considered, we obtained SAR images of higher resolution with enhanced prominent scatterers and reduced speckle. We also found that the same statistical model (Weibull) can be used in the CFAR detector of the target region on both the original and the SR image. In this paper, we derived and use a 2-D version of AWNE to reduce the peak location mismatch, attenuate better the sidelobes, and produce overall better images because the local 2-D dependencies are taken into account. In addition, the 2-D AWNE version can be compared more fairly to other 2-D superresolution methods [1], [3], [4], [5], [6] used for SAR image formation.

2.3.2. *Two-Dimensional AWNE*

Following the discussion of the previous sections we can see that the extension of AWNE to a 2-D version for discrete-time signals can be done starting from the minimum weighted norm extrapolation procedure so that Eq. (1) now takes the form

$$\hat{x}(n,m) = \sum_{i=1}^{L} \sum_{j=1}^{L} b_{i,j} q(n - n_i, m - m_j), \tag{4}$$

where we now have $L \times L$ extrapolation coefficients, the $b_{i,j}$'s, matching the $L \times L$ number of given data samples of $x(n, m)$, which are the constraints that make $\hat{x}(n, m)$ a valid extrapolation. For contiguous uniformly spaced data samples, a vector $\underline{y}$ containing the given data samples and a matrix G are used to solve for the vector $\underline{b}$ of $b_{i,j}$'s in Eq. (4). The Gram matrix G has $L \times L$ blocks denoted $g(k)$ so that

$$G = \begin{bmatrix} g(0) & g(-1) & \cdots & g(-L+1) \\ g(1) & g(0) & \cdots & g(-L+2) \\ \vdots & \vdots & \vdots & \vdots \\ g(L-1) & g(L-2) & \cdots & g(0) \end{bmatrix}, \tag{5}$$

and each block $g(k)$ is also of dimension $L \times L$, having the form

$$g(k) = \begin{bmatrix} q(0,k) & q(-1,k) & \cdots & q(-L+1,k) \\ q(1,k) & q(0,k) & \cdots & q(-L+2,k) \\ \vdots & \vdots & \vdots & \vdots \\ q(L-1,k) & q(L-2,k) & \cdots & q(0,k) \end{bmatrix}. \tag{6}$$

Clearly, the block matrix G is Hermitian and has a block-Toeplitz structure since $q(n, m) = q^*(-n, -m)$. However, the blocks are not in general Hermitian. Regarding the iterative part of the process, the window $p(n)$ used in the 1-D AWNE version has various extensions to 2-D including the circularly symmetric 2-D window [23]. The rest of the components defined in Figure 1 can be extended in a simple way because they all have 2-D counterparts.

We can summarize next all the computations of the algorithm whose given items are the $L \times L$ samples of $x(n, m)$, the window diameter (length) $J \geq L$ centered on the data, a maximum number of iterations K_0 or a stopping tolerance ε:

Step 0. Form the 2-D circularly symmetric window $p(n, m)$ of diameter J and zero-pad it to size $N_E \times N_E$ where $N_E = 2J + L - 2$.

Step 1. Extend $x(n, m)$ to a zero-padded version $x_z(n, m)$ of size $N_E \times N_E$ preserving the original samples in the center of $x_z(n, m)$.

Step 2. The initial extrapolation for $k = 0$ is given by $\hat{x}_0(n, m) = x_z(n, m)$.

Step 3. Increment k and replace the previous extrapolation with a windowed version

$$\hat{x}_{k-1}(n,m) = \hat{x}_{k-1}(n,m) \cdot p(n,m).$$

Step 4. Obtain the 2-D autocorrelation

$$q_k(n,m) = \hat{x}_{k-1}(n,m) * \hat{x}^*_{k-1}(-n,-m).$$

Step 5. Form the block-Toeplitz matrix G according its definition in Eqs. (5) and (6).

Step 6. Solve for $\underline{b}$ from

$$G \cdot \underline{b} = \underline{y},$$

where $\underline{b}$ is the vector of ordered extrapolation coefficients and $\underline{y}$ is a similarly ordered set of data samples.

Step 7. Compute the new extrapolation as

$$\hat{x}_k(n,m) = \sum_{i=1}^{L} \sum_{j=1}^{L} b_{i,j} q_k(n-i, m-j).$$

Step 8. Stop if $\|\hat{x}_k(n, m) - \hat{x}_{k-1}(n, m)\| < \varepsilon$ or $k = K_0$, otherwise go to Step 3.

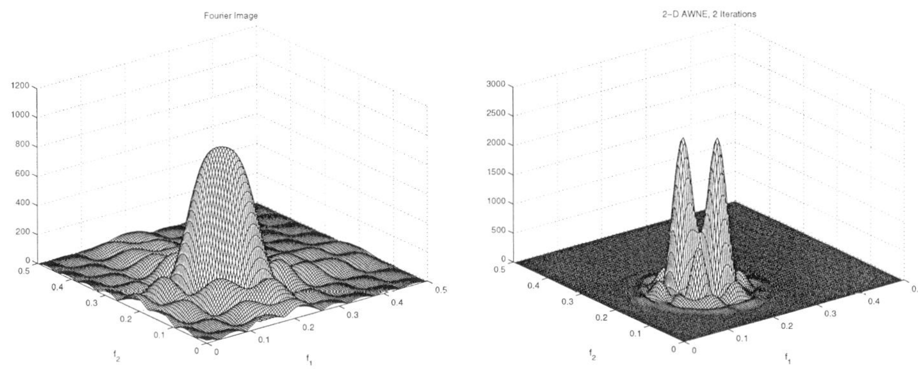

Figure 3. a) Fourier image (rectangular window), b) 2-D AWNE image (high resolution).

2.4. *Implementation Issues and Simulated Point Scatterer Examples*

The AWNE procedure involves the use of a window which can be chosen by the user. Throughout most of this paper, we will restrict our choice of window type and length to simplify our comparisons. Given the length $L \times L$ set of data (L odd) to extrapolate, a circularly symmetric [23] Hamming window of diameter $J = L$ is used in most cases in the application of the 2-D AWNE. A window length $J = L$ is used for the 1-D AWNE and for the Taylor window (1-D or 2-D separable) that is used when the Fourier imaging procedure is applied. This window length that matches the data size is a conservative choice that incorporates only original data samples and thus produces a convergence at iteration 2 due to the operation in *Step 3* and the form of $p(n, m)$ defined in *Step 0*. It also imposes a maximum extrapolation length of $N_E \times N_E$, where N_E is approximately three times the given data sample length L, see Eq.(3). Figure 3a) shows the spectrum of a square of 17×17 samples of the 2-D signal given by

$$x(n, m) = e^{j2\pi(0.2n+0.16m)} + e^{j2\pi(0.15n+0.2m)}$$

using a rectangular window where it is clear that the sinusoids cannot be resolved. Figure 3b) shows the spectrum of the 51×51 samples that result after the application of the 2-D AWNE to the given data. The high resolution characteristics of the resulting spectrum can be observed.

In order to contrast the performance of the 1-D and 2-D versions of AWNE with the Fourier method, Figure 4 shows the results of the three methods on a collection of simulated point scatterers configured to spell Σ*ERIM* used by DeGraaf in [1]. The data is corrupted by additive white Gaussian noise with a target-to-noise ratio of 33 dB. Figure 4 shows a) the true points; b) the Fourier image of the 24×24 samples of original phase history with rectangular windowing; c) the Taylor windowed Fourier image; d) the spectrum of the 72×72 samples of extrapolated phase history computed

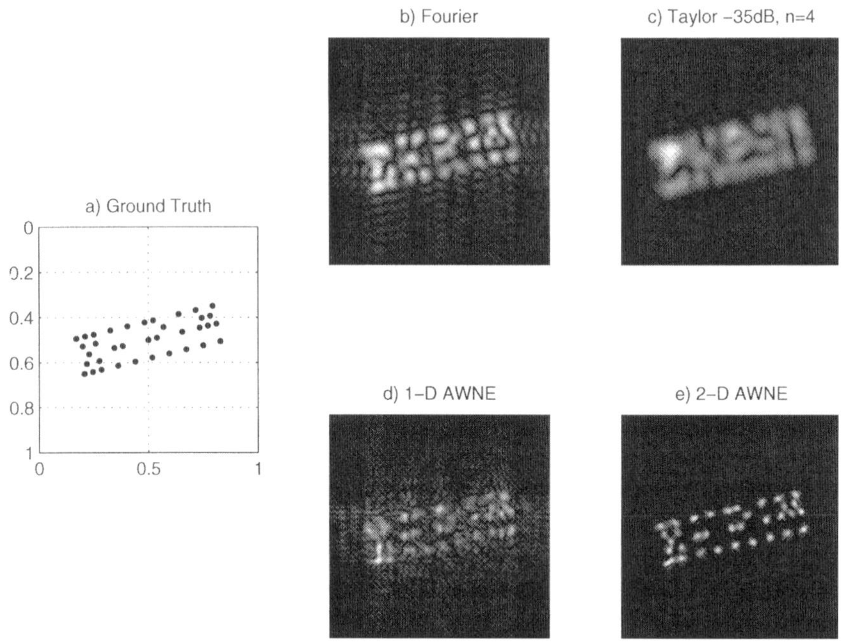

Figure 4. Simulated point scatterer Images.

with the 1-D AWNE method in a row-column separable fashion; and e) the spectrum of the 72 × 72 phase history computed with the 2-D AWNE method. Note that all the magnitude images presented in this paper are plotted on a square root scale. It is interesting to see the good localization of the point scatterers produced by the 2-D AWNE, the absence of sidelobes in the image, and the improvement over the results produced by its 1-D counterpart.

3. Processing MSTAR SAR Images with 2-D AWNE

One of the reasons for implementing and presenting the 2-D AWNE method in this paper is to improve our previous results obtained when the 1-D version [7] was applied to actual SAR imagery. The images selected from the MSTAR public target database [12] are formed using the Sandia method [24] which uses an overlapped-subaperture algorithm based on FFT and vectorial operations to allow for real-time and fine resolution operation. Among the tasks that the algorithm performs are motion compensation, phase correction, Taylor windowing and DFT computation. This database includes SAR images of different time-critical targets at different depression and aspect angles. The two images used for experimentation throughout this paper include a T72 tank.

3.1. *Obtaining the Phase History*

To produce a new image, the AWNE method works on the phase history, which is obtained by inverting the last two steps of the MSTAR image formation process. The size of an MSTAR target image is 128 × 128 complex pixels. After performing the DFT, the phase history revealed is approximately 100 × 100 in dimension. From the header file of the MSTAR image, we know that a −35 dB Taylor window is used and we assume that $\bar{n} = 4$ is used. Thus, we form the corresponding Taylor window and divide it out from the 100 × 100 phase history samples to obtain the unweighted frequency (phase history) samples. These phase history samples are ready for the application of the 1-D AWNE [7] or the 2-D AWNE algorithms. It is important to mention that we take advantage of all the previous processing done by the Sandia image formation method to focus our attention exclusively on the effects of the extrapolation procedure.

3.2. *Applying the 2-D AWNE*

The application of the 2-D AWNE to the MSTAR images is intended to show the improvement in some of the characteristics of the SAR imagery when this superresolution method is used in lieu of the traditional Fourier method. There are several experiments defined to evaluate the performance of the AWNE algorithm. The first experiment is done to evaluate the two versions of AWNE. In the previous section, we have shown that the 2-D version gives better results than the 1-D version (see Figure 4) when the methods were applied to simulated point scatterers. The application of both algorithms is presented in Figure 5 to contrast their performance with the Fourier (MSTAR) method as a basis for comparison under a handicapped situation. For fairness purpose, extrapolations that go beyond the original data size are truncated to this size. Figure 5 shows in a) the original MSTAR image (hb03793.015); in b) the image formed after the extrapolation of the middle 49 × 49 phase history samples to 100 × 100 samples using 1-D AWNE (1 iteration) in a row/column fashion; and in c) the image formed by the application of the 2-D AWNE

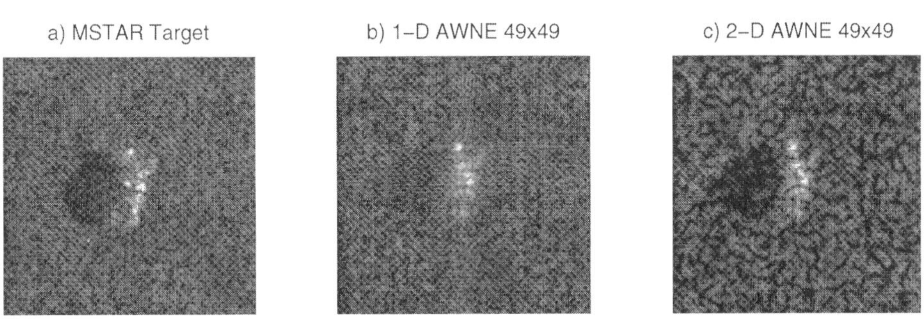

Figure 5. a) Original MSTAR image, b) 1-D and c) 2-D handicapped SR images, obtained by extrapolating a subset of 49 × 49 phase history samples to 100 × 100.

(1 iteration) to extrapolate from 49 $\times$ 49 to 100 $\times$ 100 phase history samples. Two important things can be observed by evaluating the 2-D AWNE result of Figure 5c): the preservation of the target's shadow present in Figure 5a) and the absence of the central sidelobes present in Figure 5b). The image in Figure 5c) is closer to the MSTAR image in Figure 5a) than its counterpart in Figure 5b). The scatterers are well defined even though the image has been generated with only a quarter of the phase history samples that produced the original MSTAR image. Some objective metrics were presented in [7] and [8] to evaluate quantitatively the performance of both SR versions of the AWNE method in this handicapped way using the original MSTAR image as a basis for comparison.

There is an inherent drawback to the 2-D AWNE method imposed by the need to solve L^2 linear equations per iteration of the algorithm. Motivated by the decimation/mosaicing strategy employed by DeGraaf [1], we propose two schemes for the application of the 2-D AWNE version: a *zooming* on the targets or region of interest to reduce the amount of data, and similarly a *partitioning* of the image in the spatial domain that renders a small set of phase history to extrapolate for each block. For the zooming scheme, before applying the 2-D AWNE method, the phase history is modulated to baseband by image-domain shifting, lowpass filtered, and then decimated by 3 in both range and cross-range directions. These operations result in a zooming of the target part of the MSTAR SAR image (see Figure 6). Then the decimated phase history can be used to perform *full* rather than handicapped extrapolation experiments (i.e. extrapolate the available data to a larger size than the given original phase history size in order to produce a higher resolution image). For the partition scheme, we break the complex image into blocks and recover the phase history of each one in an independent way. Then, the 2-D AWNE is applied to every block of phase history to obtain an extrapolated block of phase history and then a new image block that can be reassembled with the other blocks as shown in the block diagram in Figure 7. The following section will discuss the results from these experiments in further detail.

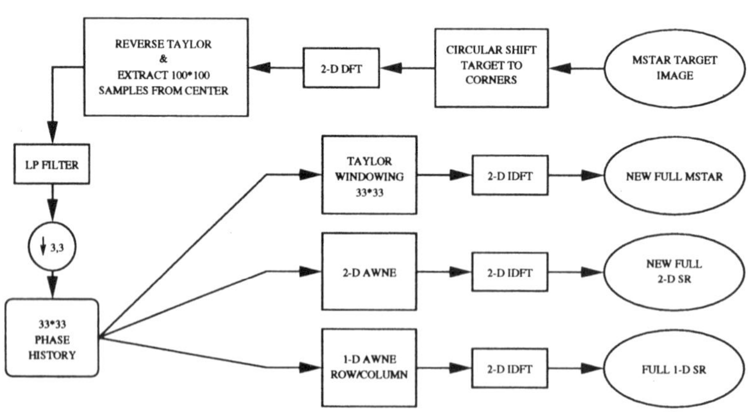

Figure 6. Block diagram for the zooming MSTAR experiments.

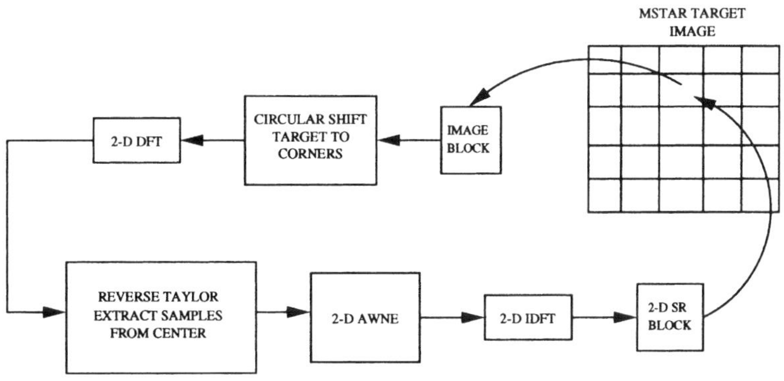

Figure 7. Block diagram for the partitioning MSTAR experiments.

4. Experimental Comparison of Fourier vs. Superresolved Images

We present some experiments to evaluate the quality of the superresolved (SR) images in comparison to Fourier images. Figure 6 shows a block diagram which summarizes the set of experiments carried out on the target chip or Region Of Interest (ROI) which covers one-ninth of the original image area. Using the *zooming* procedure defined in the previous section, we obtain the new set of 33×33 Phase History (PH) samples of the target ROI. Using this new phase history, we form the *new full MSTAR* (Fourier) image which is used as the reference or ground truth in the comparison done for the evaluation of the SR images.

4.1. Target Zooming Scheme

The full resolution experiments explore the possibility of going beyond the resolution limits implied by the given phase history data size. An example of extrapolated phase history samples obtained with the 2-D AWNE method applied to the set of 33×33 samples to extrapolate to 100×100 samples using only one iteration (a conservative choice) corresponding to the target chip can be seen in Figure 8a). Recall that the original 33×33 block of phase history data is preserved perfectly (centered at the 50th location) in the middle of the extrapolated set. Note that the AWNE method gives a natural amplitude decay away from the given data region which prevents sidelobe interference in the resulting image. Figure 9 shows: a) the new full MSTAR Fourier image (hb03793.015); b) the full 1-D (row/column separable) SR image; and c) the full 2-D SR image. Higher resolution in the point scatterers can be observed in Figures 9 b) and c) which can be better appreciated in the 3-D mesh plots presented in Figures 10 a) and b). The comparison of these with the 3-D plot of the new full MSTAR presented in Figure 8b) clearly favors the SR images in terms of resolution although the extrapolation lengths are minimal (a 3 fold increase).

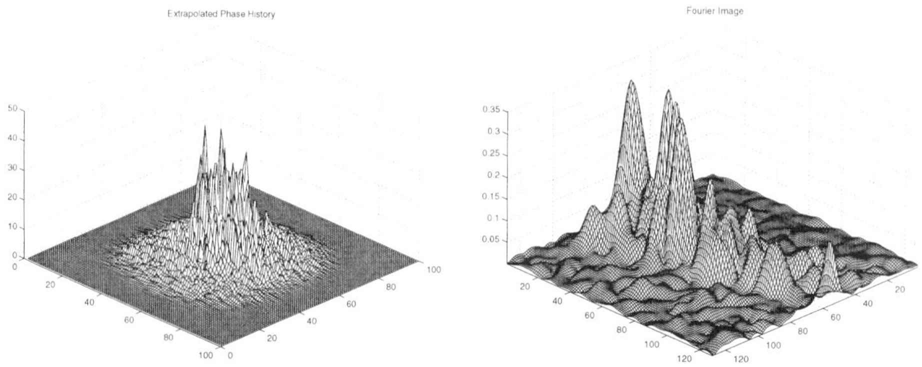

Figure 8. a) Magnitude of the extrapolated phase history from 33 × 33 to 99 × 99, b) New Full MSTAR 3-D plot.

Further analysis of the effects of extrapolation using the 2-D AWNE method can be done on Figure 11 where structural information is presented in terms of the contour plots and the convex hulls of the prominent point scatterers for each case. Figure 11 shows: a) the full new MSTAR image; and b) the full 2-D SR image. In Figure 11, the convex hull of the prominent point scatterers represents the constellation of the target. This is a geometric representation that can be used in peak-based target detection [25] by comparing as a means of constellations of peaks. The general shape of the constellation of the peaks of Figure 11a) for the new full MSTAR image is preserved in the 2-D SR image (Figure 11b)), where a better definition of this constellation can be observed. Furthermore, the contour plots in the same Figure show how the resolution of the point scatterers is superior in the full 2-D SR image and how the process also suppresses clutter around the target. From a subjective point of view, the conclusion that going beyond the original limit of phase history samples produces better images is easily established. However, objective comparison metrics cannot be used with these full SR images, as in previous work

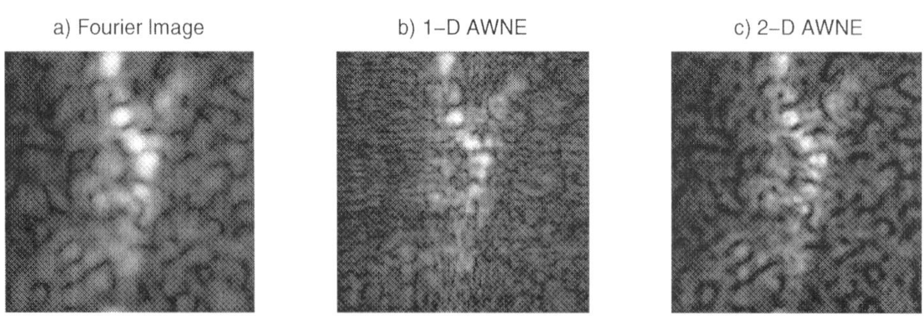

Figure 9. a) New Full MSTAR, b) Full 1-D and c) Full 2-D SR images.

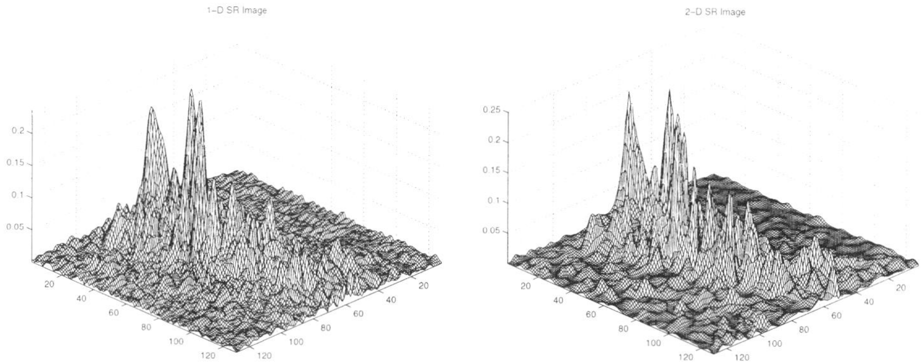

Figure 10. a) Full 1-D SR 3-D plot, b) Full 2-D SR 3-D plot.

reported in [7], [8] because of the lack of a ground truth image to compare with. An alternative approach is the objective evaluation of the full SR images based on their impact on the performance of an ATR system [2], [26]. Also, general image quality metrics that allow a comparison among various superresolution generated images coming from different methods [1], [3], [4], [5], [6] can be defined and used.

4.2. Block Partition Scheme

Here we present the results of the scheme to save computational effort described by the block diagram in Figure 7. It is an approach to reduce the amount of data to extrapolate at any given time while still processing the complete phase history data available. Figures 12

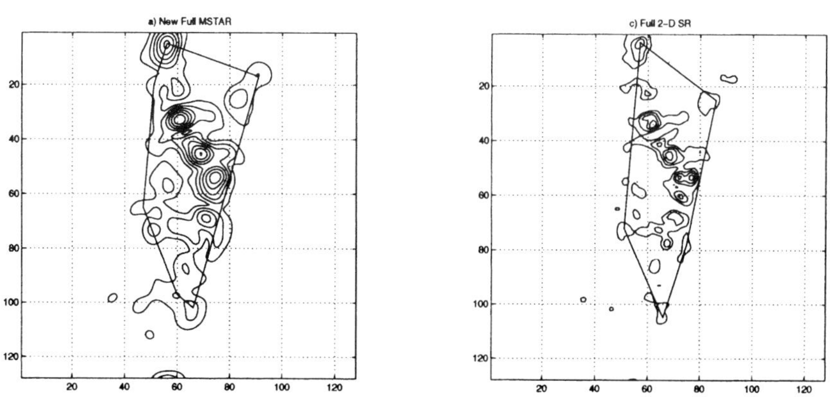

Figure 11. a) New Full MSTAR, and b) Full 2-D SR contour and convex hull plots.

Figure 12. Extrapolated image using block partitioning with extrapolation length of 3 times the original block size.

and 13 show two images processed in a partition scheme of blocks of size 16 by 16 pixels each from which we obtain 12×12 phase history samples. The images are processed with the 2D AWNE using parameters that lead to extrapolated blocks that are 3 times and 5 times larger. Both images show the preservation of the scatterer centers and the attenuation of clutter. In addition, both images present the shadow of the target as seen before in Figure 5c) and in the original MSTAR image, see Figure 5a). However, we can appreciate higher resolution in the images of Figure 12 and Figure 13 than in the ones of Figure 5 because the former ones make use of the full set of phase history data while the latter ones use only a subset of the data in a handicapped situation. An interesting behavior in the performance of our algorithm is the sparse result that is produced when large extrapolation ratios and several iterations are used. The image in Figure 13 formed with 5 fold extrapolation has better concentration of the scatterers and shows more background clutter suppression. Further experimental results with larger extrapolation lengths showed a very sparse image with single peaks representing the scatterer centers that can be used as features for an ATR system.

Clearly, the independent processing of the superresolved blocks introduces blocking artifacts as it can be observed in Figures 12 and 13. The artifacts are better appreciated in Figure 12. These artifacts are expected to affect CFAR detection applied to the image by skewing the image pixel distribution and generating false-positives outside the target region. Approaches to deal with the artifacts can be postprocessing at the boundaries, or block overlapping and averaging as in [1].

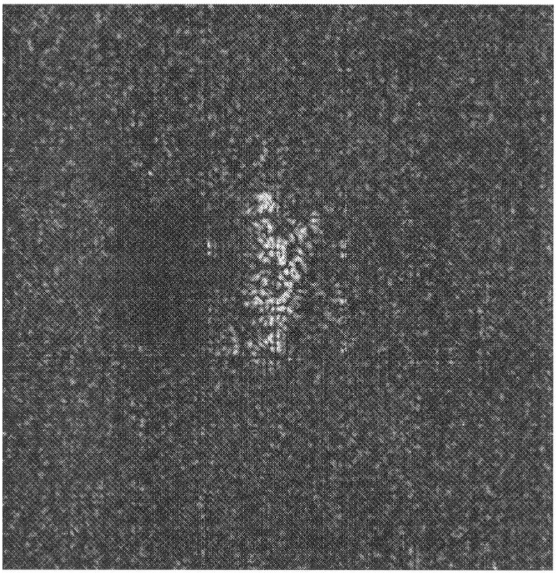

Figure 13. Extrapolated image using block partitioning with extrapolation length of 5 times the original block size.

5. Experimental Comparison to Other Superresolution Methods

In this section, Capons' minimum variance method (MVM) and the 2-D linear prediction or AR method are used to form superresolved SAR images. They are compared visually with the 2-D AWNE images under *handicapped* situations using a different MSTAR image than before. The main idea is to contrast particular features of the superresolution SAR formation methods. The superresolution methods process a subset of the full phase history (PH) samples to form SR SAR images of equivalent spectral resolution and size. As noted before, the handicapped situation can be used to stress the benefits of using superresolution methods when only limited data is available.

The 2-D linear prediction or autoregressive (AR) method [27], [28] is implemented using a quarter-plane support modified covariance version. In particular, the method uses a fast Cholesky-decomposition type of algorithm [29] and the 2-D AR power spectral density is produced by the combination of the results for quadrants 1 and 4 [28]. For this experiment the PH signal subset used is 55×55 and the order of the 2-D AR parameter array chosen is 22×22. Figure 14 illustrates the 2-D AR SR image which presents an unusual character for the target although a few prominent point scatterers are well defined. In addition, it can be observed that the shadow of the target is not well preserved. Furthermore, the introduction of spurious point scatterers throughout the background region can be observed. These are undesirable image characteristics for ATR systems based on prominent peak processing [26].

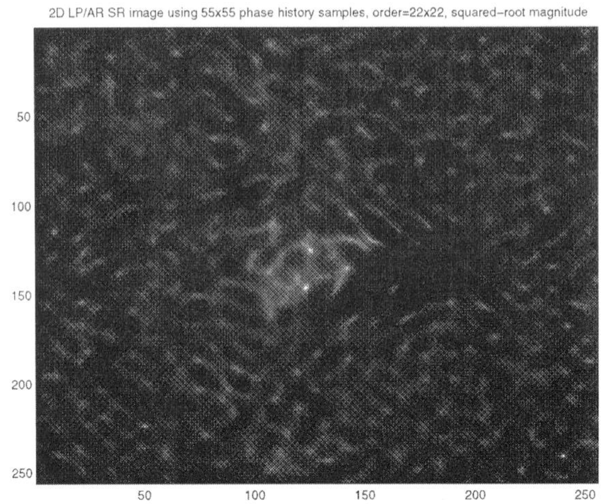

Figure 14. 2-D (AR) linear prediction SR image (hb03787.015) using 55 × 55 phase history samples and a filter order of 22 × 22.

The 2-D version of Capon's MVM [27], [28] is a popular choice in SAR imaging [1], [2], [3], [29], [30] and recently there has been increasing interest in its computational aspects. The interest comes from the argument that the main burden of the MVM does not come from the correlation estimate formation and inversion but from the computation of the estimator over all the needed frequencies. Therefore, fast algorithms to overcome this problem have been proposed in [29], [30]. For the calculation of the spectral estimate, the forward-backward averaged correlation matrix is formed using all possible $M_1 \times M_2$ subapertures, which are available from the selected PH sample subset that forms a matrix of size $N_1 \times N_2$, leading to a number of subapertures given by $A = (N_1 - M_1 + 1)(N_2 - M_2 + 1)$. Empirical results [1] suggest that the MVM produces satisfactory results when the relation between subaperture size and aperture size is kept between the following bounds

$$0.4 \leq \sqrt{\frac{M_1 M_2}{N_1 N_2}} \leq 0.5. \tag{7}$$

For the computation of the experimental results shown next, a direct implementation of the MVM is used and the empirical bound in Eq.(7) is satisfied. The next set of figures show the results for the case when approximately a quarter (55 × 55) of the full phase history data set is used to generate the superresolved images using a 256 × 256 2-D DFT. Figure 15 and Figure 16 show the corresponding superresolved images for the 2-D AWNE and Capon methods. In this case, the SR MVM image has better definition of the main scatterer center but the smoothness presented in the rest of the target area may be an undesirable feature. In addition, the shadow for the SR MVM image is almost perfectly preserved. On the other hand, the SR 2-D AWNE image has a better definition of the prominent peaks, is comparable in terms of the shadow preservation, but is inferior as far

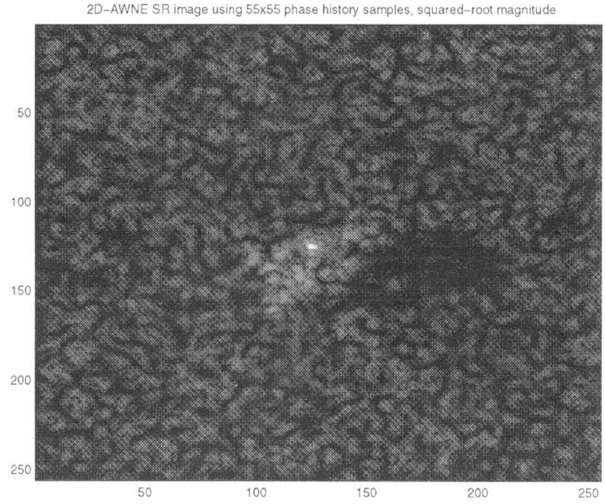

Figure 15. 2-D AWNE SR image (hb03787.015) extrapolating 55 × 55 to 163 × 163 phase history samples.

as speckle reduction and clutter suppression. Overall, the SR images are superior in comparison with the Fourier images (not shown) using the same subset of phase history samples.

A hybrid scheme between the 2-D AWNE and the MVM is proposed to produce a SR image with the best features of both methods. The main goal is to achieve the high resolution of the prominent scatterer centers produced by the 2-D AWNE and the better

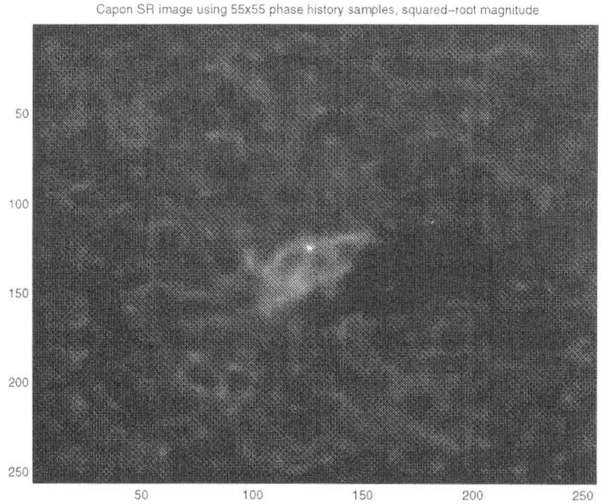

Figure 16. Capon SR image (hb03787.015) using 55 × 55 phase history samples.

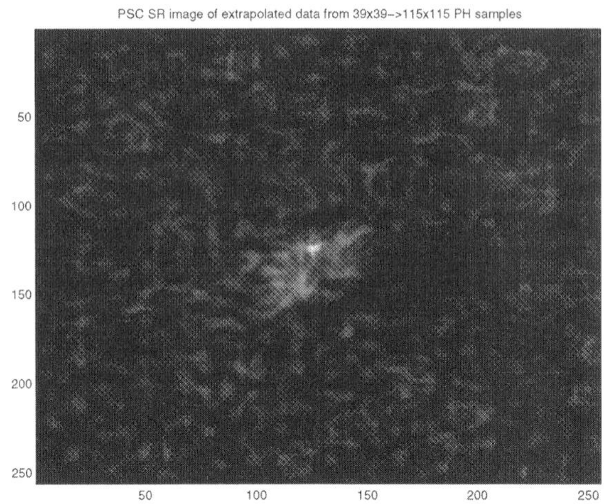

Figure 17. Hybrid Scheme. Capon via AWNE SR image (hb03787.015) using 39 × 39 phase history samples.

clutter suppression and speckle reduction produced by the MVM. To illustrate the viability of this scheme a 39 × 39 subset of phase history samples is extrapolated using the 2-D AWNE to 115 × 115 and then these samples are used by the MVM to produce a SR image. In the resultant SR image shown in Figure 17, it can be observed that the scatterer centers have better resolution than in the MVM SR image (Figure 16) and the clutter suppression and speckle reduction is better than in the 2-D AWNE SR image (Figure 15) even when the subset of original PH samples used is about 50% of the 55 × 55 set. This leads us to believe in the potential of complementing the hybrid scheme to produce superior SR SAR imagery than either method alone at expense of higher complexity.

6. Conclusions

We have presented a detailed two-dimensional (2-D) extension of the adaptive weighted norm extrapolation method and proposed its application to achieve superresolution in SAR imaging by showing its benefits on simulated and MSTAR images. The use of the 2-D AWNE procedure instead of the 1-D version [7] results in a better image with less undesirable artifacts. The 2-D AWNE proved to have better performance in the handicapped experiments realized. First, it preserved the shadow of the target and produced a closer image to the MSTAR reference (see Figure 5). The other focus of our experiments was the illustration of the potential advantages of going beyond the traditional limits of resolution by extrapolating the full aperture of phase history to a larger size. We proposed two schemes for the formation of images that allow the full use of phase history data to produce an extrapolation using the 2-D AWNE algorithm. We analyzed performance by visual comparison and by comparison of the geometric constellation of prominent point scatterers of the targets present in the images. The 2-D AWNE method proved to be superior to its 1-D counterpart and it was established that going beyond the original limit of phase history samples produces better images (see Figure 12 and 13).

Acknowledgments

The authors would like to thank Dr. Stuart DeGraaf and Dr. Armin Doerry for suggestions and valuable discussions. We also thank Dr. DeGraaf for providing the data for the simulated point scatterers of Figure 4 and S. Lawrence Marple, Jr. for providing the code for the 2-D (AR) linear prediction method.

References

1. S.R. DeGraaf, "SAR Imaging via Modern 2-D Spectral Estimation Methods," *IEEE Trans. Image Process.*, vol. 7, May 1998, pp. 729–761.
2. L.M. Novak, G.R. Benitz, G.J. Owirka, and L.A. Bessette, "ATR Performance Using Enhanced Resolution SAR," in *SPIE*, vol. 2757, Orlando, FL, April 1996, pp. 332–337.
3. J. Li and P. Stoica, "An Adaptive Filtering Approach to Spectral Estimation and SAR Imaging," *IEEE Trans. Signal Process.*, vol. 44, June 1996, pp. 1469–1484.
4. Z. Bi, J. Li, and Z.S. Liu, "Super Resolution SAR Image Formation via Parametric Spectral Estimation Methods," in *SPIE*, vol. 3370, Orlando, FL, April 1998, pp. 238–249.
5. J. Li, P. Stoica, Z. Bi, R. Wu, and E.G. Zelnio, "A Robust Hybrid Spectral Estimation Algorithm for SAR Imaging," in *32nd Asilomar Conf. on Signal, Systems and Computer*, vol. 1, Pacific Grove, CA, October 1998, pp. 1322–1326.
6. F.M. Candocia and J.C. Principe, "A Method Using Multiple Models to Superresolve SAR Imagery," in *SPIE*, vol. 3370, Orlando, FL, April 1998, pp. 197–207.
7. A.E. Brito, S.H. Chan, and S.D. Cabrera, "Application of AWNE Superresolution to MSTAR SAR Images," in *Proc. of IASTED SIP'98*, vol. 1, Las Vegas, NV, May 1998, pp. 495–499.
8. A.E. Brito, S.H. Chan, and S.D. Cabrera, "SAR Image Formation Using 2-D Re-Weighted Minimum Norm Extrapolation," in *SPIE Algorithm for SAR Radar Imagery VI*, vol. 3721, Orlando, FL, April 1999, pp. 78–91.
9. S.D. Cabrera and T.W. Parks, "Extrapolation and Spectral Estimation with Iterative Weighted Norm Modification," *IEEE Trans. Signal Process.*, vol. 39, April 1991, pp. 842–851.
10. M.D. Sacchi, T.J. Ulrych, and C.J. Walker, "Interpolation and Extrapolation Using a High-Resolution Discrete Fourier Transform," *IEEE Trans. Signal Process.*, vol. 46, January 1998, pp. 31–38.
11. A.E. Brito, "Iterative Adaptive Extrapolation Applied to SAR Image Formation and Sinusoidal Recovery," PhD thesis, Department of Electrical and Computer Engineering, The University of Texas at El Paso, El Paso, TX, 2001.
12. "Sensor Data Management System Page," *http://www.mbvlab.wpafb.af.mil/public/MBVDATA.*
13. A.K. Jain and S. Ranganath, "Extrapolation Algorithms for Discrete Signals with Application in Spectral Estimation," *IEEE Trans. Acoust. Speech Signal Process*, vol. ASSP-29, September 1981, pp. 830–845.
14. A. Papoulis, "A New Algorithm in Spectral Analysis and Band-Limited Extrapolation," *IEEE Trans. Circuits Syst.*, vol. CAS-22, September 1975, pp. 735–742.
15. C.L. Byrne and R.M. Fitzgerald, "Reconstruction from Partial Information with Applications to Tomography," *SIAM Journal on Applied Mathematics*, vol. 42, August 1982, pp. 933–940.

16. C.L. Byrne and R.M. Fitzgerald, "Spectral Estimators that Extend the Maximum Entropy and Maximum Likelihood Method," *SIAM Journal on Applied Mathematics*, vol. 44, April 1984, pp. 425–442.
17. L. Potter and K. Arun, "Energy Concentration in Band-Limited Extrapolation," *IEEE Trans. Acoust. Speech Signal Process.*, vol. 37, July 1989, pp. 1027–1041.
18. E. Diethorn and J. Munson, D.C., "A Linear, Time-Varying System Framework for Noniterative Discrete-Time Band-Limited Signal Extrapolation," *IEEE Trans. Signal Process.*, vol. 39, January 1991, pp. 55–68.
19. S. Dharanipragada and K. Arun, "Bandlimited Extrapolation Using Time-Bandwidth Dimension," *IEEE Trans. Signal Process*, vol. 45, December 1997, pp. 2951–2966.
20. S.H. Chan, A.E. Brito, and S.D. Cabrera, "Characterization of Windowing Effects in Adaptive Extrapolation of Sinusoids," in *33rd Asilomar Conf. on Signals, Systems and Computers*, vol. 2, Pacific Grove, CA, October 1999, pp. 1092–1097.
21. S.H. Chan, A.E. Brito, and S.D. Cabrera, "Imposed Window Estimate for Adaptive Extrapolation of Complex Sinusoids in Gaussian Noise," in *9th IEEE DSP Workshop*, Hunt, TX, October 2000.
22. A.E. Brito and S.D. Cabrera, "Analysis of Autocorrelation Pursuit via Non-Parametric Iterative Extrapolation," in *35th Annual Conference on Information Sciences and Systems,* vol. 2, Baltimore, MD: Johns Hopkins University, March 2001, pp. 633–638.
23. J.S. Lim, *Two-Dimensional Signal and Image Processing*, Englewood Cliffs, New Jersey: Prentice-Hall, 1990.
24. B.L. Burns and J.T. Cordaro, "A SAR Image-Formation Algorithm that Compensates for the Spatially-Variant Effects of Antenna Motion," in *SPIE,* vol. 2230, Orlando, FL, April 1994, pp. 14–24.
25. W.W. Irving, R.B. Washburn, and W.E.L. Grimson, "Bounding Performance of Peak-Based Target Detectors," in *SPIE,* vol. 3070, Orlando, FL, April 1997, pp. 245–257.
26. B. Bhanu and G. Jones III, "Performance Characterization of a Model-Based SAR Target Recognition System Using Invariants," in *SPIE*, vol. 3070, Orlando, FL, April 1997, pp. 305–321.
27. S.M. Kay, *Modern Spectral Estimation*, Englewood Cliffs, NJ: Prentice Hall, 1988.
28. S.L. Marple, *Digital Spectral Analysis with Applications*, Englewood Cliffs, NJ: Prentice Hall, 1987.
29. A. Jakobsson, S.L. Marple, and P. Stoica, "Computationally Efficient Two-Dimensional Capon Spectrum Analysis," *IEEE Trans. Signal Process.*, vol. 48, September 2000, pp. 2651–2661.
30. E.G. Larsson and P. Stoica, "Fast Implementation of Two-Dimensional APES and Capon Spectral Estimators," in *Proc. IEEE ICASSP*, vol. V, Salt Lake City, UT, March 2001, pp. 3069–3072.

Multidimensional Systems and Signal Processing, 14, 105–118, 2003

Multi-Channel Multi-Variate Equalizer Design☆

RAVIKIRAN RAJAGOPAL
Department of Electrical Engineering, Ohio State University, Columbus, OH 43210

LEE POTTER
Department of Electrical Engineering, Ohio State University, Columbus, OH 43210

Received April 14, 2001; Revised February 5, 2002; Accepted February 12, 2002

Abstract. The polarimetric calibration of synthetic aperture radar (SAR) imagery requires the equalization of a multiple-input, multiple-output (MIMO) distortion system. For wide-band, wide-angle SAR systems, the distortion is frequency and angle dependent and can be accurately modelled as a two-dimensional finite impulse response (FIR) filter bank. This paper presents a design algorithm, using a Gröbner basis, to compute a FIR filter bank that exactly inverts the multi-channel distortion. The results presented for polynomial inversion of MIMO FIR systems hold for sampled data signals of arbitrary dimension.

Key Words: MIMO polynomial inverse, Gröbner basis, filter orders, channel equalization, polarimetric radar calibration, wide-band radar

1. Introduction

Polarimetric diversity in synthetic aperture radar (SAR) measurements yields improved detection and characterization of man-made targets in clutter [5], [14], [19]. Two linearly independent polarizations are used on transmit and receive to observe the 2×2 polarimetric scattering matrix describing reflectivity at each scene location. However, physically realizable radar systems exhibit channel cross-talk and gain imbalances which distort the polarimetric measurements; therefore, polarimetric calibration is required to equalize this distortion.

Established calibration techniques [21], [23] adopt a system model for distortion that is independent of frequency and radar aspect angle. Accordingly, equalization of the "flat fading" model is achieved by a matrix multiply directly applied, pixel by pixel, to the four coregistered polarimetric images.

However, emerging wide-band radar systems transmit a fractional bandwidth[1] exceeding 25% and tens of degrees of antenna beam width, thereby rendering the flat fading model inadequate. For example, 8 cm resolution SAR imagery is obtained by an X-band system

☆ Prepared through collaborative participation in the Advanced Sensors Consortium sponsored by the U.S. Army Research Laboratory under Cooperative Agreement DAAL01-96-2-0001. The views and conclusions contained in this document are those of the authors and should not be interpreted as presenting the official policies, either expressed or implied, of the Army Research Laboratory or the U.S. Government.

with 10 GHz center frequency, 40% fractional bandwidth, and a 12 degree aperture. Similarly, foliage and ground penetrating SAR imagery with 60 cm resolution can be obtained at UHF-band wavelengths using 50% fractional bandwidth and a 35 degree aperture. For these wide-band systems, the magnitude response, phase response and polarization rotation of antennas varies significantly as a function of frequency and angle. Consequently, frequency and angle dependent, or "frequency selective," equalization must be used for calibration.

For wide-band, wide-angle radar systems, the combination of data collection, polarimetric antenna response, and SAR image processing is well modelled as a linear shift-invariant (LSI) distortion acting on the desired images of scene reflectivity [4], [16], [18]. The physical sources of distortion include nonideal transmit pulse, antenna response, phase center displacement, cable delays and amplifier gains. As a result of SAR image processing, these nonideal effects essentially disperse energy only in a neighborhood of a pixel. Therefore, we model the multiple-input, multiple-output (MIMO) LSI operator as a finite impulse response (FIR) filterbank.

In this paper, we seek a FIR filterbank to calibrate, or equalize, the multi-channel images. A FIR equalizer is desirable because of inherent stability, robustness against quantization effects, limited error propagation, and ease of implementation. While the flat fading approach can be applied independently to thousands of samples in the frequency-Doppler plane [4], [17], a FIR parameterization of the equalization filterbank provides a large and physically meaningful dimensionality reduction. In particular, this paper addresses three questions: When does there exist a FIR filterbank to equalize a MIMO FIR distortion? What is the required order of the equalization filters? How can the equalizer be computed without prior knowledge of the filter mask size and shape?

The answer for existence is intuitive for single-input, multiple-output systems in one variable: an equalizer exists if and only if all the distortion filters share no common factor. The algebraic result invoked here generalizes this intuition to multiple-inputs and multi-dimensional signals. For computation, we propose the use of Gröbner bases for an algebraic design procedure. In contrast to least-squares linear algebraic techniques, the numerically stable Gröbner procedure has small memory requirement, adaptively selects the FIR filter mask shapes, and yields small filter orders. Significantly, the smaller filter orders can potentially permit system calibration from a limited set of known calibration reflectors, such as a trihedral and rotated dihedral used in narrow-band radiometric calibration [8].

In section 2, we define notation and briefly introduce Gröbner bases. Section 3 presents the MIMO FIR distortion model and establishes the necessary and sufficient condition for invertibility using a FIR filter bank. The constructive proof gives the foundation for the proposed algebraic design procedure. A bound for filter orders is given in section 4. Section 5 presents a direct, yet memory intensive and potentially high order, linear algebraic design approach for computing an equalizing filter bank. The proposed algebraic procedure, using Gröbner bases, is described in section 6. A computed example is presented in section 7 for an ultra wide-band (UWB) polarimetric radar to illustrate the algebraic procedure. A summarizing discussion concludes the paper in section 8.

2. Preliminaries

2.1. Notation

Let **k** denote the n-tuple $(k_1, \ldots, k_n)$ of nonnegative integers $\mathbb{Z}^+$. For any n-dimensional (n-D) sampled signal, $\mathbf{x} = \{\mathbf{x}(\mathbf{k})\}$, $\mathbf{k} \in \mathbb{Z}_n^+$, the support of **x** is the set of points **k** for which **x** is nonzero. Define $|\mathbf{k}| = k_1 + k_2 + \ldots + k_n$. The degree of a finite support signal **x**, denoted by deg(**x**), is the maximum $|\mathbf{k}|$ such that $\mathbf{x}(\mathbf{k}) \neq 0$. With each finite support signal, we associate a polynomial (z-transform) in the variables $z_1, \ldots, z_n$:

$$\sum_{k_i \in \mathbb{Z}^+, 1 \leq i \leq n} \mathbf{x}(k_1, \ldots, k_n) z_1^{k_1} z_2^{k_2} \ldots z_n^{k_n} = \sum_{\mathbf{k} \in \mathbb{Z}_n^+} \mathbf{x}(\mathbf{k}) \mathbf{z}^{\mathbf{k}}. \tag{1}$$

The term $\mathbf{x}(\mathbf{k})$ is called the coefficient of the monomial $\mathbf{z}^{\mathbf{k}}$. We denote both the n-D signal and its associated polynomial by **x** and suppress the variables $z_1, \ldots, z_n$. Likewise, we write the n-D convolution of sequences $\mathbf{x}_1$ and $\mathbf{x}_2$ as the multiplication of the associated polynomials. Finally, the set of all polynomials with coefficients in a field $\mathbb{F}$ is denoted by $\mathbb{F}[z_1, \ldots, z_n]$; the complex and real fields are denote by $\mathbb{C}$ and $\mathbb{R}$, respectively.

For the purposes of this paper, the polynomials $f_1, \ldots, f_s$ are said to be *coprime* if there are no points on $\mathbb{C}^n$ where all the polynomials $f_1, \ldots, f_n$ vanish simultaneously. For polynomials in several variables, the lack of a common factor does not imply coprimeness. For definitions of various types of coprimeness of multivariate polynomials, and their implications towards factorizability of polynomials, see [7, section 2].

2.2. Gröbner Bases

Gröbner bases are an essential ingredient of the algebraic technique presented in section 6. Gröbner bases generalize the concepts of greatest common divisor and division algorithm for univariate polynomials to multivariate polynomials. Details can be found in [2], and computational issues are addressed in [3]. The problem solved by Gröbner bases is the following:

Given $f_1, \ldots, f_k$ and h, all in $\mathbb{C}[z_1, \ldots, z_n]$, can h be expressed as

$$h = \sum g_i f_i, \quad g_i \in \mathbb{C}[z_1, \ldots, z_n] \tag{2}$$

and if so, find such a representation.

For the univariate case, $n = 1$, the solution requires only the computation of the greatest common divisor (gcd) of the f_i and a check that the gcd divides h, where the division yields a unique quotient and a unique remainder. The multivariate division algorithm can be used to compute remainders on division of h by $f_1, \ldots, f_k$. However, the remainder is not unique and depends on the ordering of $f_1, \ldots, f_k$. In particular, the remainder may be nonzero even if h can be expressed as in equation (2). For example, let $f_1 = x_2^2 - 1$ and $f_2 = x_1x_2 - 1$ with $h = x_1^2x_2 + x_1x_2^2 + x_2^2$. Then, h can be written

$$
\begin{aligned}
h &= (x_1 + 1)f_1 + (x_1)f_2 + (2x_1 + 1) \\
&= (1)f_1 + (x_1 + x_2)f_2 + (x_1 + x_2 + 1)
\end{aligned}
$$

Note that neither of the remainders, $2x_1 + 1$ and $x_1 + x_2 + 1$, is divisible by f_1 or f_2, because the remainders have smaller total degrees.

A Gröbner basis F of $\langle f_1, \ldots, f_k \rangle$ is a set of polynomials such that the multivariable division algorithm yields a *unique* remainder, regardless of the order. In particular, if h allows a representation of the form of equation (2), then, on division by F, the remainder is guaranteed to be zero. Further, with respect to a particular ordering of the monomials in $\mathbb{C}[z_1, \ldots, z_n]$, the remainder is of minimal degree.

3. System Model

A system model for polarimetric wide-band SAR is given in [4], [16]. For range much larger than image diameter, the image is related to the desired scene reflectivity by a shift-invariant two-dimensional filtering operation

$$\mathbf{y}(u, r) = \mathbf{h}(u, r) *_u *_r \mathbf{x}(u, r) \tag{3}$$

where (u, r) denotes the cross-range and down-range scene location. The distortion, $\mathbf{h}$, incorporates system effects including antenna response, image processing, transmit pulse, phase center displacements, cable delays, and amplifier gains. For reciprocal scattering, the desired image $\mathbf{x}$ is vector-valued with $N = 3$ values per pixel (see Figure 1). The uncalibrated polarimetric image, $\mathbf{y}$, represents $M = 4$ pixels per location. For example, in a horizontal-vertical polarization basis, images are formed from each combination of transmit and receive polarization to yield $\mathbf{y}_{HH}$, $\mathbf{y}_{HV}$, $\mathbf{y}_{VH}$, and $\mathbf{y}_{VV}$. Thus, 12 two-dimensional FIR filters comprise the filterbank $\mathbf{h}(u, r)$ in equation (3).

This system model and equalization task may be more generally described as follows. We represent finite support n-D signals and channel responses as polynomials in $\mathbb{F}[z_1, \ldots, z_n]$. Denote the N input signals as $\mathbf{x}_i$, $i = 1, \ldots, N$. Let the M output signals $\mathbf{y}_j$ be defined as

$$\mathbf{y}_j = \sum_{i=1}^{N} \mathbf{h}_{ij}\mathbf{x}_i \quad j = 1, \ldots, M \tag{4}$$

where $\mathbf{h}_{ij}$ is the FIR channel from input i to output j. We seek polynomials $\mathbf{g}_{ij}$ such that

$$\sum_{j=1}^{M} \mathbf{g}_{ij}\mathbf{y}_j = \sum_{j=1}^{M} \sum_{k=1}^{N} \mathbf{g}_{ij}\mathbf{h}_{kj}\mathbf{x}_k = \mathbf{x}_i, \quad i = 1, \ldots, N. \tag{5}$$

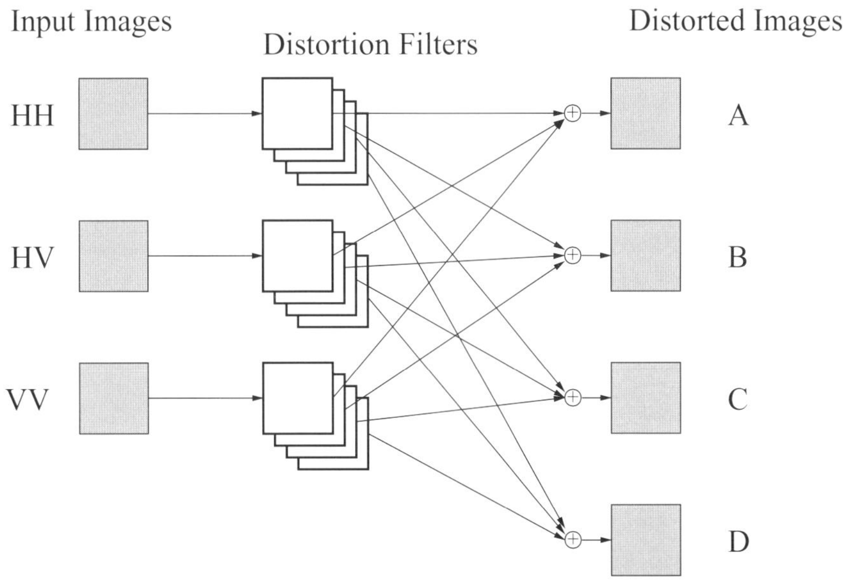

Figure 1. Example forward system model ($N = 3$ inputs, $M = 4$ outputs, images: $n = 2$).

In matrix notation, equation (5) can be written as

$$\mathcal{G}\mathcal{H} = I_N \tag{6}$$

where

$$\mathcal{H} = \begin{pmatrix} \mathbf{h}_{11} & \mathbf{h}_{21} & \dots & \mathbf{h}_{N1} \\ \mathbf{h}_{12} & \mathbf{h}_{22} & \dots & \mathbf{h}_{N2} \\ \vdots & \vdots & \ddots & \vdots \\ \mathbf{h}_{1M} & \mathbf{h}_{2M} & \dots & \mathbf{h}_{NM} \end{pmatrix}, \quad \mathcal{G} = \begin{pmatrix} \mathbf{g}_{11} & \mathbf{g}_{12} & \cdots & \mathbf{g}_{1M} \\ \mathbf{g}_{21} & \mathbf{g}_{22} & \cdots & \mathbf{g}_{2M} \\ \vdots & \vdots & \ddots & \vdots \\ \mathbf{g}_{N1} & \mathbf{g}_{N2} & \cdots & \mathbf{g}_{NM} \end{pmatrix} \tag{7}$$

and I_N denotes the $N \times N$ identity matrix. For simplicity in the sequel, we use a nonstandard indexing in equation (7) for the elements of $\mathcal{H}$. A matrix $\mathcal{H}$ is said to be polynomial if each of its elements $\mathbf{h}_{ij} \in \mathbb{F}[z_1, \dots, z_n]$. Thus, we seek a polynomial left inverse $\mathcal{G}$ of the FIR filterbank $\mathcal{H}$.

We present a necessary and sufficient condition for the existence of $\mathcal{G}$. Although the result is known [1], [7], [15], the proof is constructive and is reviewed here both to fix notation and to present the design steps implemented in Sections 5 and 6.

THEOREM 1: *A matrix $\mathcal{H}$, whose elements are multivariate polynomials with coefficients in a complete field, has a polynomial left inverse $\mathcal{G}$ iff*

1. $M \geq N$ and
2. the $N \times N$ minors of $\mathcal{H}$ are coprime, i.e., $\mathcal{H}$ is *unimodular.*

Proof: Let $\mathcal{H}_i$ denote the $N \times N$ submatrices of $\mathcal{H}$. Let D_i denote the determinant of $\mathcal{H}_i$. If the $N \times N$ minors of $\mathcal{H}$ are coprime, by the weak form of the *Nullstellensatz* [2, p. 169], there exist polynomials a_i, $1 \leq i \leq Q, Q = \binom{M}{N}$, such that

$$\sum_{i=1}^{Q} a_i D_i = 1. \tag{8}$$

Because finite products and finite sums of polynomials are also polynomials, for those $\mathcal{H}_i$ such that $D_i \neq 0$, by Cramer's rule there exist square matrices $\mathcal{G}_i$ (adjoints of $\mathcal{H}_i$) satisfying

$$\mathcal{G}_i \mathcal{H}_i = D_i I_N. \tag{9}$$

Thus, there exist $N \times M$ matrices $\tilde{\mathcal{G}}_i$ (which are $\mathcal{G}_i$ with additional zero rows corresponding to those columns in $\mathcal{H}$ which are not in $\mathcal{H}_i$) such that

$$\tilde{\mathcal{G}}_i \mathcal{H} = D_i I_N. \tag{10}$$

From (8) and (10) it follows that

$$\sum_{i=1}^{Q} (a_i \tilde{\mathcal{G}}_i) \mathcal{H} = \sum_{i=1}^{Q} a_i D_i I_N = I_N \tag{11}$$

whence

$$\mathcal{G} = \sum_{i=1}^{Q} (a_i \tilde{\mathcal{G}}_i). \tag{12}$$

Conversely, suppose that the $N \times N$ minors of $\mathcal{H}$ share a common zero w, i.e., $D_i(w) = 0$, $1 \leq i \leq Q$. Then $\text{rank}(\mathcal{H}(w)) < N$. Then, there does not exist $\mathcal{G}(w)$ such that $\mathcal{G}(w)\mathcal{H}(w) = I_N$. Thus, a polynomial left inverse of $\mathcal{H}$ does not exist. ■

If the coefficients of $\mathbf{h}_{ij}$ are real, then $\mathcal{G}$ can be chosen such that the coefficients of $\mathbf{g}_{ij}$ are also real, so long as the unimodularity condition is satisfied in $\mathbb{C}$ [12, section 1.6]. Hence, in the sequel, we will assume that the base field is $\mathbb{C}$, with the understanding that restriction to $\mathbb{R}$ can be made to obtain analogous results. Thus, results are applicable both to complex baseband SAR images formed from bandpass transmit pulses and to real-valued imagery formed from baseband sampling of received echoes.

4. Filter Orders

The proof of Theorem 1 answers the question of existence of a polynomial left inverse, and the constructive proof leads to algorithms for computing $\mathcal{G}$ given $\mathcal{H}$. In this section, we present bounds for the polynomials, $\mathbf{g}_{ij}$, in the equalization filter. Bounds may be obtained by application of a result by Kollar [12, sections 1.5, 1.7], which uses tools from algebraic geometry; for a more accessible proof of this result, see [6].

THEOREM 2: *Given $f_1, \dots, f_k$ and $h \in \mathbb{C}[z_1, \dots, z_n]$, assume that h vanishes on all the common zeros of $f_1, \dots, f_k$. Let $d_i = \deg(f_i)$ and assume that atmost three of the d_i are equal to 2. Then, one can find $g_1, \dots, g_k \in \mathbb{C}[z_1, \dots, z_n]$ and a natural number s satisfying*

$$\sum_{i=1}^{k} g_i f_i = h^s \tag{13}$$

such that $s \leq N'(n, d_1, \dots, d_k)$ and $\deg(g_i f_i) \leq (1 + \deg(h))\, N'(n, d_1, \dots, d_k)$ where

$$N'(n, d_1, \dots, d_k) = \begin{cases} d_1 \cdots d_k & \text{if } k \leq n \\ d_1 \cdots d_n & \text{if } k > n > 1\,. \\ d_1 + d_2 - 1 & \text{if } k > n = 1 \end{cases} \tag{14}$$

Kollar's result applies to arbitrary fields and is therefore more general than Theorem 2; this version for $\mathbb{C}$ will suffice here. The technical restriction that atmost three of the d_i are equal to 2 is required in the proof technique, but is not restrictive in practice because multidimensional distortion filters are typically larger than degree 2. Further, for time signals ($n = 1$), the restriction can be removed [15]. As an aside, absolute values of the coefficients of g_i can be bounded [12] given bounds on the absolute values of the coefficients of f_i, $1 \leq i \leq k$.

Theorem 2 may be used to bound the degree of the equalizing filterbank, $\mathcal{G}$. Suppose the conditions of Theorem 1 are satisfied and that each channel of the distortion filterbank is bounded in degree by d

$$d = \max_{i,j} \deg(\mathbf{h}_{ij}). \tag{15}$$

Then, the determinants D_i are bounded in degree by Nd, and the degree of every element of the adjoint, $\mathcal{G}_i$, is bounded by $(N - 1)d$. Application of Theorem 2 to equation (8) yields $\deg(a_i) \leq (Nd)^N$ whence

$$\xi \triangleq \max_{i,j} \deg(\mathbf{g}_{ij}) \leq (Nd)^N + (N - 1)d. \tag{16}$$

For most cases of interest, $Q > n$, because $\mathcal{H}$ is generically not invertible [15] otherwise. Then, (with $k > n$ in Theorem 2), $\xi = (Nd)^n + (N - 1)d$. For example in polarimetric calibration, with $N = 3$ desired images, $n = 2$ dimensions, and $d = 7$ point response spread, we have $\xi = 21^2 + 14$, which is large, but not doubly exponential [13]. Unfortunately, this bound is independent of the number of outputs, M; yet, this order bound holds for *all* invertible $\mathcal{H}$. The authors conjecture that for *generic* invertible $\mathcal{H}$, a bound can be obtained that is decreasing in M; such a result for was proven in [9] for the special case of univariate SIMO systems.

5. Linear Algebra for Computation

Because the equalizer is a linear system with finite impulse response, the solution of the equalizer coefficients, $\mathbf{g}_{ij}$, can be written as a large system of linear equations; this provides a direct, "brute force" solution for $\mathcal{G}$. In section 6, we present an alternative algebraic solution technique and discuss its advantages.

The convolution of two sequences x and y can be written as

$$x \circledast y = T(x)\mathrm{vec}(y) \tag{17}$$

where $T(x)$ is a Toeplitz matrix if x is a time signal, or a Toeplitz-block Toeplitz matrix if x is an image signal [10], and vec(y) denotes the vectorized version of y. This process can be extended to signals of arbitrary dimension with nested Toeplitz matrices with nesting level equal to the dimension of the underlying signal space. Then, solving equation (6) is equivalent to solving the following system of linear equations:

$$\begin{bmatrix} T(\mathbf{h}_{11}) & T(\mathbf{h}_{12}) & \cdots & T(\mathbf{h}_{1M}) \\ T(\mathbf{h}_{21}) & T(\mathbf{h}_{22}) & \cdots & T(\mathbf{h}_{2M}) \\ \vdots & \vdots & \ddots & \vdots \\ T(\mathbf{h}_{N1}) & T(\mathbf{h}_{N2}) & \ddots & T(\mathbf{h}_{NM}) \end{bmatrix} \begin{bmatrix} \mathrm{vec}(\mathbf{g}_{i1}) \\ \mathrm{vec}(\mathbf{g}_{i2}) \\ \vdots \\ \mathrm{vec}(\mathbf{g}_{iM}) \end{bmatrix} = \begin{bmatrix} \mathbf{0} \\ \mathrm{vec}(\delta^{(\mathbf{n})}) \\ \vdots \\ \mathbf{0} \end{bmatrix}, \quad i = 1, \ldots, N \tag{18}$$

where $\delta^{(\mathbf{n})}$ is in the ith block row and denotes the appropriately zero-padded n-dimensional Kronecker delta function.

Alternatively, equation (8) in the proof of Theorem 1 converts the MIMO inversion problem to the SIMO problem of computing a_i from the determinants, D_i. Therefore, the approach in equation (18) can be applied to solve for a_i, from which $\mathcal{G}$ can be computed.

Equation (18) provides for solution of $\mathcal{G}$ from knowledge of $\mathcal{H}$. Suppose instead that only input and output pairs, $\{\mathbf{x}, \mathbf{y}\}$, are known. Such would be the case with known calibration targets placed in a scene. The inverse (which equalizes for sources in the subspace spanned by the test targets) is then found by solving, for each i, the equation

$$\sum_j \mathbf{g}_{ij}\mathbf{y}_j = \mathbf{x}_i. \tag{19}$$

Equations (18) and (19) require prior knowledge of the support, and hence order, of each $\mathbf{g}_{ij}$. This is a deficiency in the linear algebraic approach. For an overdetermined system in equation (19), a bound on $\deg(\mathbf{g}_{ij})$ may be used. As an aside, we note that for $n = 1$ and $N = 1$, the block nested Toeplitz matrix (called the Sylvester resultant matrix in this special case) is rank deficient when the order of the inverse filters is overestimated and/or when the forward filters are not coprime. This property is exploited in [20], where the order of the inverse filters are estimated by computing the rank of the Sylvester matrix for increasing orders of the filters until it becomes rank deficient.

A second deficiency with the linear algebraic approach is the size of the coupled system of equations to be solved. Even for reasonably sized filters, the left-hand side matrix is very large and may not fit in memory. While Szegö polynomials have been used to derive fast algorithms for square nested Toeplitz matrices with square blocks [11], [22], such algorithms have not yet been extended to rectangular matrices. A third drawback is the lack of an existence test; the rank of $T(\mathcal{H})$ is affected both by coprimeness and by any selection for the unknown support of the $\mathbf{g}_{ij}$ filters, and the separate effects cannot be discerned.

6. Algebra for Computation

In this section, we propose an algebraic procedure, using Gröbner bases, for computing an equalizing filterbank. The algebraic coprimeness condition for existence in Theorem 1 and the algebraic nature of the constructive proof both suggest algebraic techniques for computation of $\mathcal{G}$. Indeed, the proposed procedure provides a memory efficient computational approach that, significantly, does not require prior knowledge of the inverse filter orders.

We propose the following algorithm for computation of $\mathcal{G}$, given $\mathcal{H}$ and $M \geq N$.

1. Compute the determinants and adjoints, D_i and $\tilde{\mathcal{G}}_i$, as in the proof of Theorem 1.
2. Compute a Gröbner basis F for $\langle D_1, \ldots, D_Q \rangle$.
3. Check whether $1 \in F$. If not, $\mathcal{H}$ is not left invertible.
4. Determine a_i in equation (8); these polynomials are computed while forming the basis F in step 2.
5. Substitute a_i from step 4 into equation (12) to obtain $\mathcal{G}$.

This procedure is not as memory intensive as the linear algebra approach, because no more than three polynomials ever need be stored in memory at once. In addition, by permutation of the reduced Gröbner basis polynomials, an inverse can be computed with either minimal order or minimum number of nonzero coefficients. Note that nonrectangular support regions may be returned for $\mathbf{g}_{ij}$, and no prior knowledge of the support is required. Construction of the Gröbner basis in step 2 is the most computationally intensive element of the algorithm.[2]

This Gröbner basis technique can be used to compute the inverse given an input-output description rather than given $\mathcal{H}$ directly, i.e., to solve equation (19).

As an additional benefit of the algebraic approach, the algorithm may be used to produce a physically meaningful filterbank when no exact inverse exists. Suppose that step 3 of the algorithm above fails, i.e., $1 \notin F$. In this case, we can still recover filtered versions of the input signals, i.e., we can find $\mathcal{G}$ such that

$$\mathcal{G}\mathcal{H} = \text{diag}(q_1, \ldots, q_M)I_N \tag{20}$$

where $q_i \in \mathbb{C}[z_1, \ldots, z_n]$, That is, we can recover $\mathbf{x}_i$ filtered by q_i. The single-channel filters q_i can come from the common zeros of D_i, or may be constructed from prior physical considerations, such as frequency notches in every channel or the desire to null cross-polarized returns. The following algorithm can be used to compute such a $\mathcal{G}$:

1. Compute D_i, $\tilde{\mathcal{G}}_i$ as in the proof of Theorem 1.
2. Compute a Gröbner basis F for $\langle D_1, \ldots, D_Q \rangle$.
3. Compute a reduced [2] Gröbner basis $\tilde{F}$ and its representation in terms of the D_i.
4. For each i from 1 to M, do
   (a) Check whether the remainder of q_i on division by $\tilde{F}$ is 0. If not, the required $\mathcal{G}$ does not exist.
   (b) The division in step 4(a) and the representation of $\tilde{F}$ gives quotients a_k such that

$$\sum_{k=1}^{Q} a_k D_k = q_i. \tag{21}$$

   (c) Compute the ith row of $\mathcal{G}$ as

$$(i\text{th row of } \mathcal{G}) = \sum_{k=1}^{Q} a_k (i\text{th row of } \tilde{\mathcal{G}}_k). \tag{22}$$

This procedure may be viewed as producing an algebraic "pseudo-inverse" that inverts $\mathcal{H}$ at those points in $\mathbb{C}^n$ where an inverse exists. From step 4(c), the authors conjecture that an algebraic technique may be used to computing the MIMO equalizer directly without recourse to the determinants D_i. Step 4(c) implies that the rows of $\mathcal{G}$ are independent of each other. Hence, it may be possible to split the inversion problem into N subproblems which can be solved separately.

Suppose that the desired q_i are monomials $\mathbf{z}^{\mathbf{k}}$, which would correspond to "delays" or shifts. Thus, if the reduced Gröbner basis computed in step 3 divides the delays, it is possible to recover shifted versions of the output. One can show that this is equivalent to requiring in Theorem 1 that all common zeros of the minors D_i lie at the origin. A reformulation of Theorem 1 in terms of Laurent polynomials would provide a more succint description [7] of this result, without providing any additional insight, and while requiring an increase in the complexity of the proof.

7. Example: Radar Calibration

Consider a polarimetric SAR system which transmits and receives both horizontally and vertically polarized waves. By reciprocity [21], this system can be modeled as a

3-input 4-output filterbank of the form in equation (7). For illustrative purposes, suppose $\mathcal{H}$ is given by

$$\mathcal{H} = \begin{bmatrix} 3x^2+2y^2-1 & x^2+y+4 & y^2+2 \\ x & x & xy+1 \\ 1 & 1 & y \\ x^2+y^2-3 & y+3 & y^2+x+4 \end{bmatrix}. \tag{23}$$

First, from the `gb(generators(minors(3, H)),ChangeMatrix=>true)` and `getChangeMatrix` commands of `Macaulay2`, we find that the Gröbner basis of the determinants D_i is given by

$$F = \left\{ y+7, \quad x^2+y^2-\frac{1}{2}y-\frac{5}{2} \right\} \tag{24}$$

$$y+7 = 2D_4 - D_1 \tag{25}$$

$$x^2+y^2-\frac{1}{2}y-\frac{5}{2} = \frac{1}{2}D_1. \tag{26}$$

Note that the Gröbner basis has only two elements; hence, any polynomial which is expressible as a combination of $D_1, \ldots, D_4$ can be expressed as a combination of these two polynomials and conversely.

Second, since $1 \notin F$, $\mathcal{H}$ is not invertible.

Third, we decide to compute $\mathcal{G}$ such that $\mathcal{G}\mathcal{H} = (y+7)I_3$. The corresponding inverse is

$$\mathcal{G} = 2\tilde{\mathcal{G}}_4 - \tilde{\mathcal{G}}_1 = \begin{bmatrix} 1 & 2x+6-2y+yx^2 & -3x^2-6x+2xy+y+2-x^3y & -2 \\ -1 & -y^2-2x-6-yx^2-5y & xy^2+6x+x^3y+5xy+5+3x^2 & 2 \\ 0 & y+7 & -xy-7x & 0 \end{bmatrix} \tag{27}$$

where the $\tilde{\mathcal{G}}_i$ are the adjoints of $\mathcal{H}_i$ with appropriate zero columns added:

$$\tilde{\mathcal{G}}_1 = \begin{bmatrix} -1 & -yx^2-4y+2 & x^3y+x^2+y+4xy+4-2x & 0 \\ 1 & 3yx^2+2y^3-y-y^2-2 & -3x^3y-3x^2-2y^3x-2y^2+xy+1+xy^2+2x & 0 \\ 0 & -2x^2-2y^2+5+y & 2x^3+2xy^2-5x-xy & 0 \end{bmatrix} \tag{28}$$

$$\tilde{\mathcal{G}}_4 = \begin{bmatrix} 0 & x+4-3y & -x^2-4x+3xy+y+3 & -1 \\ 0 & -y^2-x-4+yx^2+y^3-3y & xy^2+4x-x^3y-y^3x+3xy-y^2+3 & 1 \\ 0 & y+6-x^2-y^2 & -xy-6x+x^3+xy^2 & 0 \end{bmatrix}. \tag{29}$$

For a simpler case, consider the 2-by-3 forward system where a second image is summed with filtered versions of the first:

$$\mathcal{H} = \begin{bmatrix} 2x - y + 3 & 1 \\ xy & 1 \\ x + 3 & 1 \\ x + y + 2 & 1 \end{bmatrix}. \tag{30}$$

In this case, the Gröbner basis of the minors contains 1, and an exact equalizer exists. Following the procedure above, the inverse is found to be

$$\begin{aligned} \mathcal{G} &= \frac{1}{3}\begin{bmatrix} 1 & -1 & 0 & 0 \\ -xy & 2x - y + 3 & 0 & 0 \end{bmatrix} - \frac{1}{3}\begin{bmatrix} 1 & 0 & -1 & 0 \\ -x - 3 & 0 & 2x - y + 3 & 0 \end{bmatrix} \\ &\quad - \frac{x}{3}\begin{bmatrix} 0 & 0 & 1 & 1 \\ 0 & 0 & -x - y - 2 & x + 3 \end{bmatrix} \\ &= \begin{bmatrix} 0 & -1 & -x + 1 & x \\ x - xy + 3 & 2x - y + 3 & x^2 + xy + y - 3 & -x^2 - 3x \end{bmatrix}. \end{aligned} \tag{31}$$

Note that many of the inverse filters do not have rectangular support; the support regions found here could not have been predicted in advance. When the equalizer computed here is applied on a set of images, it yields computational savings over the linear algebra solution which typically yields filters with maximum support.

8. Summary

The problem of designing calibration filters for polarimetric SAR systems was formulated as computing FIR MIMO equalizers for multi-dimensional systems. A necessary and sufficient condition was presented for the existence of an equalizing filterbank, and bounds were given for the order of the filters. An algebraic procedure was proposed for the construction of the filterbank, either from knowledge of the MIMO distortion system or from a sufficient set of input-output measurements. Built on Gröbner bases, the design algorithm alleviates two deficiencies in a linear least-squares design procedure: the search for appropriate, non-rectangular filter support regions and the large memory requirements. Further, the new algorithm provides a useful equalization filterbank when no exact inverse exists.

Notes

1. Fractional bandwidth is defined as system bandwidth divided by center frequency.
2. The computations can be performed using freely available packages such as `Macaulay2` from `http://www.math.uiuc.edu/Macaulay2`. The relevant commands are `gb`, `minors`, `generators`, `getChangeMatrix`.

References

1. C.A. Berenstein and A. Yger, "On Explicit Solutions to the Bezout Equation," *Systems & Control Letters*, vol. 4, February 1984, pp. 33–39.
2. D.A. Cox, J. Little, and D. O'Shea, *Ideals, Varieties, and Algorithms*, Springer-Verlag, 1992.
3. J.H. Davenport, Y. Siret, and E. Tournier, *Computer Algebra*, Academic Press, 1988.
4. E. Ertin and L.C. Potter, "Polarimetric Calibration for Wide Band Synthetic Aperture Radar Imaging," *IEEE Proceedings - Radar, Sonar and Navigation*, vol. 145, no. 5, October 1998, pp. 275–280.
5. E. Ertin and L.C. Potter, "Polarimetric Classification of Scattering Centers Using M-ary Bayesian Decision Rules," *IEEE Transactions on Aerospace and Electronic Systems*, vol. 36, no. 3 Part: 1, July 2000, pp. 738–749.
6. N. Fitchas and A. Galligo, "Nullstellensatz Effectif et Conjecture de Serre (Théorème de Quillen-Suslin) pour le Calcul Formel," *Math. Nachr.*, vol. 149, 1990, pp. 231–253.
7. E. Fornasini and M.E. Valcher, "Multidimensional Systems with Finite Support Behaviors: Signal Structure, Generation, and Detection," *SIAM Journal on Control and Optimization*, vol. 36, no. 2, 1998, pp. 760–769.
8. A. Freeman and J. van Zyl, J. Klein, H. Zebker, and Y. Shen, "Calibration of Stokes and Scattering Matrix Format Polarimetric SAR Data," *IEEE Transactions on Geoscience and Remote Sensing*, vol. 30, no. 3, May 1992, pp. 531–539.
9. G. Harikumar and Y. Bresler, "FIR Perfect Signal Reconstruction from Multiple Convolutions: Minimum Deconvolver Orders," *IEEE Transactions on Signal Processing*, vol. 46, no. 1, January 1998, pp. 215–218.
10. G. Harikumar and Y. Bresler, "Exact Image Deconvolution from Multiple FIR Blurs," *IEEE Transactions on Image Processing*, vol. 8, no. 6, June 1999, pp. 215–218.
11. J.H. Justice, "A Levinson-Type Algorithm for Two-Dimensional Wiener Filtering Using Bivariate Szegö Polynomials," *Proceedings of the IEEE*, vol. 65, no. 6, June 1977, pp. 882–886.
12. J. Kollar, "Sharp Effective Nullstellensatz," *Journal of the American Mathematical Society*, vol. 1, no. 4, October 1988, pp. 963–975.
13. E. Mayr and A. Meyer, "The Complexity of the Word Problem for Commutative Semigroups and Polynomial Ideals," *Advances in Mathematics*, vol. 46, 1982, pp. 305–329.
14. L. Novak, S. Halversen, G. Owirka, and M. Hiett, "Effects of Polarization and Resolution on SAR ATR," *IEEE Transactions on Aerospace and Electronic Systems*, vol. 33, no. 1, January 1997, pp. 102–116.
15. R. Rajagopal and L. Potter, "MIMO FIR Equalizers and Orders," in *ICASSP 2001 Proceedings*, 2001.
16. R. Rau and J.H. McClellan, "Analytic Models and Postprocessing Techniques for UWB SAR," *IEEE Transactions on Aerospace and Electronic Systems*, vol. 36, no. 4, October 2000, pp. 1058–1074.
17. G.A. Showman and J.H. McClellan, "Blind Polarimetric Equalization of UWB SAR Imagery," in *Proceedings of 2000 International Conference on Image Processing*, 2000, pp. 673–676.
18. M. Soumekh, "Reconnaissance with Ultra Wideband UHF Synthetic Aperture Radar," *IEEE Signal Processing Magazine*, vol. 12, no. 4, July 1995, pp. 21–40.
19. W. Steedly and R. Moses, "High Resolution Exponential Modeling of Fully Polarized Radar Returns," *IEEE Transactions on Aerospace and Electronic Systems*, vol. 27, no. 3, May 1991, pp. 459–469.

20. L. Tong, G. Xu, and T. Kailath, "Blind Identification and Equalization Based on Second-Order Statistics: A Time Domain Approach," *IEEE Transactions on Information Theory*, vol. 40, no. 2, March 1994, pp. 340–349.
21. J.J. van Zyl, "Calibration of Polarimetric Radar Images Using Only Image Parameters and Trihedral Corner Reflector Responses," *IEEE Transactions on Geoscience and Remote Sensing*, vol. 28, no. 3, 1990, pp. 337–347.
22. M. Wax and T. Kailath, "Efficient Inversion of Toeplitz-Block Toeplitz Matrix," *IEEE Trans. on Acoustics, Speech and Signal Processing*, vol. 31, no. 5, October 1983, pp. 1218–1221.
23. H. Zebker, J. Van Zyl, S. Durden, and L. Norikane, "Calibrated Imaging Radar Polarimetry: Technique, Examples, and Applications," *IEEE Transactions on Geoscience and Remote Sensing*, vol. 29, no. 6, November 1991, pp. 942–961.

Multidimensional Systems and Signal Processing, 14, 119–137, 2003

Signal Processing for Large Bandwidth and Long Duration Waveform SAR

ZHIPING LIN[1] ezplin@ntu.edu.sg
School of EEE, Nanyang Technological University, Singapore

YONGHONG ZENG eyhzeng@ntu.edu.sg
School of EEE, Nanyang Technological University, Singapore
National University of Defense Technology, P.R. China

GUOAN BI egbi@ntu.edu.sg
School of EEE, Nanyang Technological University, Singapore

JOCELYN YEO
DSO National Laboratories, Singapore

Received April 11, 2001; Revised September 28, 2001; Accepted December 7, 2001

Abstract. We first show that for some kinds of signals a "bandwidth and time duration reduction technique" can be used to simulate waveform distortions caused by moving targets, that is, it is correct to measure the waveform distortions at very large TB with relatively small v by reducing TB and increasing v while keeping TBv unchanged, where T is the duration of the transmitted signal, B is the bandwidth and v is the relative speed of targets. We then study the waveform distortions in SAR signals caused by moving antenna. Based on the bandwidth and time duration reduction technique, a lot of time and memory are saved in simulations. We then confirm by simulations that waveform distortions do pose problems when processing very large bandwidth and long duration SAR data using conventional SAR processing methods. Finally we propose the concepts of wideband and narrowband processing of SAR data. Models are set up for wideband and narrowband SAR data processing, and new methods are presented for reconstructing targets using the proposed models. Simulations show that the methods can improve the quality of the simulated SAR images.

Key Words: large bandwidth, long duration waveform, SAR data processing, SAR model, SAR image

1. Introduction

The common mode of operation for conventional SAR is like this: while an aircraft carrying a SAR sensor is flying along a particular path, an antenna transmits a short duration waveform (pulse), then receives an echo corresponding to this pulse, transmits another pulse and so on. Due to the motion of antenna, relative motion of targets during transmission and reception of a pulse always exists. This relative motion causes distortions (dilations) in the received waveforms. In conventional SAR, since the pulse duration is very short, this distortion can be ignored. This makes traditional SAR signal processing relatively easy and explains the fact that currently most (if not all) practical SAR systems use very short pulses as their waveforms for transmission [13]. On the other hand, using large bandwidth and long duration waveforms do have some advantages. First, low probability of interception (LPI) may be achieved. It is relatively easy for an intercept receiver to detect the radiated signal of conventional radars at long ranges [1], [10], [19]. To reduce a radar's detectability to a hostile intercept receiver, its peak power should be made as low as possible. The radiated energy should be spread over a wide angular, over a long time interval, and over a wide frequency band [12]. So, if a SAR system can transmit a long duration waveform, the instantaneous power of the transmitted signal will be reduced and LPI be achieved. Second, large bandwidth and long duration waveform leads to large time-bandwidth product and may improve the resolution of SAR imaging [5], [14]. Ultra high carrier frequency waveform can be used to improve the cross-range resolution. The Twin-Otter SAR developed at the Sandia National Laboratories can operate on frequency band as high as in the Ka (14 to 16 GHz) and Ku (32.6 to 37 GHz) bands [15]. Its spotlight resolution can reach higher than 1 meter. The central carrier frequency of the currently used millimeter-wave radar can be larger than 10GHz [7], [9], [18]. Such kind of radars have been widely used in many areas [7], [9], [18]. We believe large bandwidth and ultra high carrier frequency could be used in future SAR systems.

For a SAR system with large bandwidth, ultra high carrier frequency and large time duration, the waveform distortion (dilation) due to relative motion of targets during transmission and reception of a pulse cannot be ignored anymore, especially when the speed of the platform is very high such as in spaceborne radar [3], [4], [6], [8]. Since traditional signal processing methods for SAR ignore such distortions, they may not be valid anymore here. There are many auto-focus and motion compensation algorithms [2], [13], [16]. However, since each of them has its own purpose and is based on its own conditions, they cannot be used for our case directly. For example, most known motion compensation algorithms have the following assumptions [2], [13] (for example, see [13], section 6.7, pp. 465–477): (a) the speed of the targets are constant; (b) the speed is so small that the targets can be treated as motionless during transmission and reception of a pulse. These two assumptions are obviously not valid in our case. In SAR, the relative speeds of targets with respect to the antenna are dependent on the location of the antenna and the location of the targets. In large bandwidth and long duration case, dilations in received SAR signal cannot be ignored based on our simulations. This means that targets cannot be treated as motionless during transmission and reception of a pulse. So the problem discussed by us is much more difficult and new methods must be proposed. The purpose of this paper is to investigate the waveform distortion in SAR and seek methods to improve the distorted images.

The rest of paper is organized as follows. We first propose a "bandwidth and time duration reduction technique" in section 2. In order to reduce the computational complexity for simulation purpose, we show that for some kinds of signals it is correct to measure waveform distortions for very large TB with relatively small υ by reducing TB and increasing υ while keeping $TB\upsilon$ unchanged, where and henceforth T is the duration of the transmitted signal, B is the bandwidth and υ is the relative speed of the target. We then study the waveform distortion due to the moving antenna in section 3. Simulations are done to show waveform distortions in very large bandwidth and long duration SAR data by using conventional SAR processing methods. The concepts of wideband and narrowband processing of SAR data are proposed in section 4. We present new methods for reconstructing targets using the proposed models. Simulations show that the methods can improve the quality of the simulated SAR images.

2. Bandwidth and Time Duration Reduction Technique

2.1. Waveform Distortions Due to Moving Targets

It is well known that when a transmitter/receiver and an object (target) reflecting the transmitted waveform have relative motion, the received waveform will be distorted. The amount of distortion depends on the waveform, the relative motion and the speed of propagation of the signal [17]. Assume that the transmitted signal is $p(t)$ and the target is moving at a constant velocity v. It has been shown [17] that the echoed signal of a single point-target at distance R_0 (the distance of the target at time 0) is

$$s(t) = p(a(t - \tau)), \tag{1}$$

where

$$\tau = 2R_0/c, \ a = \frac{c - v}{c + v}, \tag{2}$$

and c is the speed of the light. When the transmitted signal is a narrowband signal, the time scaling can be further approximated by the Doppler shift. More accurately, if the "narrowband condition" [11], [17] is satisfied, that is,

$$TB \le \frac{0.1c}{v}. \tag{3}$$

or

$$TBv \le 0.1c, \tag{4}$$

the echoed signal can be approximated by

$$s(t) \approx p(t - \tau)e^{-j2\pi f_d}, \tag{5}$$

where $f_d = \frac{2v}{c} f_c$ is the Doppler frequency shift and f_c is the central carrier frequency. However, for wideband signals we have to consider the more general case of time scaling, that is, the time scaling (dilation) factor a must be considered and it should not be approximated by a Doppler shift.

2.2. Bandwidth and Time Duration Reduction Technique

Before carrying out actual simulations, it is noticed that for signals with large time bandwidth product, the amount of data to be processed will be huge. For wideband signal of frequency bandwidth B and time duration T, at least 2TB discrete samples are required to represent the signal. If $T = 10^{-3}$ (s) and $B = 1000$ (MHz), at least 2 million samples are required! In SAR applications, we also need to sample the synthetic aperture domain. The number of samples depends on the size of the synthetic aperture, the radar frequency and the size of the cross range in the illuminated area. If 1000 samples are chosen in the synthetic aperture, we have totally 2×10^9 samples. This will require considerable amount of processing time as well as huge memory. In order to reduce the complexity for simulating the waveform distortion in SAR, we propose a "bandwidth and time duration reduction technique" in the following to simplify the simulation process.

It is observed that for certain waveforms such as chirps, the shape and properties of the waveforms will be preserved after the bandwidth of waveforms is reduced while the time duration keeps unchanged. Since the main concern here is waveform distortion, it suffices to show that waveform distortion remains unchanged for both signals with the original bandwidth and with a reduced bandwidth. Notice the factors causing waveform distortion are given in expression (3). Hence if the bandwidth B is reduced 10 times, and the velocity of aircraft is at the same time increase 10 times, condition (3) remains unchanged and the amount of waveform distortion, if any, will remain unchanged. We show this in the following when the transmitted signal is a chirp.

Let the transmitted signal be an up-chirp, $p(t) = e^{j\alpha t + j\beta t^2}$,where α is the carrier frequency and β is the chirp rate. When the speed of the target is v, the received signal $s(t)$ is described in (1). If the target is stationary, the received signal is then $s_1(t) = p(t - \tau)$, where τ is the same as that in (1). For the investigation of waveform distortion, τ can be assumed to be 0. So, the difference between $s(t)$ and $s_1(t)$ is then

$$s(t) - s_1(t) = e^{ja\alpha t + ja^2\beta t^2} - e^{j\alpha t + j\beta t^2}. \tag{6}$$

Since $v \ll c$, we have $a = \frac{c-v}{c+v} \approx 1 - \frac{2v}{c}$. Furthermore, $a^2 \approx 1 - \frac{4v}{c}$. Therefore,

$$s(t) - s_1(t) \approx e^{j\alpha t + j\beta t^2}\left(e^{-j\frac{2v}{c}\alpha t - j\frac{4v}{c}\beta t^2} - 1\right). \tag{7}$$

$$|s(t) - s_1(t)| \approx |e^{-j\frac{2v}{c}\alpha t - j\frac{4v}{c}\beta t^2} - 1|. \tag{8}$$

We know that the bandwidth B of the signal is approximately βT and the central carrier frequency is nearly $\alpha + \beta T/2$. If we keep the ratio of the bandwidth and carrier frequency to be constant, that is, the relative bandwidth $\frac{\beta T}{\alpha + \beta T/2}$ is a constant q, and also keep the product of the speed of the target and bandwidth, Bv, to be constant, we know that the difference in (8) remains unchanged. In fact, since $\alpha = \left(\frac{1}{q} - \frac{1}{2}\right)\beta T = \left(\frac{1}{q} - \frac{1}{2}\right)B$ and $\beta v = \frac{B}{T}v$, (8) can be turned to

$$|s(t) - s_1(t)| \approx |e^{-j\frac{2Bv}{c}\left(\frac{1}{q} - \frac{1}{2}\right)t - j\frac{4Bv}{cT}t^2} - 1|, \tag{9}$$

where Bv, q and T are fixed. So, the difference is constant. That is, we can measure the waveform distortion by keeping Bv to be constant and allow B to be small while v very large.

Table 1. Mean square error of the distorted signal (Chirp).

Bandwidth	Duration	Speed of targets	Error
20MHz	$10\mu s$	3000Km/s	1.4161
200MHz	$10\mu s$	300Km/s	1.4046
2GHz	$10\mu s$	30Km/s	1.4035
20GHz	$10\mu s$	3Km/s	1.4034
20MHz	$100\mu s$	300Km/s	1.4234
200MHz	$100\mu s$	30Km/s	1.4026
2GHz	$100\mu s$	3Km/s	1.4033
20GHz	$100\mu s$	300m/s	1.4140

Simulations are conducted to test this observation, and it works even when the bandwidth is reduced as much as 1000 times! There are some small deviations due to the finite data length effect, which are insignificant for the investigation on waveform distortion. For example, to investigate waveform distortion for a signal with bandwidth 20GHz, and the velocity of target being 300 m/s, it is equivalent to do the same for a signal with bandwidth 20MHz, and the velocity of target being 300 Km/s! Of course there is no aircraft in the real world can travel at such a high speed. But it works well if we just want to know the effect of time-bandwidth product on waveform distortion. We use the up-chirps for our simulations. The simulations result are listed in Table 1. In our simulations, the center carrier frequency is the same as the bandwidth, that is, $q = 1$. The table gives the mean square error of the distorted signal with respect to the undistorted signal ($\upsilon = 0$), that is,

$$\sqrt{\sum_i |s_1(t_i) - s(t_i)|^2} \Big/ \sqrt{\sum_i |s_1(t_i)|^2}. \tag{10}$$

We assume that we have one target at distance 2000 meters. As can be seen from Table 1, the relative errors caused by waveform distortions are significant and hence cannot be ignored.

This observation is also valid for step chirp signals. We can prove this theoretically similarly as we did for chirp before. For other signals, this observation must be checked carefully. A major problem is that for most other kinds of signals, bandwidths depend on time durations. That is, when the bandwidth increase the time duration usually must be shortened. For example, for a biphase signal, its bandwidth depends on the width of each subpulse, say T_s. If the time duration of the signal is fixed, say T, then to keep with the increase of the bandwidth, the number of subpulses, say N, must also increase so that the reciprocal of the width of each subpulse, N/T, is proportional to the increased bandwidth. When the bandwidth increase from B to $10B$, the number of subpulses must also increase form N to $10N$. However, in practice, in order to achieve good auto-correlation function only certain types of biphase signals are useful such as Barker code. If we use the Barker code, when the time duration of the signal T is fixed, the length-N Barker code is very different from the length-$10N$ Barker code since the Barker code is chosen randomly. So, how can we compare the distortions of the received signals if the transmitted signals

Table 2. Mean square error of the distorted signal (Frank code).

Bandwidth	Duration	Speed of targets	Error
20MHz	10μs	3000Km/s	1.4050
80MHz	10μs	750Km/s	1.4029
320Hz	10μs	187.5Km/s	1.3906
1.28GHz	10μs	46.875Km/s	1.3908

are very different? It is meaningless to compare the distortions on such cases. Furthermore, it is very difficult to get the Barker code for large number of subpulses. Frank code is easy to be used since we can easily produce any length Frank code. We have done some simulations on this code (see Table 2). Here we use the width of the mainlobe as the bandwidth of the signal. Therefore, if the time duration is fixed to be $10\mu s$, $20MHz$ bandwidth corresponds to $N = 100$ subpulses. For the step frequency code, such as Costas code, this observation also seems nearly correct. Table 3 gives the simulation results, where the length-24 Costas code is used.

Another observation is that for certain waveforms such as chirps, the impact of waveform distortion on the reconstructed image quality is almost unchanged if we fix B and the product Tv respectively while letting T and v vary. We have done some simulations to support this observation in one-dimensional range radars. We assume the following scene. The transmitted signal is an up-chirp. There are four targets in the area. The distance of the center of targets is 2000m (meter) and all targets are located in $[2000 - 50, 2000 + 50]$ m. Three targets are at distance 2000m, 2035m and 1975m respectively, which are independent on the radar parameters. Another targets is at distance $2035 + 2 * dx$ m, where dx is the sampling interval in the range domain which depends on the bandwidth. The bandwidth and the central carrier frequency are 200MHz. In Figure 1-a, we assume that the targets are stationary and the time duration of transmitted signal is 10μs. In Figure 1-b to 1-d, we assume the speed of the targets is 300Km/s, 30Km/s and 3Km/s respectively, and the time duration of the signals are 10μs, 100μs and 1000μs respectively. From Figure 1-b to 1-d, we see that the images of the reconstructed targets are almost unchanged. So, for simulation purpose, in order to save time and memory, we can replace long time duration by high speed. For example, to investigate the impact of waveform distortion on the quality of reconstructed image when the time duration is 1000μs and the velocity of target is 3 Km/s, it is equivalent to do the same for a signal with time duration of 10μs and the velocity of target being 300 Km/s!

Table 3. Mean square error of the distorted signal (Costas code).

Bandwidth	Duration	Speed of targets	Error
20MHz	10μs	3000Km/s	1.3796
80MHz	10μs	750Km/s	1.4136
320Hz	10μs	187.5Km/s	1.4148
1.28GHz	10μs	46.875Km/s	1.4151

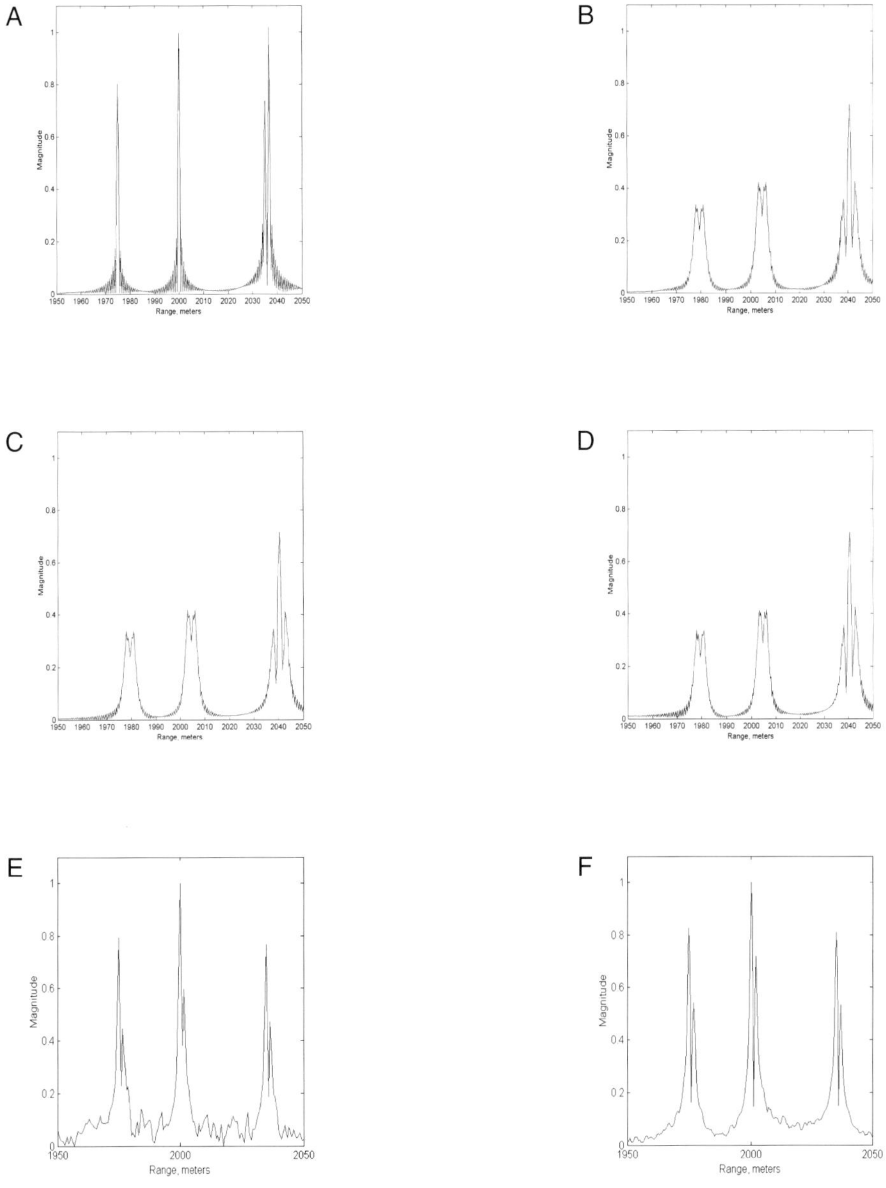

Figure 1. (a) Range reconstruction via matched filtering ($B = 200MHz$, $T = 10us$, $\upsilon = 0$). (b) Range reconstruction via matched filtering ($B = 200MHz$, $T = 10us$, $\upsilon = 300Km/s$). (c) Range reconstruction via matched filtering ($B = 200MHz$, $T = 100us$, $\upsilon = 30Km/s$). (d) Range reconstruction via matched filtering ($B = 200MHz$, $T = 1000us$, $\upsilon = 3Km/s$). (e) Range reconstruction via matched filtering (Frank code) ($B = 200MHz$, $T = 0.25us$, $\upsilon = 1200Km/s$). (f) Range reconstruction via matched filtering (Frank code) ($B = 200MHz$, $T = 16us$, $\upsilon = 18.75Km/s$).

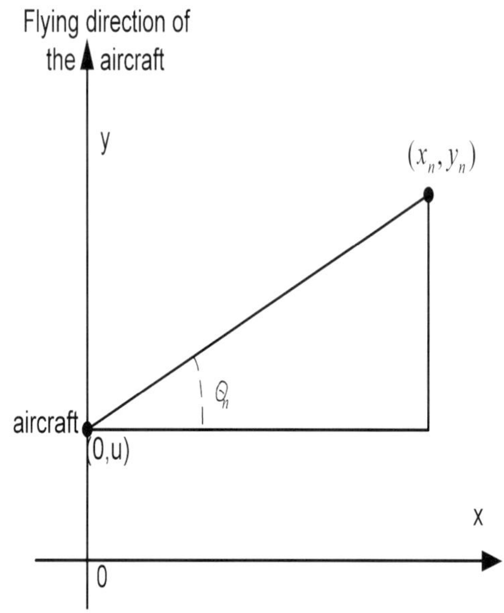

Figure 2. System Model for SAR.

This observation is also true for step chirp signal. For other types of signals, this observation must be checked carefully. For Frank code, this observation seems nearly correct. To keep the bandwidth unchanged, we keep the ratio N/T to be constant. In our simulations, the bandwidth is fixed at $200MHz$. When $T = 0.25\mu s$, N is chosen to be 5. Figures 1-*e* to 1-*f* are the simulation results.

In general, for some signals like chirps the impact of waveform distortion on the quality of reconstructed image is nearly unchanged if we keep TBv constant while letting v be very large and TB be relatively small. This bandwidth and time duration reduction technique will enable us to save a large amount of time and memory for simulations in large bandwidth and long duration SAR signal processing.

3. Waveform Distortion in SAR Due to Moving Antenna

For large bandwidth and ultra high carrier frequency SAR applications, the waveform distortion due to the moving antenna may not be ignored. We consider the following model for SAR [13]: the transmitted signal is $p(t)$, the aircraft is moving at a constant velocity v, and the speed of propagation of the signal is c. The targets are located at (x_n, y_n) $(n = 1, 2, \cdots, N)$, where x_n is the slant-range and y_n is the cross-range (see Figure 2). For simplicity, all the targets are assumed motionless. The aircraft is flying along the y axis and its location is then $(0, u)$, where $u \in [-L, L]$. $2L$ is the length of synthetic aperture.

From Figure 2, we see that

$$\sin\theta_n = \frac{y_n - u}{\sqrt{x_n^2 + (y_n - u)^2}}. \tag{11}$$

The reflected signal by target (x_n, y_n) is then

$$s_n(t, u) = \sigma_n p(\lambda_n(t - \tau_n)), \tag{12}$$

where σ_n is the reflective coefficient of target (x_n, y_n),

$$\lambda_n = \frac{c + \upsilon\sin\theta_n}{c - \upsilon\sin\theta_n}, \tag{13}$$

$$\tau_n = 2\sqrt{x_n^2 + (y_n - u)^2}/c. \tag{14}$$

So, the received signal should be

$$s(t, u) = \sum_{n=1}^{N} s_n(t, u) = \sum_{n=1}^{N} \sigma_n p(\lambda_n(t - \tau_n)). \tag{15}$$

A major difference between the dilation λ_n here and the dilation a in one dimensional case in (1) is that λ_n not only depends on the speed υ but also on the direction of the target. This causes the wideband processing method for SAR which will be discussed in the following section much more difficult.

In current SAR practice, the scale factors λ_n are ignored, that is, the received signal is approximated by [13]

$$\bar{s}(t, u) = \sum_{n=1}^{N} \bar{s}_n(t, u) = \sum_{n=1}^{N} \sigma_n p(t - \tau_n). \tag{16}$$

We will show that such a simplification may lead to big errors sometimes when the bandwidth of the transmitted waveform is very large and the time duration of the waveform is long. We have done some simulations for spotlight SAR by using the wavefront reconstruction method and M. Soumekh's Matlab programs ([13], chapter 5, pp. 319–372) and the above received signal formulae (15) and (16). As discussed above, we know that the waveform distortion is determined by the time bandwidth product TB and the relative speed υ. Furthermore, based on the derivation in section 2, we know that for some waveforms such as chirps, the impact of waveform distortion on the reconstructed image remains almost unchanged if we keep $TB\upsilon$ constant while letting υ be very large and TB be relatively small. To save computational time, we would like TB to be as small as possible.

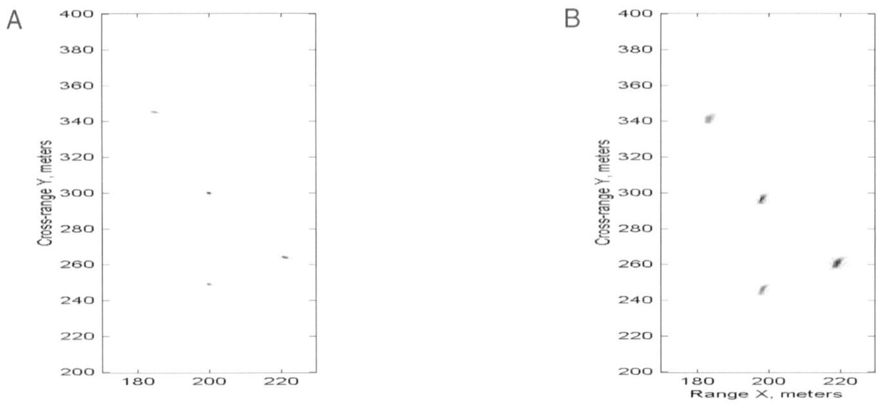

Figure 3. (a) Reconstructed SAR image ($B = 200MHz$, $T = 0.1us$, $v = 0$). (b) Reconstructed SAR image ($B = 200MHz$, $T = 0.1us$, $v = 40{,}000Km/s$).

So, in our simulations, we keep *TB* to be small while letting the speed v be very large. This is of course not the case in practice. However, the results are almost the same when *TB* is very large while v is relatively small. We give three examples in the following.

(I) For Figures 3-a to 3-b, the basic parameters are as follows:

$B = 200MHz$ (bandwidth)
$f_c = 200MHz$ (central carrier frequency)
$X_c = 200m$ (target center in slant range)
$X_0 = 30m$ ($2X_0$ is the length of slant range)
$Y_c = 300m$ (target center in cross range)
$Y_0 = 60m$ ($2Y_0$ is the length of cross range)
$L = 200m$, $T = 0.1\mu s$.

The results for $v = 0$, $v = 40{,}000Km/s$ are shown in the Figures 3-a to 3-b respectively. When $v = 40{,}000Km/s$, the TBv is then 8×10^8 m/s. If we consider the ordinary speed for aircraft, that is, $v = 200m/s$, then, to keep the TBv unchanged, we need *TB* to be 4×10^6. We can choose *B* to be $4GHz$ and *T* to be $1ms$, or other pairs of values for *B* and *T*. This is practical.

(II) For Figures 4-a to 4-b, the basic parameters are the same as those in Figure 3-a and Figure 3-b except that the speed of the aircraft is reduced to $v = 400Km/s$ and the time duration is raised to $T = 10\mu s$. The results for $v = 0$, $v = 400Km/s$ are shown in the Figures 4-a to 4-b respectively. It is clearly seen that the reconstructed image in Figure 4-b is of the same quality as that in Figure 3-b. The TBv here is the same as that in the previous example. We can imagine that similar results occur when the aircraft flies at the ordinary speed, $v = 200m/s$, while the bandwidth *B* is $8GHz$ and *T* is 0.5ms. In spaceborne radar [3], [4], the speed can be as high as $v = 10Km/s$. Hence we can expect similar results occur

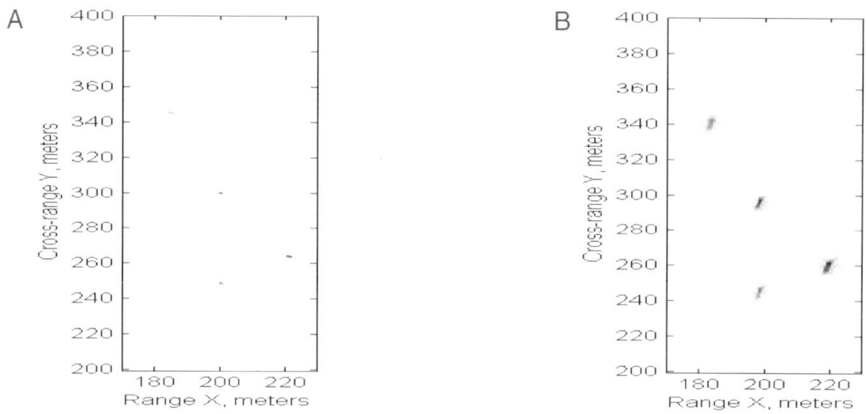

Figure 4. (a) Reconstructed SAR image ($B = 200MHz$, $T = 10us$, $\upsilon = 0$). (b) Reconstructed SAR image ($B = 200MHz$, $T = 10\mu s$, $\upsilon = 400Km/s$).

for spaceborne radar when $B = 800MHz$ and $T = 0.1ms$, which is reasonable in the near future.

(III) For Figures 5-a to 5-b, the basic parameters are as follows:

$$B = 2GHz,\ f_c = 2GHz,\ X_c = 200m,\ X_0 = 20m$$
$$Y_c = 300m,\ Y_0 = 30m,\ L = 50m,\ T = 0.1\mu s.$$

The results for $\upsilon = 0$, $\upsilon = 20{,}000Km/s$ are shown in the Figures 5-a to 5-b respectively. When $\upsilon = 20{,}000Km/s$, the $TB\upsilon$ is then 4×10^9 m/s. If we consider the ordinary speed for

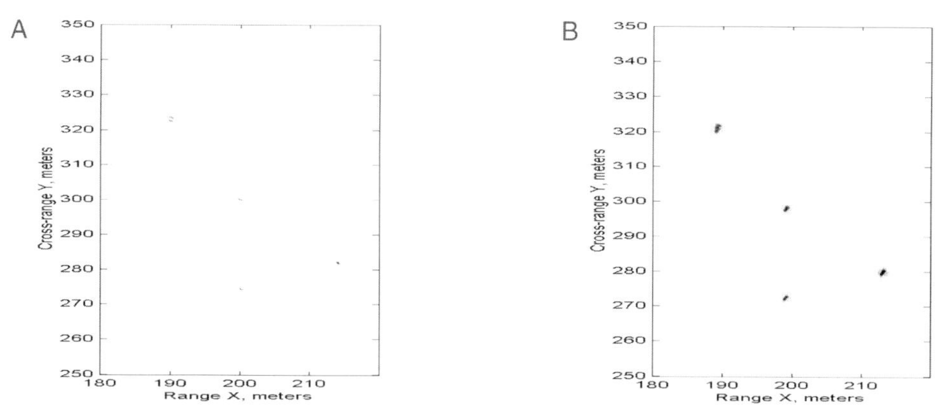

Figure 5. (a) Reconstructed SAR image ($B = 2GHz$, $T = 0.1\mu s$, $\upsilon = 0$). (b) Reconstructed SAR image ($B = 2GHz$, $T = 0.1\mu s$, $\upsilon = 20{,}000Km/s$).

aircraft, that is $v = 200m/s$, then, to keep the TBv unchanged, we need TB to be 2×10^7. We can choose B to be $20GHz$ and T to be $1ms$, which are reasonable in SAR.

The simulations show that the waveform distortion cannot be ignored when the bandwidth of the transmitted waveform is very large and the time duration of the waveform is long. We should develop methods to deal with the problem.

4. Wideband and Narrowband Processing for SAR Signals

As discussed in the previous section, we need new methods to deal with the waveform distortion in SAR signals. Unlike ordinary range radars, SAR detect targets in two- or three-dimensional space. This leads to much higher complexity in signal processing algorithms for SAR signals. The wideband and narrowband processing methods for one-dimensional signals are no longer valid for SAR [11], [17]. Although there are many auto-focus and motion compensation algorithms, as explained in the introduction they cannot be used here directly. New methods should be proposed. In this section, based on the analysis of the received SAR signal, we will propose new methods for the problem. Like one-dimensional case, we call a method which considers the dilation of the received SAR signal a **wideband processing method** for SAR signals.

We now further analyze the received SAR signal. For $v \ll c$,we have

$$\lambda_n \approx 1 + \frac{2v}{c}\sin\theta_n, \quad \frac{1}{\lambda_n} \approx 1 - \frac{2v}{c}\sin\theta_n. \tag{17}$$

So, by using (17) and the first order approximation in the Taylor series expansion, we have

$$\begin{aligned} s(t,u) &= \sum_{n=1}^{N} \sigma_n p(\lambda_n(t-\tau_n)) \\ &\approx \sum_{n=1}^{N} \sigma_n p\left[\left(1+\frac{2v}{c}\sin\theta_n\right)(t-\tau_n)\right] \\ &\approx \sum_{n=1}^{N} \sigma_n p(t-\tau_n) + \sum_{n=1}^{N} \sigma_n p'(t-\tau_n)\frac{2v\sin\theta_n}{c}(t-\tau_n). \end{aligned} \tag{18}$$

Let $S(\omega, u)$ be the Fourier transform of $s(t, u)$ with respect to fast time t and $P(\omega)$ be the Fourier transform of $p(t)$. We get

$$\begin{aligned} S(\omega,u) &= \sum_{n=1}^{N} \sigma_n/\lambda_n P(\omega/\lambda_n) e^{-j\omega\tau_n} \\ &\approx \sum_{n=1}^{N} \sigma_n P(\omega/\lambda_n) e^{-j\omega\tau_n} \end{aligned}$$

$$\approx \sum_{n=1}^{N} \sigma_n P\left(\omega - \frac{2v\sin\theta_n}{c}\omega\right) e^{-j\omega\tau_n}$$

$$\approx \sum_{n=1}^{N} \sigma_n \left[P(\omega) - P'(\omega)\frac{2v}{c}\frac{y_n - u}{\sqrt{x_n^2 + (y_n - u)^2}}\omega \right] e^{-j\omega\tau_n}$$

$$\approx P(\omega)\sum_{n=1}^{N} \sigma_n e^{-j\omega 2\sqrt{x_n^2+(y_n-u)^2}/c}$$

$$-\omega P'(\omega)\sum_{n=1}^{N} \sigma_n \frac{2v}{c}\frac{y_n - u}{\sqrt{x_n^2 + (y_n - u)^2}} e^{-j\omega 2\sqrt{x_n^2+(y_n-u)^2}/c}. \quad (19)$$

Since the currently available methods (see [13], e.g.) for detecting targets (x_n, y_n) all ignore the last term in the equation (18) or (19), errors are inevitable. Suitable methods for compensating the impact of the term must be found.

Like narrowband processing in the one-dimensional case [6], [8], we can also replace the dilation by Doppler frequency shifts. However, in SAR the Doppler frequency shifts are not only related to the speed of the vehicle, but also related to the direction of the targets. This makes the problem much more difficult. When $TB \le \frac{0.1c}{v}$, a **narrowband processing for SAR** signals is as follows.

Let f_d^n be the Doppler frequency shift, that is,

$$f_d^n = \frac{2v\sin\theta_n}{c} f_c, \quad (20)$$

where f_c is the central carrier frequency. If the narrowband condition (3) is satisfied, we have

$$s(t, u) \approx \sum_{n=1}^{N} \sigma_n p(t - \tau_n) e^{j2\pi f_d^n (t-\tau_n)}. \quad (21)$$

In frequency domain, we have,

$$S(\omega, u) \approx \sum_{n=1}^{N} \sigma_n P(\omega + 2\pi f_d^n) e^{-j\omega\tau_n}$$

$$\approx \sum_{n=1}^{N} \sigma_n P\left(\omega + \frac{4\pi v f_0}{c}\frac{y_n - u}{\sqrt{x_n^2 + (y_n - u)^2}}\right) e^{-j\omega 2\sqrt{x_n^2+(y_n-u)^2}/c}. \quad (22)$$

As discussed above, either wideband processing or narrowband processing for SAR signals is very different from that in the one-dimensional case. The known methods for the one-dimensional case (see [17], e.g.) cannot be used directly. In the following, we give two iterative methods which can be used to improve the quality of reconstructed SAR images.

From (16) and (18) we see that

$$\bar{s}(t,u) = \sum_{n=1}^{N} \sigma_n p(t-\tau_n)$$
$$= s(t,u) - \sum_{n=1}^{N} \sigma_n p'(t-\tau_n)\frac{2\upsilon\sin\theta_n}{c}(t-\tau_n). \tag{23}$$

If we can get $\bar{s}(t, u)$, we can use the currently known methods such as the wavefront reconstruction method [13] to locate the targets (x_n, y_n) and find the reflective coefficient σ_n. In practice, what we get is the received signal $s(t, u)$. Due to the impact of the second term in (18), using known methods such as the wavefront reconstruction method on the received signal $s(t, u)$ cannot get images in good quality, which are shown in Figures 3–5. However, when we get a rough estimation of $(x_n, y_n)$and σ_n, we can use them to improve the quality of the image. If the rough estimation of the targets and the reflectivities are $(\tilde{x}_n, \tilde{y}_n)$ and $\tilde{\sigma}_n$ respectively ($n = 1, 2, \cdots, N$), we get an approximation for $\bar{s}(t, u)$ by

$$\bar{s}(t,u) \approx s(t,u) - \sum_{n=1}^{N} \tilde{\sigma}_n p'(t-\tilde{\tau}_n)\frac{2\upsilon\sin\tilde{\theta}_n}{c}(t-\tilde{\tau}_n), \tag{24}$$

where

$$\sin\tilde{\theta}_n = \frac{\tilde{y}_n - u}{\sqrt{\tilde{x}_n^2 + (\tilde{y}_n - u)^2}}, \tag{25}$$

$$\tilde{\tau}_n = 2\sqrt{\tilde{x}_n^2 + (\tilde{y}_n - u)^2}/c. \tag{26}$$

We call this method “**Wideband Compensation**”.

We have done simulations based on the above method. An example is shown below. We first use the wavefront reconstruction method [13] on the received signal $s(t, u)$ to get a rough image of the targets (Figure 4-b). From the rough image, we get an estimation of the location of the targets and the reflectivities. Then we use (24) to get an approximation for $\bar{s}(t, u)$. Finally, the wavefront reconstruction method is used again on $\bar{s}(t, u)$ to get an image (Figure 6-a). The process can be done iteratively until we get a satisfactory image. Simulations have shown that this method can improve the quality of the image. The radar parameters here are described in the example (I) of section 3. Figure 4-a is the reconstructed image when the speed of the aircraft $\upsilon = 0$. Figures 7-a to 7-c are the corresponding 3D images of Figure 4-a, Figure 4-b and Figure 6-a respectively, where the values are normalized so that they are in the interval [0,255]. The mean square error (MSE) between Figure 7-a and Figure 7-c is 0.0207. The mean square error (MSE) between Figure 7-a and Figure 7-b is 0.0398. So after one iteration, the MSE is reduced by nearly one half.

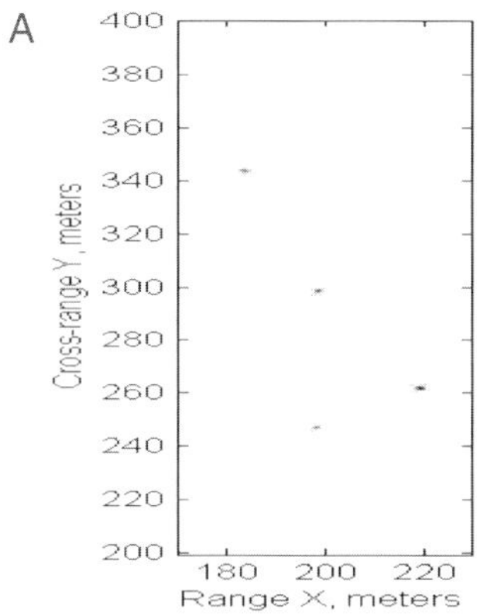

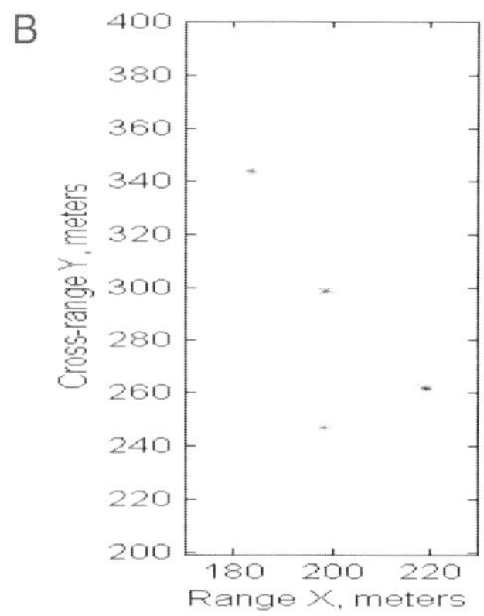

Figure 6. (a) Wideband compensation SAR image ($v = 400Km/s$). (b) Narrowband compensation SAR image ($v = 400Km/s$).

When the narrowband condition is satisfied, we can also improve the image quality using the approximation (21). Based on (21) and the first order approximation of the Taylor series expansion, we have

$$s(t,u) \approx \sum_{n=1}^{N} \sigma_n p(t-\tau_n) + \sum_{n=1}^{N} \sigma_n j2\pi f_d^n (t-\tau_n) p(t-\tau_n). \tag{27}$$

So,

$$\begin{aligned} \bar{s}(t,u) &= \sum_{n=1}^{N} \sigma_n p(t-\tau_n) \\ &\approx s(t,u) - \sum_{n=1}^{N} \sigma_n j2\pi f_d^n (t-\tau_n) p(t-\tau_n). \end{aligned} \tag{28}$$

As discussed above, we can use the wavefront reconstruction method to reconstruct the image if we have a good approximation of $\bar{s}(t, u)$. When we get a rough estimation of (x_n, y_n) and σ_n, we can use them to improve the quality of the image. If the rough estimation of the targets and the reflectivities are $(\tilde{x}_n, \tilde{y}_n)$and $\tilde{\sigma}_n$ respectively ($n = 1, 2, \cdots, N$), we get an approximation for $\bar{s}(t, u)$ as

$$\bar{s}(t,u) \approx s(t,u) - \sum_{n=1}^{N} \tilde{\sigma}_n j2\pi \tilde{f}_d^n (t-\tau_n) p(t-\tilde{\tau}_n), \tag{29}$$

where

$$\tilde{f}_d^n = \frac{2v \sin \tilde{\theta}_n}{c} f_c, \tag{30}$$

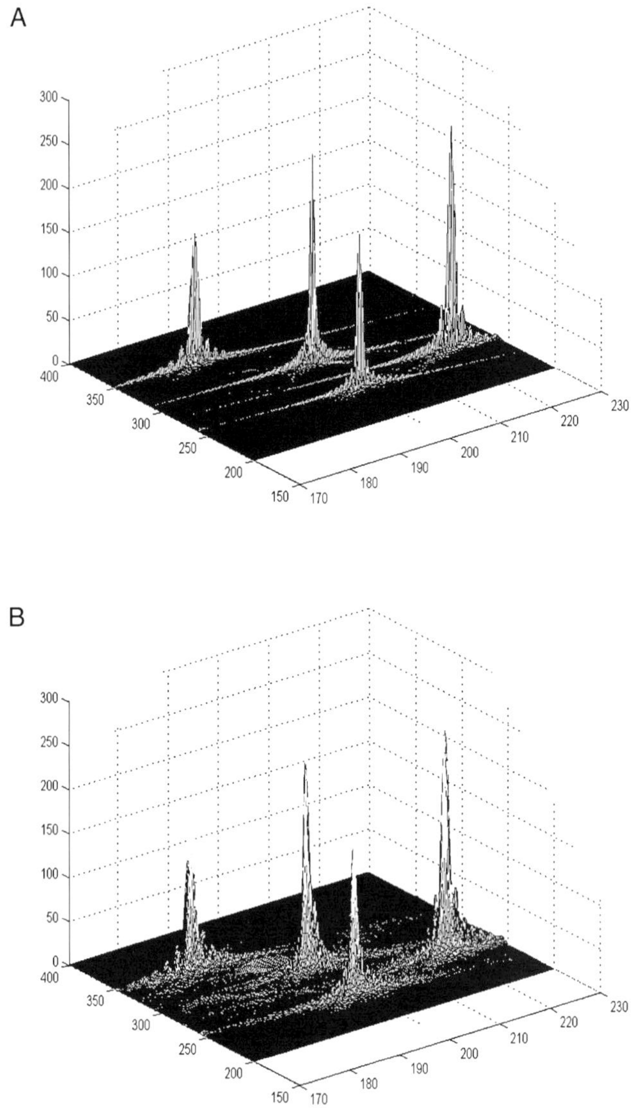

Figure 7. (a) Reconstructed SAR image ($\upsilon = 0$). (b) Reconstructed SAR image ($\upsilon = 400Km/s$).

and $\sin\tilde{\theta}_n$ and $\tilde{\tau}_n$ are defined in (25) and (26). We call this method "**Narrowband Compensation**".

We have also done simulations based on the "**Narrowband Compensation**" method. A simulation result is given in Figure 6-b, where the signal is chirp. The case used here is the same as the example discussed above for "wideband compensation" method. Figures 7-d

C

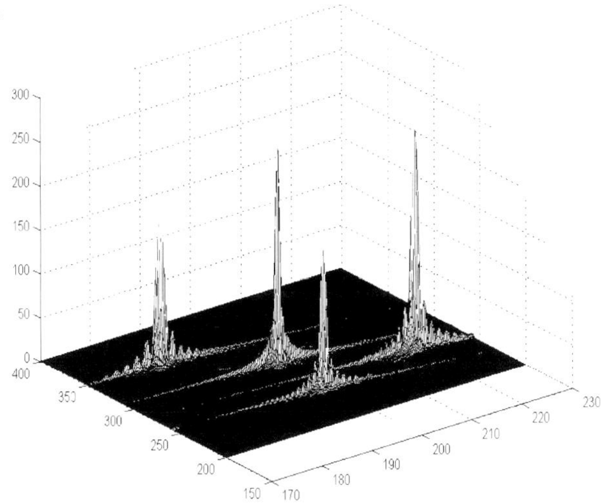

D

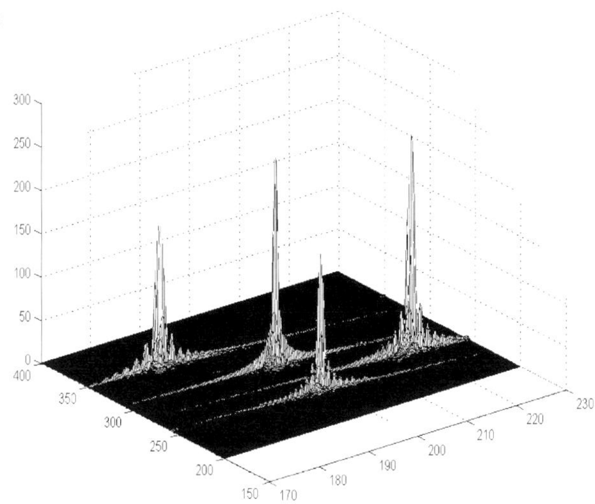

Figure 7. (c) Wideband compensation SAR image ($v = 400Km/s$). (d) Narrowband compensation SAR image ($v = 400Km/s$).

is the corresponding 3D images. For this example, the "**Narrowband Compensation**" is sufficient. But for some other examples, the "**Narrowband Compensation**" is not efficient. To save space, we will not show them here.

From the simulation, we see that for some wideband SAR signals, sometimes the "**Narrowband Compensation**" method is valid even if the narrowband condition (3) is violated. This may be due to the fact that the dilation caused by the speed of the vehicle is different in different directions. The narrowband condition should also be changed according to different directions. Narrowband condition (3) is only the worst case.

An advantage of "**Narrowband Compensation**" over "**Wideband Compensation**" is that the "**Narrowband Compensation**" can be used for any type of signals even if the signals are non-continuous, while the "**Wideband Compensation**" requires that the signals are smooth (have at least first-order derivative). At present time, our method is only valid for a small number of separated point-targets. For large number of closely located targets, the method needs to be improved further.

5. Conclusions

It has been shown that in SAR, due to the motion of the aircraft, waveform distortion always exists. For large bandwidth and long duration SAR systems, which we believe could be the future SAR systems, waveform distortions due to moving antennas cannot be ignored. Huge amount of time and memory are needed to simulate the processing of large bandwidth and long duration SAR signals. In order to reduce the computational complexity and memory for simulation purpose, we have shown that for some kinds of signals it is correct to measure waveform distortions for very large TB with relatively small υ by reducing TB and increasing υ while keeping $TB\upsilon$ constant. Based on this observation, we are able to run simulations on SAR waveform distortions using general personal computers. The concept of wideband and narrowband process of SAR data are proposed. New method are presented for reconstructing targets using the proposed wideband and narrowband SAR model. Simulations have shown that the methods can improve the quality of the simulated SAR images.

The study is only concerned with the signal processing aspect. There are additional problems associated with large bandwidth and long duration SAR systems, such as the leakage of transmitted signals into received signals. The investigation of these problems and the search of more efficient wideband SAR processing algorithms will be our future works.

Acknowledgments

Thank the reviewers for their invaluable comments which help us to improve the paper.

Note

1. Corresponding author: Zhiping Lin, Nanyang Technological University, School of EEE, Block S2, Singapore 639798.

References

1. M. Burgos-Garcia, J. Sanmartin-Jara, F. Perez-Martiniz, and J.A. Retamosa, "Radar Sensor Using Low Probability of Interception SS-FH Signals," *IEEE AES Systems Magazine*, vol. 15, April 2000, pp. 23–28.
2. W. Carrara, R.S. Goodman, and R.M. Majewski, *Spotlight Synthetic Aperture Radar: Signal Processing Algorithms*, Artech House, 1995.
3. C. Elachi, "Space Imaging Radar in Planetary Exploration and Earth Observation," *AIAA Journal*, vol. 39, no. 4, 2001, pp. 553–563.
4. B. Huneycutt and J. Zuzek, "Frequency Use and Needs of Spaceborne Active Sensors," *International Geoscience and Remote Sensing Symposium (IGARSS)*, vol. 6, 2000, pp. 2457–2463.
5. M.G.M. Hussain, "Ultra-Wideband Impluse Radar - an Overview of the Principles," *IEEE AES Systems Magazine*, Sept. 1998, pp. 9–14.
6. Q. Jin and K.M. Wong, "The Estimation of Time Delay and Doppler Stretch of Wideband Signals," *IEEE Transactions on Signal Processing*, vol. 43, no. 4, April 1995, pp. 904–916.
7. D. Klugmann, "A Compact 94-GHz FM-CW Doppler Cloud Radar Using Semiconductor Devices for Power Generation," *Proceedings of IEEE International Conference on Geoscience and Remote Sensing Symposium (IGARSS)*, vol. 5, 2000, pp. 1801–1803.
8. X.X. Niu, P.C. Ching, and Y.T. Chan, "Wavelet Based Approach for Joint Time Delay and Doppler Stretch Measurements," *IEEE Transactions on Aerospace and Electronic Systems*, vol. 35, no. 3, July 1999, pp. 1111–1119.
9. S. Noda, K. Inomata, M. Watanabe, T. Fukae, and M. Tobioka, "A Millimeter-Wave Radar for Train Application," *Proceedings of IEEE International Conference on Vehicle Electronics Conference (IVEC)*, vol. 1, 1999, pp. 153–158.
10. J.F. Pietrasinski, T.W. Brenner, and C.J. Lesnik, "Selected Tendencies of Modern Radar and Radar Systems Development," *Proceedings of 12th International Conference on Microwave and Radar*, vol. 1, 1998, pp. 133–137.
11. A.W. Rihaczek, *Principles of High Resolution Radar*, McGraw-Hill, 1969.
12. M.I. Skolnik, *Introduction to Radar System*, 3rd ed., McGraw-Hill Higher Education, 2001.
13. M. Soumekh, *Synthetic Aperture Radar Signal Processing with MATLAB Algorithms*, John Wiley & Sons, INC. 1999.
14. J.D. Taylor, *Ultra-Wideband Radar Technology*, CRC Press, 2000.
15. B. Walker, G. Sander, M. Thompson, B. Burns, R. Fellerhoff, and D. Dubbert, A High-Resolution, Four-Band SAR Testbed with Real-Time Image Formation, http://www.sandia.gov/radar/files/igarss96.pdf.
16. L.M.H. Ulander, P.O. Frolind, and T. Martin, Processing and Calibration of Ultra-Wideband SAR Data from CARABAS-II, http://www.estec.esa.nl/ceos99/papers/p143.pdf.
17. L.G. Weiss, "Wavelets and Wideband Correlation Processing," *IEEE Signal Processing Magazine*, no. 1, Jan. 1994, pp. 13–32.
18. X.B. Yang, J.Y. Yang, and C.G. Jing, "A Research on Millimeter-Wave LFMCW Radar for Air-Field Object Imaging," *Conference Digest of International Conference on Infrared and Millimeter Waves*, 2000, pp. 413–414.
19. M.C. Zari, A.F. Zwilling, D.A. Hess, J. Jo, C.S. Anderson, and D. Chiang, "Soldier Identification System Utilizing Low Probability of Intercept (LPI) Techniques," *IEEE AES Systems Magazine*, vol. 12, July 1997, pp. 21–26.

Multidimensional Systems and Signal Processing, 14, 139–159, 2003

Target-Centered Models and Information-Theoretic Segmentation for Automatic Target Recognition

MICHAEL D. DEVORE mdd2@cis.wustl.edu
Electronic Systems and Signals Research Laboratory, Dept. of Electrical Engineering, Washington University, St. Louis, MO 63130

JOSEPH A. O'SULLIVAN jao@ee.wustl.edu
Electronic Systems and Signals Research Laboratory, Dept. of Electrical Engineering, Washington University, St. Louis, MO 63130

Received May 9, 2001; Revised May 2, 2002; Accepted May 10, 2002

Abstract. We present an approach to automatic target recognition (ATR) from synthetic aperture radar (SAR) imagery which combines advantages of both model-based and template-based approaches. Prior observations are used to estimate the statistical properties of reflectance over regions in the training scene. These target-centered statistical models can then be used to estimate the statistical properties of sensor output for arbitrary pose. Two-sided hypothesis tests which are maximally powerful at the most likely alternative are developed in a information-theoretic framework to address target model segmentation and confuser rejection. Segmentation of target from clutter is performed in the target-centered coordinate system using all prior observations to produce a consistent segmentation over all poses. We present performance and computation complexity results as a function of segmentation threshold, confuser-rejection threshold, and operating conditions for publicly available SAR data.

Key Words: automatic target recognition, synthetic aperture radar, statistical hypothesis testing, performance-complexity, resource consumption rate, information-based segmentation, information-based confuser rejection, MSTAR

1. Introduction

Automatic target recognition (ATR) systems are generally based on some characterization of sensor output conditioned on the class, state, and pose of objects which may be in a scene. In the problem of ATR based upon synthetic aperture radar (SAR) images, several approaches to object characterization have been reported in the literature. The family of approaches generally referred to as model-based methods represent various attempts to characterize the physical structure of the target classes of interest which are then used to predict the SAR sensor output for arbitrary combinations of target class, pose, etc. Classification of an object in some observation is performed by searching over the hypothesis space for the parameter combination that yields a predicted observation closest to the actual observation. The physical structure of the targets can be represented by CAD models as in the Moving and Stationary Target Acquisition and Recognition (MSTAR) ATR system [19] with features or signatures predicted from electromagnetic modeling tools such as described by Keydel and Lee [9]. A variation on this approach is to extract information about primitive shapes comprising a target class from sets of training images of those targets as in the work of Richards, et al. [13] and Chiang, et al. [1].

The family of approaches to ATR often referred to as template-based methods attempt to build a database of typical sensor outputs from training images of a wide range of target classes and parameterizations such as reported by Owirka, et al. [12] and Ross, et al. [14]. Such databases represent a direct model of the sensor output for each target class bypassing a model of the target structure itself. The database is typically searched for the sensor output which best resembles an actual observation, and the corresponding target class and parameters are declared as the ATR system output. In what may be viewed as a hybrid between model-based and template-based approaches, some have sought to start with a model describing the physics of SAR imaging and from this model determine a statistical model for the sensor output [2], [11]. Unknown parameters of the statistical model are estimated from training data and stored as a function of target class, pose, etc. Similar to the template-based approaches, a target in an observation is classified by searching the parameter database for the set of parameters which, when applied to the statistical sensor model, best describe the observation.

Model-based approaches have the advantage of being able to predict sensor output at arbitrary target pose, but construction of the models can be problematic given limited training data. Template-based methods often bin sample data into some partitioning of pose-space and determine average templates over each pose subset. Such partitioning has the advantage of not requiring large amounts of training data but suffers from an inability to describe the likely sensor output at an exact target pose. In this paper, we consider methods which combine the ability of model-based approaches to predict signatures of targets in previously unseen poses with the statistical sensor modeling and ability to employ training images enjoyed by template-based approaches. The idea is to estimate the statistical properties not of pixels in an image but of patches in the scene as a function of viewing angle. The result is a collection of distribution parameter functions defined in a target-centered coordinate system. Verbout, et al. [17] build three-dimensional templates which they use to predict SAR images under previously unseen squint angles but still maintain templates for each combination of depression and azimuth angles. By comparison, we develop target models in which each location on the target surface has an associated function of depression and azimuth that characterizes the region of a SAR image corresponding to the target location. These target models and associated statistical functions comprise deformable templates which can accommodate an infinite variety of target pose and radar power fluctuation.

Given an estimated model of a training scene, it is desirable to segment the model into target and clutter regions. Also, given a SAR image, it is desirable to infer whether or not the image is of a target for which a model exists. Both of these problems can be phrased as a two-sided hypothesis test $\theta = \theta_0$ versus $\theta \neq \theta_0$, where θ_0 characterizes a known distribution. No uniformly most powerful test, which has maximum power for any given significance across all possible alternatives, exists for this problem. Often, researchers impose restrictions on the set of tests under consideration and select the test which is most powerful across the subset. For example, the restriction to unbiased tests, for which the power is larger than the significance level, yields the uniformly most powerful unbiased

(UMPU) test (cf. [15, p. 349]). By contrast, we employ the test which maximizes power at the most likely alternative, yielding a decision based upon the empirical relative entropy between distributions parameterized by $\hat{\theta}$ and θ_0. Segmentation affects the way target classification is performed because the physical dimensions of segmented target models varies with target class, and the size of a target's projection onto the slant plane varies with pose. We discuss modifications to the generalized likelihood ratio test that accommodate varying lengths of observation vectors and relate segmentation to the problem of dimensionality reduction.

In Section 2, we describe conditional probabilistic models for SAR imagery, two-sided hypothesis testing, and the use of variable-sized target images in classification and confuser rejection. Section 3 describes the use of training data to estimate target-centered models of the training scene, segmentation of the training scene to obtain target models, and prediction of image statistical properties from arbitrary pose. Section 4 contains an analysis of the results obtained by applying these methods to a variety of multi-class ATR problems using SAR data from the Moving and Stationary Target Acquisition and Recognition (MSTAR)[1] program. Conclusions follow in Section 5.

2. Model-Based ATR

In this section we briefly review the conditionally complex Gaussian model for SAR imagery and its implications for ATR. We discuss similarities between image segmentation and dimensionality reduction, and we present a classification algorithm motivated by this connection. Finally, we address the problem of deciding whether an image is of a target class not represented among the available models.

2.1. SAR Image Model

The scene being imaged by the sensor platform represents a source of information which may be regarded as probabilistic in nature with some *a priori* probability mass function on the class A of objects in the scene, $P_A(a)$, and some conditional probability density on the pose $\mathbf{\Psi}$ that a given object class is likely to present, $p_{\mathbf{\Psi}|A}(\psi|a)$. In the most general case, $\mathbf{\Psi}$ is a vector with continuous-valued components that capture the three rotational degrees of freedom for a target, the three translational degrees of freedom, and the states of articulated components. Information from the scene is collected via the sensing platform which yields an observation $\mathbf{R}$ that has a probabilistic dependence $p_{\mathbf{R}|A,\mathbf{\Psi}}(\mathbf{r}|a, \psi)$ on the scene. The observation $\mathbf{R}$ represents a measurement in vector form, and for the SAR imaging scenario we can take $\mathbf{R}$ to be an mn length vector with components taken row-wise from an $m \times n$ SAR image.

The complex Gaussian distribution for $p_{\mathbf{R}|A,\mathbf{\Psi}}$ is introduced for high resolution radar in [7] and is extended to SAR in [11]. According to this model, a resolution cell in the scene contains a number of point-like scatterers with complex return signals that combine coherently to produce a complex value R_i in the corresponding pixel. If the number of scatterers is large and no one dominates within a resolution cell, the real and imaginary components of R_i follow an approximate Gaussian distribution. If the phase of the scatterers is uniformly distributed, the pixel value will have a complex mean of zero. Assuming that the pixels in a SAR image represent nonoverlapping resolution cells in the scene, the complex pixel values are independent from one another and the conditional density function for an entire SAR image is a product over all pixel i of the conditional density for each pixel. We denote by $\sigma_i^2(a, \psi)$ the variance of pixel i when target a presents itself with pose ψ.

The power in the radar echo is a function of the distance to the target. Additionally, it is known that the power transmitted by a radar platform can fluctuate from one device to another within the same model line and for any given unit the power output can drift from a nominal value over time. This power fluctuation has the effect that the pixels in a reconstructed SAR image may be uniformly scaled by some constant c. We incorporate this constant into the model by assuming that the received SAR image $\mathbf{R} = c\mathbf{Y}$, where $\mathbf{Y}$ represents the image received under some nominal power. Then R_i is a complex, conditionally Gaussian random variable with variance $c^2\sigma_i^2(a, \psi)$ and the conditional density for $\mathbf{R}$ becomes

$$p_{\mathbf{R}|A,\mathbf{\Psi}}(\mathbf{r}|a, \psi) = \prod_i \frac{1}{\pi c^2 \sigma_i^2(a, \psi)} e^{-|r_i|^2/c^2\sigma_i^2(a,\psi)}. \tag{1}$$

This conditional density is a continuous-valued function defined over the discrete parameter A and the continuous pose parameter $\mathbf{\Psi}$. In Section 3 we consider target models that allow us to approximate $p_{\mathbf{R}|A,\mathbf{\Psi}}$ for arbitrary $\mathbf{\Psi}$.

2.2. *Two-Sided Hypothesis Testing*

It is well known that the likelihood ratio test maximizes the power over all other tests for any given significance level for the case of a simple hypothesis $\mathcal{H}_0$ against a simple alternative $\mathcal{H}_A$. For an observation $\mathbf{r}$ with independent components, this test is equivalent to choosing $\mathcal{H}_A$ if $\sum_{i=1}^N \ln[p_A(r_i)/p_0(r_i)] > \gamma$. In many cases, we need to test a simple $\mathcal{H}_0$, $\mathbf{r} \sim p_0$, against a compound alternative $\mathcal{H}_A$, $\mathbf{r} \not\sim p_0$. This is a two-sided hypothesis testing problem for which there is no uniformly most powerful test against all alternatives. We therefore seek a test based upon the most likely alternative. The resulting test will form the basis for both confuser rejection and target model segmentation.

In the case of zero-mean complex Gaussian distributions, $\mathcal{H}_0$ is an assertion that $r_i \sim \mathcal{CN}(0, \sigma_i^2)$. For any given alternative, $r_i \sim \mathcal{CN}(0, \xi_i^2)$, the likelihood ratio test becomes

$$\sum_{i=1}^N \ln\left(\frac{\sigma_i^2}{\xi_i^2}\right) + \frac{|r_i|^2}{\sigma_i^2} - \frac{|r_i|^2}{\xi_i^2} > \gamma. \tag{2}$$

From a single observation component r_i, the most likely alternative is $r_i \sim \mathcal{CN}(0, |r_i|^2)$.

Substituting $\xi_i^2 = |r_i|^2$ into the log-likelihood ratio above yields

$$\sum_{i=1}^{N} \ln\left(\frac{\sigma_i^2}{|r_i|^2}\right) + \frac{|r_i|^2}{\sigma_i^2} - 1 > \gamma. \tag{3}$$

The left side of this inequality is the sum of relative entropies between the distributions $\mathcal{CN}(0, \sigma_i^2)$ and $\mathcal{CN}(0, |r_i|^2)$. This relative entropy is defined in terms of the natural logarithm and so has units of nats.

2.3. *Hypothesis Testing Algorithms*

It is pointed out in [11] that in experiments with SAR data the likelihood function $p_{\mathbf{R}|A,\Psi}$ is often very sharply peaked in the vicinity of the true target and pose. In this case, the generalized likelihood ratio test (GLRT) for target classification and maximum *a posteriori* (MAP) estimates for pose yield results nearly identical to the minimum probability of error classifier and minimum mean-square error estimator, respectively. The algorithm yielding the GLRT inferred target $\hat{a}_{\mathrm{GLRT}}$ and MAP estimate for pose $\hat{\psi}_{\mathrm{MAP}}$ can be represented in a single equation as

$$\begin{bmatrix} \hat{a}_{GLRT} \\ \hat{\psi}_{MAP} \\ \hat{c}^2 \end{bmatrix} = \underset{[a,\psi,c^2]^T}{\operatorname{argmax}}\; p(\mathbf{r}|a,\psi,c^2)p(\psi|a)P(a). \tag{4}$$

Given that $\mathbf{r}$ is hypothesized to have come from a target a at pose ψ, the maximum-likelihood estimator for c^2 is

$$\hat{c}^2(a,\psi) = \frac{1}{N}\sum_i \frac{|r_i|^2}{\sigma_i^2(a,\psi)}, \tag{5}$$

where N is the number of components in $\mathbf{r}$.

These inference methods assume that all components of the observation vector $\mathbf{R}$ are relevant to the recognition problem. In practice, however, the components of the observation vector that are germane may vary with the hypothesis considered. This is the case if a model dependent segmentation is employed. The set of pixels corresponding to target regions in the scene clearly varies with target class a, due to varying target sizes and shapes, and with pose ψ, due to target location in the image, etc. Model segmentation is very closely related to the subject of dimensionality reduction in which one seeks a lower dimension subset of statistics to characterize an observation vector for use in a recognition system. Dimensionality reduction can result in lower error rates in cases where the class statistics are estimated from a finite set of training data. As the clutter distribution is estimated from training samples, its inclusion can drastically hinder recognition accuracy and we seek a method of reducing the dimensionality of the estimated statistics to reduce the impact of inaccurately known clutter.

Schmid and O'Sullivan [16] consider an approach to dimensionality reduction based on hard thresholding of the information of each measurement component relative to an assumed "null-hypothesis." This null-hypothesis can be interpreted as representing the distribution of an uninformative component, and selection is performed based upon whether the empirical relative entropy between the component value and the null-hypothesis exceeds some threshold. Components of an observation vector which are used in computing likelihoods varies with target class. To accommodate this variation, likelihoods of the dimensionality reduced statistics under each class are computed relative to likelihoods under the null-hypothesis. They derive asymptotic analytical bounds on the probability of error in a two-class case and demonstrate lower error bounds than without segmentation for a two-class problem with simulated high resolution radar profiles.

DeVore, Schmid, and O'Sullivan [4] apply this method to ATR from SAR imagery and show that as few as 250 components from each of four target classes can be retained while decreasing the classification error rate by several percentage points. For individual target classes, the error rate was drastically reduced for classes with small training sets and was not increased even for those with large training sets. Their dimensionality reduction method is equivalent to the hypothesis testing problem in (3). We apply this method to target model segmentation and select a varying number of components from each class rather than some fixed-small number. Details and examples of this segmentation are provided in Section 3.2. Here we discuss appropriate modification of the GLRT to accommodate class dependent component selection.

Application of their method requires a definition of the null-hypothesis which we take to be a complex Gaussian distribution with zero mean and variance equal to that of the average background clutter, ξ^2. Let $\mathbf{I}(a, \psi)$ be a binary mask with the value one in each pixel location hypothesized to be on target when object a is imaged with pose ψ and zero in all other pixel locations. The modified GLRT becomes

$$\begin{bmatrix} \hat{a}_{GLRT} \\ \hat{\psi}_{MAP} \\ \hat{c}^2 \end{bmatrix} = \underset{[a,\psi,c^2]^T}{\text{argmax}}\; p(\psi|a)P(a) \prod_{i:I_i(a,\psi)=1} \frac{p(r_i|a,\psi)}{p_{R|\xi^2}(\mathbf{r}|\xi^2)}. \tag{6}$$

Assuming uniform distribution on a and ψ, substituting the complex Gaussian probability distribution in (1), and noting that maximizing a positive function is equivalent to maximizing its logarithm yields

$$\begin{aligned} \begin{bmatrix} \hat{a}_{GLRT} \\ \hat{\psi}_{MAP} \end{bmatrix} &= \underset{[a,\psi]^T}{\text{argmax}} \sum_i \left[-\ln \hat{c}^2(a,\psi)\sigma_i^2(a,\psi) - \frac{|r_i|^2}{\hat{c}^2(a,\psi)\sigma_i^2(a,\psi)} + \ln \xi^2 + \frac{|r_i|^2}{\xi^2} \right] I_i(a,\psi) \\ &= \underset{[a,\psi]^T}{\text{argmax}} \left\{ -N(a,\psi)\ln \hat{c}^2(a,\psi) + \sum_i \left[-\ln \sigma_i^2(a,\psi) - 1 + \ln \xi^2 + \frac{|r_i|^2}{\xi^2} \right] I_i(a,\psi) \right\}, \end{aligned} \tag{7}$$

where

$$\hat{c}^2(a,\psi) = \frac{1}{N(a,\psi)} \sum_i \frac{|r_i|^2}{\sigma_i^2(a,\psi)} I_i(a,\psi) \tag{8}$$

and $N(a, \psi)$ is the number of nonzero elements of the segmentation mask $\mathbf{I}(a, \psi)$.

2.4. Confuser Rejection

A common problem in ATR systems concerns the ability to properly note those received images that contain objects not represented in the model database, so called confuser targets, rather than misclassifying them. Such images may be formed from "actual" targets that have not been modeled and included in the database or they may be of clutter that appears target-like. One approach to this problem of confuser rejection can be found in classical statistical hypothesis testing. Denote the most likely combination of target class and pose determined from the previous algorithms as the null-hypothesis, $\mathcal{H}_0$, which represents an assertion that $\mathbf{r}$ is an image of target class a_0 with pose ψ_0. That is, $\mathcal{H}_0$ represents an assertion that $\text{var}(r_i) = \hat{c}\sigma_i^2(a_0, \psi_0)$, for $i = 1, 2, \ldots, N$. Denote the alternate hypothesis by $\mathcal{H}_A$ which represents an assertion that $\mathbf{r}$ is not an image of a_0 with pose ψ_0. That is, $\mathcal{H}_A$ represents an assertion that $\text{var}(r_i) \neq \hat{c}\sigma_i^2(a_0, \psi_0)$, for at least one $i = 1, 2, \ldots, N$. As discussed in Section 2.2, we conduct a likelihood ratio test against the most likely alternative. From (3), we reject the image $\mathbf{r}$ as a confuser if the relative entropy between these two distributions exceeds some threshold γ as

$$\begin{aligned} D(p(\cdot|\text{var}(r_i) &= |r_i|^2) \| p(\cdot|\text{var}(r_i) = \hat{c}^2\sigma_i^2(\hat{a},\hat{\psi}))) \\ &= \sum_i \left[\ln\left(\frac{\hat{c}^2\sigma_i^2(\hat{a},\hat{\psi})}{|r_i|^2} \right) + \frac{|r_i|^2}{\hat{c}^2\sigma_i^2(\hat{a},\hat{\psi})} - 1 \right] > \gamma. \end{aligned} \tag{9}$$

3. Model Building

For our purposes, a target model consists of both a mathematical description and a collection of parameters sufficient to characterize the output of the sensor platform given some target and pose. As mentioned in the introduction, some approaches to ATR rely on CAD representations of targets which, combined with simulation of the sensor's physical properties, allow the prediction of sensor outputs. An alternative approach of constructing target models from training data is described in this section. We first describe how the probabilistic description of sensor output from Section 2.1 can be combined with training data to estimate the radar reflection statistics over the surface of a target. Through this approach, model training becomes an issue of parameter estimation. Following this is a discussion of how these estimates can be used to characterize the sensor output for an arbitrary pose. Finally, we discuss segmentation of the estimated parameters into target and clutter regions.

3.1. Model Parameter Estimation

Given a probabilistic sensor description and a collection of training data for a target representing a wide variety of target pose, we seek estimates of the contribution to sensor output statistics from all locations on the target surface. For a given target a, the region z of the target surface that corresponds to pixel i is dependent upon the pose ψ and we can represent this dependence as a function $z_i(\psi)$. The region z corresponding to one pixel at a given pose will contribute to the variance of different pixels when viewed at some other pose. The model of a target a will consist of the model for SAR imagery in (1) and estimates of the variance function $\zeta_z^2(a, \psi)$ for the resolution cells $z_i(\psi)$ on the target. SAR imaging is inherently insensitive to translations along the radar line of sight, and translations perpendicular to the line of sight can be well modeled by simple shifts of the resulting image so that ζ_z^2 need not incorporate variation due to translation.

The variance ζ^2 is a function of the region of the target surface being imaged and its pose relative to the radar platform. Because we are considering storage and computation on digital computer systems, we must restrict the class of variance functions to those with finite representations. In practice, these functions will be estimated from a finite collection of training data so representations that strike a reasonable balance between approximation error in representation and estimation error in the function parameters is desirable. Several methods of regularization for function estimation have been reported. For an overview of many of these methods in an image formation context, see O'Sullivan, et al. [10]. Here, we consider a restriction of ζ^2 to functions that are piecewise constant over both the scene and pose. The piecewise constant regions in the scene will be the $z_i(\psi_0)$ corresponding to the pixels i in an image of the target at some nominal pose ψ_0. The piecewise constant regions in pose will follow from a uniform discretization of the components of ψ.

Suppose we collect a set of M power-normalized training images $\{\mathbf{T}'_1, \mathbf{T}'_2, \ldots, \mathbf{T}'_M\}$ from some fixed depression angle with position vectors $\{\mathbf{s}_1, \mathbf{s}_2, \ldots, \mathbf{s}_M\}$ in the slant plane and with azimuth angles $\{\phi_1, \phi_2, \ldots, \phi_M\}$. Neglecting articulated components, the transformation $U(-\phi_j, -\mathbf{s}_j) \in \mathbf{SE(2)}$ maps points (r, c) in image $\mathbf{T}'_j$ into points (x, y) in the slant plane. The image $\mathbf{T}_j = U(-\phi_j, -\mathbf{s}_j)\,[\mathbf{T}'_j]$ is a representation of the radar return from the scene registered with nominal pose $\psi_0 = [\phi_0, \mathbf{s}_0]^T = [0, \mathbf{0}]^T$ and the pixels in this image correspond to the regions in the scene over which the variance function will be approximated as piecewise constant. The collection of all such transformed data, $T = \{\mathbf{T}_1, \mathbf{T}_2, \ldots, \mathbf{T}_M\}$, constitutes a registered set of training data from which the variance function ζ^2 can be estimated.

For a fixed target position $\mathbf{s} = \mathbf{0}$, the variance ζ^2 is a function of slant plane location and azimuth angle. Let the 2π radians of target azimuth be discretized into N_w uniform intervals with the kth interval centered at $\theta_k = 2\pi/N_w$. An estimate of the variance function $\zeta_i^2(a, \theta_k)$ can be determined from the registered training images within some training interval of width d as

$$\hat{\zeta}_i^2(a, \theta_k) = \frac{1}{|\mathcal{I}_k|} \sum_{\mathbf{r} \in \mathcal{I}_\mathbf{k}} |r_i|^2, \tag{10}$$

where $\mathcal{I}_k = \{\mathbf{T}_j \in \mathcal{T} : \phi_j \in [\theta_k - d/2, \theta_k + d/2]\}$. It is possible to define $d > 2\pi/N_w$ so that training images formed from azimuth angles near the edges of the training intervals contribute to both neighboring intervals. The left panel of Figure 1 shows a SAR magnitude image in a logarithmic intensity scale of target class a = T-72 tank with azimuth angle $\phi = 135°$. The middle panel shows a registered image formed by a nearest-neighbor rotation of the data. The right panel shows an image of the variance estimate $\hat{\zeta}^2$(T-72, 135°) formed from training data with $N_w = 64$ windows and with training interval width $d = 11.25°$.

3.2. *Target Model Segmentation*

The estimation method from the previous section yields estimated variances for regions in the scene as a function of orientation relative to the radar platform. Although the corresponding pixels vary with orientation, a region in the scene that corresponds to the surface of a target remains on the target regardless of the angle from which it is viewed. Segmentation of target regions from clutter in this target-centered reference frame is a problem of target model segmentation. This is in contrast with the segmentation of SAR imagery as in [6] and [18].

Segmentation of the model can be phrased as a hypothesis testing problem which pits a hypothesis $\mathcal{H}_0$, an assertion that a given region is clutter, against the alternate $\mathcal{H}_A$, an assertion that the region is not clutter. The clutter hypothesis is defined by a zero-mean complex Gaussian distribution with variance ξ^2. As in Section 2.2, we conduct a test against the most likely alternative, which is that the pixel comes from a complex Gaussian

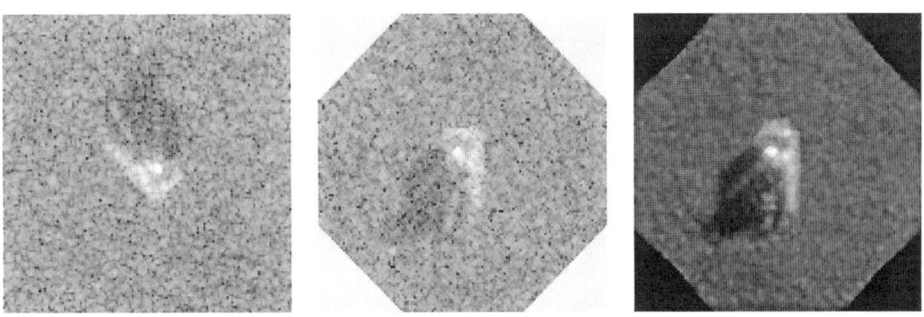

Figure 1. SAR magnitude image on a logarithmic scale of a target T-72 from the MSTAR dataset with azimuth angle 135°, the same image registered in pose, and estimated variances determined from registered images with azimuth near 135°.

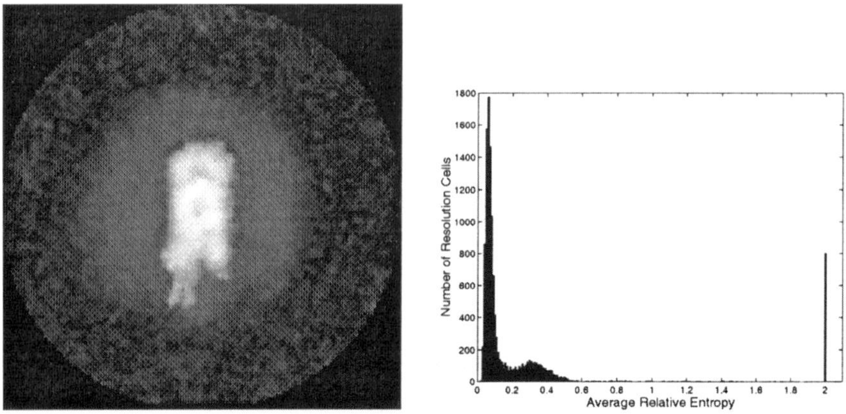

Figure 2. Average relative entropy over all orientations between the distributions parameterized by ζ_i^2 and $\xi^2 = 0.0028$ for the T-72 tank. The left panel shows an image of the relative entropy values on a logarithmic intensity scale. The right panel shows the histogram of average relative entropy values. The sharp peak at the value 2 indicates that approximately 800 scene regions have average relative entropies greater than or equal to 2.

distribution with conditional variance $\hat{\zeta}_i^2$ (a, θ)). For each target a, we reject the clutter hypothesis if the average over all orientations of the relative entropy between the distribution in (1) parameterized by $\hat{\zeta}^2$ and that parameterized by the clutter level ξ^2 surpasses some threshold η. The scene region i is taken to be on the target if

$$E_\theta\left[D(p(\cdot|\sigma^2 = \hat{\zeta}_i^2(a,\theta))\|p(\cdot|\sigma^2 = \xi^2))\right]$$
$$= \frac{1}{N_w}\sum_k \left(\frac{\hat{\zeta}_i^2(a,\theta_k)}{\xi^2} + \ln\frac{\xi^2}{\hat{\zeta}_i^2(a,\theta_k)} - 1\right) > \eta. \tag{11}$$

We denote the target model segmentation by a binary mask $\mathbf{S}(a)$ which has $S_i(a) = 1$ if region i satisfies the above inequality and zero otherwise.

The left panel of Figure 2 shows the average relative entropy on a logarithmic intensity scale for the T-72 tank assuming a background clutter level of $\xi^2 = 0.0028$ (determined experimentally from the MSTAR dataset) and with variance estimates described in the example of the previous section. In some SAR images used in estimation, the tank turret was slightly skewed resulting in the appearance of two barrels at a slight angle. This is an example of the model inaccuracies that can arise if the data are not registered in articulation as well as pose. The gray inner ring surrounding the pixels on target correspond to the regions of the scene masked by the target shadow at varying orientations and the grey-speckled outer ring corresponds to background clutter. The right panel shows the histogram of the average relative entropy values. The isolated peak at the value two indicates that approximately 800 resolution cells have an average relative entropy of two or greater.

Segmentation is a function of the threshold η and larger values result in smaller numbers of regions being included in the segmentation. Figure 3 shows segmented target models of the T-72 tank on a logarithmic intensity scale for several decreasing values of η from 125

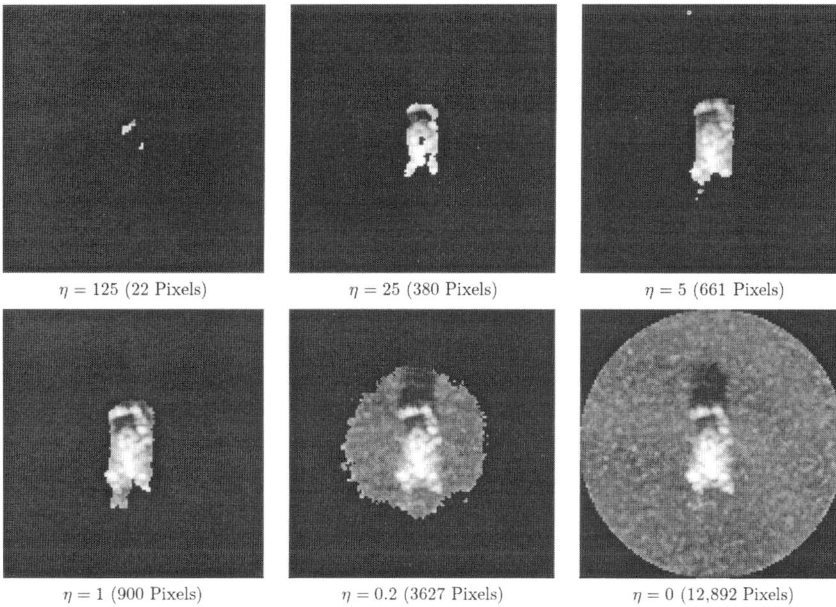

Figure 3. Model of target T72 trained over 64 windows of width 11.25° at various segmentations resulting from different threshold levels. The segmentation threshold level and the number of resolution cells with average relative entropies surpassing the threshold are given beneath each image.

to zero nats assuming a background clutter level of $\xi^2 = 0.0028$. Regions of the scene that have no corresponding SAR image pixels for some azimuth angles are explicitly chosen to represent clutter regions resulting in the circular shape in the segmentation for $\eta = 0$.

3.3. *Sensor Output Characterization*

The segmented target models derived in the previous sections represent the variability in the radar return for regions on the target as a function of azimuth angle in a coordinate system centered at the target. The output of a SAR radar platform is in a sensor-centric coordinate system, and to form a statistical characterization of the resulting data we need a transformation of the target model. SAR imaging is inherently insensitive to translations in the cross range direction and translations in the range direction affect only the power in the received data which is accommodated by the estimate $\hat{c}^2$ in (8). However, the radar return from a scene is sensitive to the orientation of the target relative to the radar platform. Thus, to estimate the variances of pixel values in an image of target a with pose consisting of location $\mathbf{s}$ and azimuth angle θ, we find the $\theta_k = 2\pi k/N_w$ closest to θ and transform the target model into the sensor coordinate system. That is, the image variance estimate is $\hat{\sigma}^2(a, \psi) = U(a, \psi)[\hat{\zeta}^2(a, \theta_k)]$ and $\mathbf{I}(a, \psi) = U(a, \psi)[S(a)]$ is a binary pixel mask in the image. Note that while $\hat{\zeta}^2$ is a piecewise constant function, the transformation U is a function of continuous ψ so $\hat{\sigma}^2$ is also a function of continuous-valued pose.

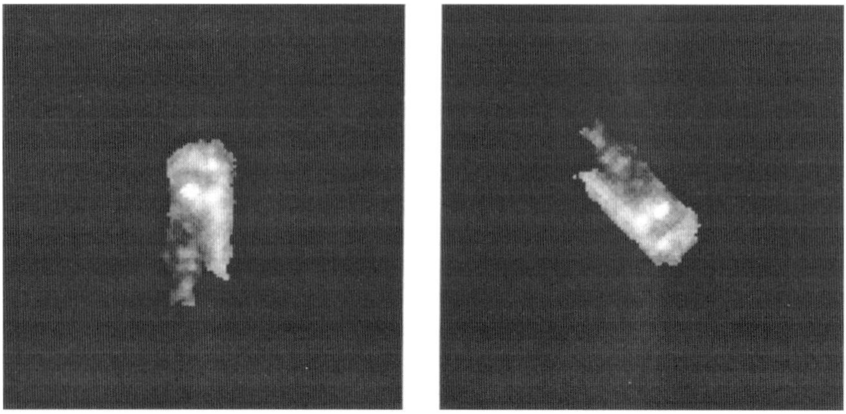

Figure 4. Left panel shows a segmented target model for a T-72 tank illuminated at 135° azimuth. Right panel shows the corresponding sensor output model.

Figure 4 shows a target model on a logarithmic scale, segmented with $\eta = 1$ for a T-72 tank when illuminated from 135° azimuth and the corresponding model for sensor output. The sensor output model is a characterization of the SAR image in the left panel of Figure 1.

3.4. *Performance Sensitivity*

We view the tradeoff between performance and complexity as the fundamental criterion for comparing alternate ATR algorithms. Each implementation delivers some level of accuracy (i.e. correct classification rate or average pose estimation error) and entails some complexity (i.e. storage or processing requirements). ATR is a difficult problem with many interrelated parameters governing the final system. In evaluating a general approach to ATR, we seek to decouple the effects of these parameters in order to understand their relative importance. DeVore, et al. [3] consider ATR performance as a function of resolution in pose space. In this section, we consider issues related to variation in segmentation.

The segmentation threshold η represents a free parameter in the design of an ATR system as described in the previous sections. Figure 3 indicates that for the T-72 data, a threshold value of 1 nat nicely separates target from clutter pixels. In the examples of the next section we consider the ATR accuracy as a function of η. This evaluation is used to determine whether this same threshold works well across a wide range of target classes and sheds light on the sensitivity of ATR to the segmentation threshold. We are also interested the encoding of estimated sensor modules $\hat{\sigma}^2(a, \psi)$ for rapid communication from the model database to the ATR processor and successive computation of the object of the argmax function in (7). The ability to refine initial classifications of SAR image content is potentially useful as operational demands on the ATR system change in response to the complexity or level of threat in the environment. Also, it offers the potential for restricting the parameter space of the search based on initial approximate calculations of the most likely target classes.

The number of nonzero pixels in the segmentation mask increases with decreasing η. By communicating from the database the pixel locations and values with average relative entropy greater than some η_1 for all possible hypotheses (a, ψ) and computing the log likelihood ratios in (7), an initial classification and pose estimation can be made. If η_1 is large so that relatively few pixels are included on average, the classification can be made quickly but with increased error rate. The initial classification can be refined by communicating only the locations and values of the pixels with average relative entropy between η_2 and η_1 for all possible hypotheses. If the subexpressions in the object of the argmax function are retained for each hypothesis and if an average of N_2 pixels are added by decreasing the entropy threshold, the estimate $\hat{c}^2(a, \psi)$ in (8) can be refined with N_2 divide and accumulate operations and a single division. The initial classification can be refined by computing the logarithm of the updated $\hat{c}^2$, a multiplication by the new $N(a, \psi)$, and a logarithm and divide accumulate operation for each of the N_2 additional pixels. This process can be continued until the all components of the sensor output models have been communicated from the database and a final classification has been reached. The additional computational cost in supporting the successive refinement from η_1 to η_2 rather than determining a classification for η_2 directly is one logarithm, multiply, divide, and subtract operation for each hypothesis. This is the same additional cost in going from η_n to $\eta_{n+1} < \eta_n$. The additional communication cost for supporting successive-refinement depends on the representation of pixel locations and values used.

4. Example

In this section we present examples of the previously described methods for actual SAR data from the MSTAR dataset. First, we present experimental results for a ten-class recognition problem with forced decisions and no confuser targets. We examine classification error rates and average pose estimation error as a function of the segmentation threshold and relate them to the average number of bits transmitted between the model database and ATR processor and to the average number of floating-point operations to compute the object of the argmax function in (7). We then divide the dataset into a four-class problem with four confuser targets and examine receiver operating characteristics, probability of correct classification versus confuser rejection rate, as a function of segmentation threshold level. Finally, we consider two different four-class problems with confuser targets which represent extended operating conditions (EOC) [8], [14], in which the test images are of targets with some configuration or version variation from the training images. In each example, the target models $\hat{\zeta}^2(a, \theta_k)$ were estimated from the training data with $N_w = 64°$ and $d = 11.25°$. From these, the sensor output models $\hat{\sigma}^2(a, \phi)$, where $\phi = [\theta, \mathbf{s}]^T$, were generated with 256 equally spaced azimuth angles θ and 25 target locations $\mathbf{s}$, corresponding to a target location unknown within 5 pixels in both range and cross-range directions.

Table 1. MSTAR dataset used in ten-class examples.

	Train			Test		
Target	Vehicles	Images	Depression	Vehicles	Images	Depression
2S1	b01	299	17°	b01	274	15°
BMP-2	9563, 9566, c21	697	17°	9563, 9566, c21	587	15°
BRDM-2	E-71	298	17°	E-71	263	15°
BTR-60	k10yt7532	256	17°	k10yt7532	195	15°
BTR-70	c71	233	17°	c71	196	15°
D7	92v13015	299	17°	92v13015	274	15°
T62	A51	299	17°	A51	273	15°
T-72	132, 812, s7	691	17°	132, 812, s7	582	15°
ZIL131	E12	299	17°	E12	274	15°
ZSU 23 4	d08	299	17°	d08	274	15°

4.1. Ten Class Recognition

In this example, SAR images of ten target classes from the MSTAR dataset were partitioned into two nonoverlapping sets, one for estimation of model parameters through the methods of Section 3 and the other for evaluation through the methods of Section 2.3. Details about the dataset used for each are given in Table 1. Note that images formed with a 17° depression angle were used for estimation and those from a 15° depression angle were used for testing. Also, the target classes T-72 and BMP-2 incorporate images from three different vehicles of the same class.

Performance results for this example are shown in Figure 5 as a function of decreasing segmentation threshold, η, logarithmically spaced from a value of 125 nats to 1 nat. The left

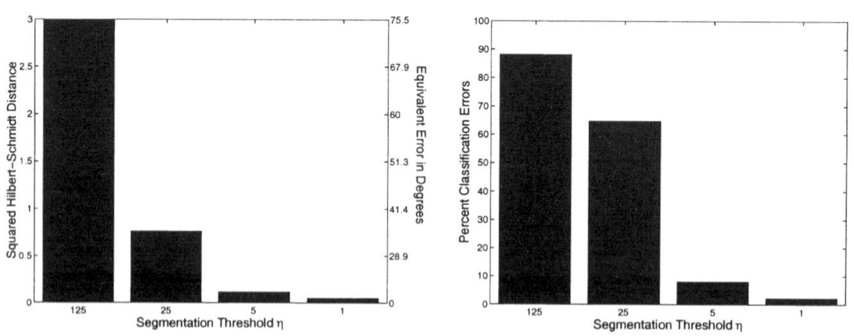

Figure 5. Error rates as a function of decreasing segmentation threshold. Left panel shows the average squared Hilbert-Schmidt distance between actual and estimated target azimuth and the equivalent error in degrees. Right panel shows the percentage of misclassified test images.

graph shows the average error between estimated target azimuth, $\hat{\theta}$, and actual target azimuth, θ, given that the true target class is known. The left vertical axis gives the error in terms of the squared Hilbert-Schmidt distance, d^2_{HS}, defined by Grenander, et al. [5] as the squared Hilbert-Schmidt norm of the difference between the rotation matrices, $O(\hat{\theta})$, $O(\theta) \in \mathbf{SO(N)}$, that describe the orientation of the target. That is, the value of $d^2_{HS}\,(O(\hat{\theta}), O(\theta))$ is given by

$$d^2_{HS}(O_1, O_2) = \mathrm{tr}[(O_1 - O_2)^T(O_1 - O_2)], \qquad (12)$$

where tr[$\cdot$] denotes the matrix trace operation and $(\cdot)^T$ denotes matrix transpose. The right vertical axis gives the equivalent error in degrees determined as $\cos^{-1}(1 - d^2_{HS}/4)$. The graph shows that the average orientation estimation error drops from over 75° with large segmentation thresholds down to 9.36° for $\eta = 1$. The right graph shows the percentage of test images incorrectly classified as a function of η. Error rates drop from 88.1% down to 2.26% over the same range of segmentation threshold.

The amount of computational resources consumed in recognizing an object in a SAR image is also related to the segmentation threshold as indicated in Figure 6. The left panel of the figure shows the average number of pixels satisfying the threshold inequality (11) for varying values of η. The right vertical axis of the same graph shows the average number of bits that must be transmitted between the model database and ATR processor for each $\hat{\sigma}^2(a, \phi)$. This value assumes that each pixel value is represented by 14 bits denoting the pixel location (sufficient for up to a 128 × 128 image) and 64 bits for the floating-point representation of the variance value. For this example, there were 10 target classes and 256 azimuth angles under consideration so the total number of bits transmitted to recognize a single target image is 2560 times this value. Since variation in target location $\mathbf{s}$ can be handled by simple image shifts in the ATR processor, no additional

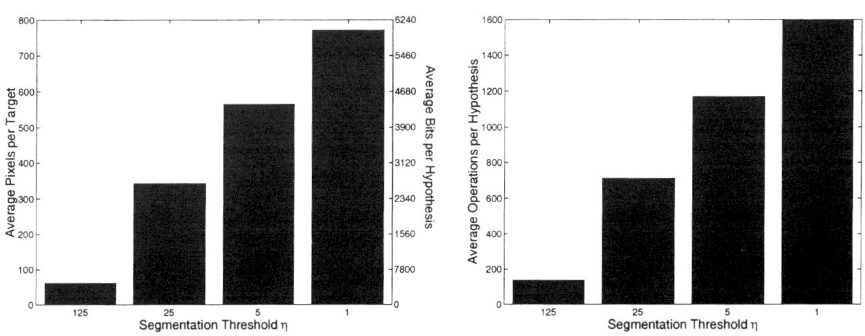

Figure 6. Consumed resources as a function of decreasing segmentation threshold. Left panel shows the average number of pixels in a segmented target model and an upper bound on the number of bits to communicate the models. Right panel shows the average number of floating-point operations to compute the object of the argmax function for each combination of target class and pose assuming η is reduced in four steps from 125 to 1.

communication is required to incorporate the 25 target locations considered in this example.

The right panel of Figure 6 shows the average number of operations required to compute the object of the argmax function in (7), assuming that η is initially set to 125 nats and is successively lowered to 25, 5, and finally 1 nat, with a classification decision being reached at each level. Assuming one operation for each divide and accumulate operation and one for the division, estimation of c^2 will take $N + 1$ operations on average, where N is the average number of pixels on target from the graph in the left panel. The remaining terms in the object of the argmax function can be computed in $N + 13$ operations assuming one operation for each divide and accumulate operation involving ξ^2, one addition for the (precomputed) term involving $\ln \sigma^2$, 10 operations for logarithm computation, one for multiplication by N, and a final addition to join the terms. This is a total of $2N + 14$ operations. An additional cost of 13 operations is incurred for each successive refinement of η as discussed in Section 3.4, assuming 10 operations for recomputing the logarithm and one each for the redundant multiplication, division, and addition operations. For this example, the number of operations indicated in this graph must be carried out for each of the 10 target classes, 256 azimuth angles, and 25 target locations, totaling 64,000 times, in the classification of a single target image.

The four graphs in these two figures demonstrate that large segmentation thresholds result in high recognition errors but can be very easy to compute. Clearly, recognition error rates near 80% are not going to be acceptable in practice but the error rates achievable with thresholds less than 5 nats may be, particularly if the decision must be made extremely quickly or with limited hardware capability. Also, it may be useful for a recognition system to deliver an initial classification quickly but suffering a relatively high error rate with the understanding that a more accurate decision is pending more exhaustive computation. Finally, it may be possible to exploit the recognition results with large segmentation threshold values by eliminating from consideration highly unlikely combinations of target class and pose while periodically lowering η.

4.2. *Four Class with Confuser Rejection*

For the experiments in this section, the MSTAR dataset was partitioned into three nonoverlapping sets consisting of training, testing, and confuser images. Table 2 details the four target classes used for estimating target models. These are a subset of the ten-class training set from the previous section. Table 3 shows the testing and confuser data. The testing data consists of images of the same vehicles used in training but taken from a different depression angle. The confuser data consists of images of four vehicles not used in training. In this example, the ability to correctly classify the testing data while rejecting the confuser data is investigated.

Figure 7 shows the percentage of correctly classified testing data and percentage of confuser targets that were not rejected, for each of four segmentation threshold levels, η. The curves are parameterized by the confuser rejection threshold γ from (9). A test image is said to be correctly classified if the recognition algorithm (7) yields the correct target class a and the image is not rejected as a confuser. The rightmost point of each curve

Table 2. MSTAR dataset used for training target models in the confuser rejection examples.

Class	Vehicle	Serial No.	Depression	Images
BMP-2	#1	9563	17°	233
	#2	9566		231
	#3	c21		233
BRDM-2	#1	E-71	17°	298
BTR-70	#1	c71	17°	233
T-72	#1	132	17°	232
	#2	812		231
	#3	s7		228

shows the best correct recognition rate achievable for each segmentation threshold if confuser rejection was not of concern. With $\eta = 1$, a correct recognition rate of 87.7% and 6.4% confuser rejection rate are achievable with $\gamma = 1.1$ nat. With $\eta = 5$, a correct recognition rate of 83.3% and 7.9% confuser rejection rate are achievable with $\gamma = 1.1$ nat.

4.3. *Extended Operation Conditions*

For the experiments in this section, the MSTAR dataset was partitioned into four nonoverlapping sets to investigate the ability to correctly recognize variants of modeled target classes while rejecting images of confuser targets. The data used for training are detailed in Table 4. Four target classes are represented, each from a single vehicle and incorporating two slightly different depression angles. Two sets of test data were constructed from the data in such a way that no vehicle used for training has images in

Table 3. MSTAR dataset used for evaluating confuser rejection capabilities. Target classes listed above the double line are represented in the target model database. Those below are confuser targets for this experiment.

Class	Vehicle	Serial No.	Depression	Images
BMP-2	#1	9563	15°	195
	#2	9566		196
	#3	c21		196
BRDM-2	#1	E-71	15°	263
BTR-70	#1	c71	15°	196
T-72	#1	132	15°	196
	#2	812		195
	#3	s7		191
2S1	#1	b01	15°	274
BTR-60	#1	k10yt7532	15°	195
D7	#1	92v13015	15°	274
T-62	#1	A51	15°	273

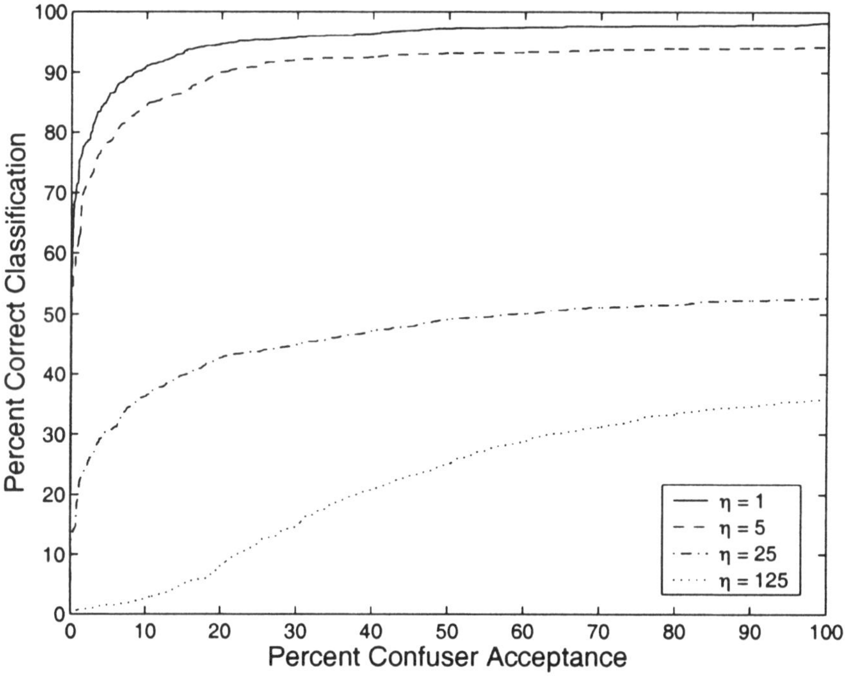

Figure 7. Correct classification rate as a function of confuser acceptance rate for each of four segmentation thresholds.

either test set and these are detailed in Table 5. The configuration variant test set consists of images of vehicles that represent minor differences in vehicle configuration from those in the training set. The version variant set consists of images of vehicles that exhibit more significant version differences from the training set. The same confuser images from the previous section were used in these experiments.

Figure 8 shows the correct classification rate and rate of confuser acceptance for both experiments. The left panel shows the results for test images that differ in configuration from the training data and the right panel shows the results for more significant version variations. The ability to simultaneously classify test images and reject confusers is

Table 4. MSTAR images used for training target models in tests under extended operation conditions.

Model	Vehicle	Serial No.	Depression	Images
BMP-2	#3	c21	15° and 17°	429
BTR-70	#1	c71	15° and 17°	429
BRDM-2	#1	E-71	15° and 17°	561
T-72	#1	132	15° and 17°	428

Table 5. MSTAR images used in the configuration variation and version variation EOC testing.

Configuration Variant Testing Set				Version Variant Testing Set			
Target	Serial No.	Depression	Images	Target	Serial No.	Depression	Images
T-72 #2	812	15° and 17°	426	T-72 #3	s7	15° and 17°	419
T-72 #4	A04	15° and 17°	573	T-72 #8	A32	15° and 17°	572
T-72 #5	A05	15° and 17°	573	T-72 #9	A62	15° and 17°	573
T-72 #6	A07	15° and 17°	573	T-72 #10	A63	15° and 17°	573
T-72 #7	A10	15° and 17°	567	T-72 #11	A64	15° and 17°	573
BMP-2 #1	9563	15° and 17°	428				
BMP-2 #2	9566	15° and 17°	427				

hindered by test images that vary from the training data as compared to the experiments of the previous section. In both the configuration and version variant tests, the target segmentation threshold $\eta = 5$ yielded better performance than $\eta = 1$ over most of the graphs. In both configuration and version variant experiments, the segmentation threshold $\eta = 125$ results in more than 20% of the confuser images being closer to training data, in the sense of (9), than any of the test images.

5. Conclusions

We have presented an approach to ATR from SAR imagery in which statistical models are estimated in a target-rather than sensor-centered reference frame. These models can be used to approximate the statistics of sensor output data at arbitrary pose relative to the scene. The algorithms account for rotations and for translations both perpendicular to and parallel to the radar line of sight. A method of addressing two-sided hypothesis tests which

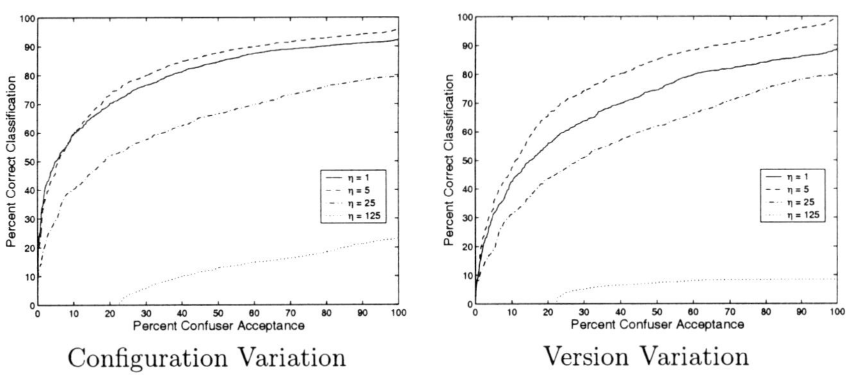

Figure 8. Correct classification rate for configuration variants and version variants as a function of confuser acceptance rate for each of four segmentation thresholds.

is maximally powerful at the most likely alternative is described and applied to both target model segmentation and confuser rejection. This hypothesis testing method yields a decision based upon the empirical relative entropy between the most likely distribution given the observed data and the distribution of the null-hypothesis.

These algorithms were demonstrated on actual SAR data in a variety of experiments designed to indicate the relative sensitivity to variation in segmentation threshold, confuser rejection threshold, and operating conditions. The experiments indicate that a segmentation threshold of 1 nat for the relative entropy between the most likely distribution and the clutter distribution yields consistently low error rates at a complexity of under 800 pixels per target and 1600 floating-point operations per hypothesized combinations of target class and orientation. This threshold yields a classification error rate of under 3% for a ten class problem under standard operating conditions and a Pcc-Pfa curve exhibiting a sharp knee in a four-class confuser rejection problem.

Acknowledgments

This work was supported in part by the US Army Research Office grant DAAH04-95-1-0494, by the Office of Naval Research grant N00014-98-1-06-06, and by the Boeing Foundation.

Note

1. The MSTAR dataset is publicly available and can be requested through the Sensor Data Management System web site at the URL http://www.mbvlab.wpafb.af.mil/public/sdms/.

References

1. H.-C. Chiang, R.L. Moses, and L.C. Potter, "Model-Based Classification of Radar Images," *IEEE Transactions on Information Theory*, vol. 46, no. 5, Aug. 2000, pp. 1842–1854.
2. M.D. DeVore, A.D. Lanterman, and J.A. O'Sullivan, "ATR Performance of a Rician Model for SAR Images," in F.A. Sadjadi (Ed.), *Automatic Target Recognition X, Proc. of SPIE*, vol. 4050, 2000.
3. M.D. DeVore, J.A. O'Sullivan, S. Anand, and N.A. Schmid, "Probabilistic Approach to Model Extraction from Training Data," in E.G. Zelnio (Ed.), *Algorithms for Synthetic Aperture Radar Imagery VIII, Proc. of SPIE*, 2001.
4. M.D. DeVore, N.A. Schmid, and J.A. O'Sullivan, "Analytical and Experimental Performance-Complexity Tradeoffs in ATR," in *Proceedings of the Thirty-Fourth Asilomar Conference on Signals, Systems, and Computers*, Oct. 2000.
5. U. Grenander, M.I. Miller, and A. Srivastava, "Hilbert-Schmidt Lower Bounds for Estimators on Matrix Lie Groups for ATR," *IEEE Transactions on Pattern Analysis and Machine Intelligence*, vol. 20, no. 8, Aug. 1998, pp. 790–802.
6. W.W. Irving, L.M. Novak, and A.S. Willsky, "Multiresolution Approach to Discriminating Targets from Clutter in SAR Imagery," *IEEE Transactions on Aerospace and Electronic Systems*, vol. 33, no. 4, Oct. 1997, pp. 1157–1169.

7. S.P. Jacobs, Automatic Target Recognition Using High Resolution Radar Range-Profiles. PhD thesis, Washington University, St. Louis, Missouri, May 1997.
8. E.R. Keydel and S.W. Lee, "Signature Prediction for Model-Based Automatic Target Recognition," in E.G. Zelnio and R.J. Douglass (Eds.), *Algorithms for Synthetic Aperture Radar Imagery III, Proc. of SPIE*, vol. 2757, 1996, pp. 306–317.
9. E.R. Keydel, S.W. Lee, and J.T. Moore, "MSTAR Extended Operating Conditions: A Tutorial," in E.G. Zelnio and R.J. Douglass (Eds.), *Algorithms for Synthetic Aperture Radar Imagery III, Proc. of SPIE*, vol. 2757, 1996, pp. 228–242.
10. J.A. O'Sullivan, R.E. Blahut, and D.L. Snyder, "Information-Theoretic Image Formation," *IEEE Transactions on Information Theory*, vol. 44, no. 6, Oct. 1998, pp. 2094–2123.
11. J.A. O'Sullivan, M.D. DeVore, V. Kedia, and M.I. Miller, "Automatic Target Recognition Performance for SAR Imagery Using a Conditionally Gaussian Model," *IEEE Transactions on Aerospace and Electronic Systems*, vol. 37, no. 1, Jan. 2001, pp. 91–108.
12. G.J. Owirka, S.M. Verbout, and L.M. Novak, "Template-Based SAR ATR Performance Using Different Image Enhancement Techniques," in *Algorithms for Synthetic Aperture Radar Imagery VI, Proc. of SPIE*, vol. 3721, 1999, pp. 302–319.
13. J.A. Richards, J.W. Fisher III, and A.S. Willsky, "An Expectation-Maximization Approach to Target Model Generation from Multiple SAR Images," in E.G. Zelnio (Ed.), *Algorithms for Synthetic Aperture Radar Imagery VII, Proc. of SPIE*, vol. 4053, 2000, pp. 652–664.
14. T.D. Ross, S.W. Worrell, J.C. Mossing, and M.L. Bryant, "Standard SAR ATR Evaluation Experiments Using the MSTAR Public Release Data Set," in E.G. Zelnio (Ed.), *Algorithms for Synthetic Aperture Radar Imagery V, Proc. of SPIE*, vol. 3370, 1998, pp. 566–573.
15. G.G. Roussas, *A Course in Mathematical Statistics*, Academic Press, 2nd ed., 1997.
16. N.A. Schmid and J.A. O'Sullivan, "Thresholding Method for Reduction of Dimensionality," *IEEE Transactions on Information Theory*, vol. 47, no. 7, Nov. 2001, pp. 2903–2920.
17. S.M. Verbout, W.W. Irving, and A.S. Hanes, "Improving a Template-Based Classifier in a SAR Automatic Target Recognition System by Using 3-D Target Information," *Lincoln Laboratory Journal*, vol. 6, no. 1, Spring 1993, pp. 53–76.
18. R.A. Weisenseel, W.C. Karl, D.A. Castañon, G.J. Power, and P. Douville, "Markov Random Field Segmentation Methods for SAR Target Chips," in E.G. Zelnio (Ed.), *Algorithms for Synthetic Aperture Radar Imagery VI, Proc. of SPIE*, vol. 3721, 1999, pp. 462–473.
19. J. Wissinger, R.B. Washburn, N.S. Friedland, A. Nowicki, D.R. Morgan, C. Chong, and R. Fung, "Search Algorithms for Model Based SAR ATR," in E.G. Zelnio and R.J. Douglass (Eds.), *Algorithms for Synthetic Aperture Radar Imagery III, Proc. of SPIE*, vol. 2757, 1996, pp. 279–293.

Multidimensional Systems and Signal Processing, 14, 161–181, 2003

Extraction of Three-Dimensional Motion and Geometric Invariants from Range Dependent Signals

MARK A. STUFF
Veridian Systems, Ann Arbor Research and Development Center, Ann Arbor, Michigan, USA

PEDRO SANCHEZ
Veridian Systems, Ann Arbor Research and Development Center, Ann Arbor, Michigan, USA

MARTIN BIANCALANA
Veridian Systems, Ann Arbor Research and Development Center, Ann Arbor, Michigan, USA

Received December 27, 2000; Revised November 20, 2001; Accepted December 7, 2001

Abstract. The theory needed to extract three-dimensional size, shape, and motion of a rigid moving body from a series of range-only measurements is explained. For a rigid body, moving with arbitrarily complex, unknown motions, the three-dimensional size and shape of a configuration of points on the body can be calculated from the range data, without any prior knowledge of the geometry of the configuration. The calculations are possible because there exist motion-invariant functions of the range data, which uniquely determine the Euclidean geometry of the points.

This theory is shown to work on synthetic range data. When the synthetic data is corrupted by noise the process is shown to still produce reasonable results.

Key Words: range measurements, remote sensing, radar, ISAR, SAR, invariants, moving targets, ATR

1. Introduction

Several important remote sensing systems collect and process signals which are functions of the ranges (or more precisely, the propagation times) to components of remote objects or scenes. Typical examples include most radar, sonar, ultrasound, and seismic systems. These sensors usually collect a series of one dimensional signals, from a remote object, despite the three-dimensional extent of the object. To use these sensors to measure two or three dimensional characteristics, they need to be deployed in spatially distributed arrays, or operated in ways that collect signals over some interval of time during which the angular relation of the sensor and object changes.

This changing angular relation, in two dimensions, is the basis for conventional synthetic aperture radar (SAR). The resulting data can be processed to form a two-dimensional projection of the three-dimensional scene, if the angular history of the collection is known. For typical SAR systems, this angular history is determined by making precise simultaneous inertial measurements to record the motion of the radar with respect to the ground. Similarly, other ranging sensor systems depend on accurate knowledge of the geometric relationship of the location of the sensor and the parts of the object being imaged. This geometric knowledge allows a processor to exploit the angular diversity needed to construct higher dimensional measurements.

If objects in the scene are moving independently, their angular motions impose additional complexity on the sensor's signal. In practical cases, this motion cannot be described in a single plane. The unmeasured motion, especially the "out of plane" motion, causes the image of the object to be smeared, geometrically distorted, and displaced with respect to the background. This is typical for moving vehicles in SAR imagery.

This paper treats the case where the motion of the object of interest is not known *a priori*. The unknown motions are permitted to be arbitrarily complex, so that no simple parametric model can describe them. (For the radar applications, we want to be able to handle realistic vehicle motion even for off-road travel and evasive manuevers.) The system has only the signals, but does not know the relative orientations of the sensor and the object during the time when the signals were collected. We want to solve this problem making no assumptions about the size, shape, motion, or composition of the object.

This is a difficult problem. No single paper such as this one can describe the complete solution. In this paper, we limit our discussion to the part of the problem which arises after the ranges to a configuration of points have been extracted from the sensor signals. This part of the theory will be the same for all types of ranging sensors. For each particular type of sensor, the geometric theory presented here needs to be preceded by a method which exploits the physics of the particular sensor to extract the required range estimates.

Extracting the ranges from the sensor signals may be simple, or may be extremely difficult. For the radar case, if the signal contains only strong returns from well resolved scattering mechanisms, then the range measurements are straight forward. However, if multiple scattering mechanisms contribute energy to a single range bin, then it becomes difficult to extract the range measurements.

We assume that the object of interest has a constant shape, during the collection of the data. The theory is developed for a single rigid body. It can be applied piecewise to articulated bodies made up of rigid parts. For an actively deforming body this theory lays the groundwork for determination of the best rigid approximation to the changing shape.

In many sensing applications, the sensor is remote; that is, the source of the radiation is far away from the objects of interest. In these cases, when the distance to the source is many times larger than the rigid object of interest, we can neglect wavefront curvature on the scale of the rigid object. We treat wavefronts from a distant source as if they were planar wavefronts passing through the local neighborhood of the rigid object. Although this approximation is used to simplify the analysis in this paper, we expect that the results can be extended to the spherical case if that is warranted.

We begin understanding the geometry by showing that there exist geometrically invariant functionals of the ranges to points on a moving object. These functionals completely characterize the size and shape of the three-dimensional configurations of points on a moving object. We then show results of processing known range histories, using the theory to reconstruct the three-dimensional configuration of the original points.

We continue by adding random noise to the range measurements and comparing the results for one particular example. This does not provide a comprehensive picture of the robustness of the method; the example only provides a preliminary impression of the stability of one implementation of the three-dimensional geometric measurement process.

The theorems which we will present in this paper are the fruits of an analytical search for the geometric invariants of far field range histories. The analysis of geometrical problems by computation and classification of invariants has a long tradition in mathematics and physics. Algebraic invariants theory is considered by some to have begun with a theorem by G. Boole in 1841 and to have been seriously applied to geometry by Felix Klein and Sophus Lie begining in 1872 [1]. This tradition has continued fruitfully for the past 150 years [2], [3], [4]. Recent interest in computer vision problems has reinvigorated interest in geometric invariants, as evidenced by many recent publications, typical of which are the volumes edited by Mundy et al [5], [6].

The recent applications of invariance theory to computer vision are primarily concerned with information collected from two dimensional optical scenes. Some research applying invariance theory to radar scenes has been done [7]. But applications to sequences of fundamentally one dimensional signals, such as radar signals, are not abundant in the literature. However, some elements of these ideas have been discussed in a few recent conference papers [8], [9], [10], [11], [12].

2. The Existence Theorem for a Size and Shape Determining Motion-Invariant Functional

Because we ignore wavefront curvature on the scale of the rigid object, we can model the range history, $\rho_q(t)$, of a moving scattering center as

$$\rho_q(t) = \begin{bmatrix} c_1(t) & c_2(t) & c_3(t) \end{bmatrix} \begin{bmatrix} x_q \\ y_q \\ z_q \end{bmatrix} + b(t) \tag{1}$$

where $\begin{bmatrix} c_1(t) & c_2(t) & c_3(t) \end{bmatrix}'$ is an unknown, time varying, unit vector which describes the illumination direction of the radar with respect to the reflecting object, $b(t)$ is an unknown, time varying translation, and $\begin{bmatrix} x_q & y_q & z_q \end{bmatrix}'$ are the coordinates of the scattering center in the inertial reference frame of the object. (See [8] for a more detailed discussion.)

First we demonstrate a particular invariant functional. This particular functional is of special interest to us because it will uniquely characterize the Euclidean geometric properties of the configuration.

THEOREM 1: (*The Existence Theorem*). *For any configuration of points*

$$\begin{bmatrix} x_1 \\ y_1 \\ z_1 \end{bmatrix} \begin{bmatrix} x_2 \\ y_2 \\ z_2 \end{bmatrix} \cdots \begin{bmatrix} x_Q \\ y_Q \\ z_Q \end{bmatrix} \tag{2}$$

in $\mathbb{R}^3$, which is not co-planar, the rank of the centered point configuration matrix

$$\boldsymbol{S} = \begin{bmatrix} x_1 - \bar{x} & y_1 - \bar{y} & z_1 - \bar{z} \\ x_2 - \bar{x} & y_2 - \bar{y} & z_2 - \bar{z} \\ \vdots & \vdots & \vdots \\ x_q - \bar{x} & y_q - \bar{y} & z_q - \bar{z} \\ \vdots & \vdots & \vdots \\ x_Q - \bar{x} & y_Q - \bar{y} & z_Q - \bar{z} \end{bmatrix} \tag{3}$$

is 3, and the range histories $\{\rho_q(t)\}_{q=1}^{q=Q}$ satisfy an invariant equation

$$\sum_{i=1}^{i=Q-1} \sum_{j=i+1}^{j=Q} \omega_{ij}\left(\rho_i(t) - \rho_j(t)\right)^2 - 1 = 0 \tag{4}$$

where

$$\omega_{ij} = \left(-\boldsymbol{S}(\boldsymbol{S}'\boldsymbol{S})^{-2}\boldsymbol{S}'\right)_{ij} \quad \forall_{1 \le i < j \le Q}. \tag{5}$$

Before we prove the theorem, we will define some convenient notation. For a column vector, $\boldsymbol{x}$, we denote its transpose as $\boldsymbol{x}'$, so that $\boldsymbol{x}' = \begin{bmatrix} x_1 & x_2 & \cdots & x_Q \end{bmatrix}$. We also let $\mathbf{1}' = [1 \;\; 1 \cdots 1]$, the constant vector of ones. We also will denote averages over certain components of vectors or matrices using the notation

$$a_{i\cdot} = \frac{1}{Q}\sum_{j=1}^{j=Q} a_{ij},\ a_{\cdot j} = \frac{1}{Q}\sum_{i=1}^{i=Q} a_{ij}, \text{ and } a_{\cdot\cdot} = \frac{1}{Q}\sum_{i=1}^{i=Q} a_{i\cdot} = \frac{1}{Q}\sum_{j=1}^{j=Q} a_{\cdot j} = \frac{1}{Q^2}\sum_{i=1}^{i=Q}\sum_{j=1}^{j=Q} a_{ij}. \tag{6}$$

To simplify the proof of the theorem, we will first prove a lemma concerning quadratic forms in centered variables. These are quantities of the form

$$\left[(x_1 - \bar{x})\ (x_2 - \bar{x}) \cdots (x_Q - \bar{x})\right] \begin{bmatrix} a_{11} & a_{12} & \cdots & a_{1Q} \\ a_{21} & a_{22} & \cdots & a_{2Q} \\ \vdots & \vdots & & \vdots \\ a_{Q1} & a_{Q2} & \cdots & a_{QQ} \end{bmatrix} \begin{bmatrix} (x_1 - \bar{x}) \\ (x_2 - \bar{x}) \\ \vdots \\ (x_Q - \bar{x}) \end{bmatrix} = \boldsymbol{x}'\boldsymbol{C}'\boldsymbol{A}\boldsymbol{C}\boldsymbol{x} \tag{7}$$

where $\boldsymbol{C}$ is the centering matrix,

$$\boldsymbol{C} = \boldsymbol{I} - \frac{1}{Q}\mathbf{1}\mathbf{1}'. \tag{8}$$

The matrix values a_{ij} are not unique in such a form. First, as with any quadratic form, we can, without loss of generality, require symmetry of the matrix entries, $a_{ij} = a_{ji}$ (see, for example, [13]). For the usual quadratic form, the symmetry requirement alone is sufficient to assure uniqueness of the matrix elements. But, because of the centering, symmetry alone is not sufficient to give us unique coefficients in expression 7. The extra condition needed for uniqueness is that each row and each column of $\boldsymbol{A}$ sum to zero, because the centering matrices, $\boldsymbol{C}$ and $\boldsymbol{C}'$ will annihilate any non-zero mean component of the rows or columns of the matrix. But the usual parameterization of the quadratic form is not the most convenient one possible. The following lemma will give us a nicer alternative parametrization.

LEMMA 1: *For all $\boldsymbol{x} \in \mathbb{R}^Q$, and any $Q \times Q$ matrix, $\boldsymbol{A}$*

$$\boldsymbol{x}'\boldsymbol{C}'\boldsymbol{A}\boldsymbol{C}\boldsymbol{x} = \left[(x_1 - \bar{x})\ (x_2 - \bar{x}) \cdots (x_Q - \bar{x})\right] \begin{bmatrix} \sum\limits_{j\neq 1} b_{1j} & -b_{12} & \cdots & -b_{1Q} \\ -b_{21} & \sum\limits_{j\neq 2} b_{2j} & \cdots & -b_{2Q} \\ \vdots & \vdots & & \vdots \\ -b_{Q1} & -b_{Q2} & \cdots & \sum\limits_{j\neq Q} b_{Qj} \end{bmatrix}$$

$$\times \begin{bmatrix} (x_1 - \bar{x}) \\ (x_2 - \bar{x}) \\ \vdots \\ (x_Q - \bar{x}) \end{bmatrix} = \sum_{i=1}^{i=Q-1} \sum_{j=i+1}^{j=Q} b_{ij}(x_i - x_j)^2 \tag{9}$$

where

$$\begin{aligned} b_{ij} &= -a_{ij} + \frac{1}{Q}\sum_{k=1}^{k=Q} a_{ik} + \frac{1}{Q}\sum_{k=1}^{k=Q} a_{kj} - \frac{1}{Q^2}\sum_{k=1}^{k=Q}\sum_{l=1}^{l=Q} a_{kl} \quad \forall_{1\leq i<j\leq Q} \\ &= -a_{ij} + a_{i\cdot} + a_{\cdot j} - a_{\cdot\cdot} \end{aligned} \tag{10}$$

Proof of Lemma 1: See Appendix.

Having completed the proof of the lemma, we can now proceed to prove the existence theorem.

Proof of Theorem 1: See Appendix.

3. Uniqueness Conditions for the Invariant Functional

The preceding theorem shows the existence of a particular invariant functional of interest. But to be able to estimate the invariant parameters, $\{\omega_{ij}\}_{1\le i<j\le Q}$, which appear in this particular invariant functional, we will need to distinguish it (using only conditions which are observable from the data, $\{\rho_i(t)\}_{i=1}^{i=Q}$) from other invariant functionals of the same functional form. The difficulty is that while we have shown that for $\omega_{ij} = -(\boldsymbol{S}(\boldsymbol{S}'\boldsymbol{S})^{-2}\boldsymbol{S}')_{ij}$, the functional equation, equation 4, $\sum_{i=1}^{i=Q-1}\sum_{j=i+1}^{j=Q} w_{ij}(\rho_i(t)-\rho_j(t))^2 - 1 = 0$, is true for all t, there exist other sets of constants, $\{\vartheta_{ij}\}_{1\le i<j\le Q} \ne \{\omega_{ij}\}_{1\le i<j\le Q}$, such that $\sum_{i=1}^{i=Q-1}\sum_{j=i+1}^{j=Q} \vartheta_{ij}(\rho_i(t)-\rho_j(t))^2 - 1 = 0$ is true for all t. (It can be shown that this non-uniqueness is always present when $Q > 4$.) Consequently, we want to state additional conditions, which can be observed from the data, such that these extraneous invariants will not be admitted.

To accomplish that, we need some definitions, and some introductory lemmas. First, we define the l-rank of a set of functions, $\{f_q\}_{q=1}^{q=Q}$ to be the dimension of the linear space spanned by the set of functions. We denote this number by $\kappa_l(f)$. Also, we define the q-rank of the set of functions to be the dimension of the linear space spanned by the products $\{f_i f_j\}_{1\le i\le Q, 1\le j\le Q}$. We denote this number by $\kappa_q(f)$.

With these definitions, we can state and prove the following lemmas, which will be used in the proof of the uniqueness theorem.

LEMMA 2: *If the q-rank, $\kappa_q(\rho)$, of the centered range data, $\{\rho_q(t)\}_{q=1}^{q=Q}$, is 6, then*

- *the q-rank, $\kappa_q(c)$, of the illumination directions, $\{c_i(t)\}_{i=1}^{i=3}$, is 6,*
- *the l-rank, $\kappa_l(\rho)$, of the centered range data is 3, and*
- *the l-rank, $\kappa_l(c)$, of the illumination directions is 3.*

Proof of Lemma 2: See Appendix.

LEMMA 3: *If the q-rank of the illumination directions is 6, and $\boldsymbol{A}$ is any symmetric three by three matrix, such that $\mathbf{c}(t)'\boldsymbol{A}\mathbf{c}(t) = 1\ \forall_t$, then $\boldsymbol{A} = \boldsymbol{I}_3$, the three by three identity matrix.*

Proof of Lemma 3: See Appendix.

LEMMA 4: *Let $\boldsymbol{P} : \mathbb{R}^Q \to \mathbb{R}^Q$ be the orthogonal projection matrix onto the linear span of the set of vectors, $\{\tilde{\rho}(t)\}$, assumed by the family of centered range histories, $\{\tilde{\rho}_q\}_{q=1}^{q=Q}$. If $\kappa_l(c) = 3$, then*

$$\boldsymbol{P} = \boldsymbol{P}_S = \boldsymbol{S}(\boldsymbol{S}'\boldsymbol{S})^{-1}\boldsymbol{S}' \tag{11}$$

where $\boldsymbol{P}_S$ is the orthogonal projection onto the range (column space) of the matrix $\boldsymbol{S}$.

Proof of Lemma 4: See Appendix.

Incidentally, this projection matrix, $\boldsymbol{P}$, precisely characterizes the affine invariants of the configuration, $\boldsymbol{S}$. But because we want a more powerfully discriminating set of Euclidean invariants, we go on, to state the uniqueness theorem.

THEOREM 2: (*The Uniqueness Theorem*). *Consider a set of centered range histories* $\{\tilde{\rho}_i\}_{i=1}^{i=Q}$ *from a configuration of* Q *points, for which* $\kappa_q(\tilde{\rho}) = 6$. *If* $\boldsymbol{\Omega}$ *is a centered symmetric matrix,*

$$\boldsymbol{\Omega} = \begin{bmatrix} \sum_{j\neq 1} \omega_{1j} & -\omega_{12} & \cdots & -\omega_{1Q} \\ -\omega_{21} & \sum_{j\neq 2} \omega_{2j} & \cdots & -\omega_{2Q} \\ \vdots & \vdots & & \vdots \\ -\omega_{Q1} & -\omega_{Q2} & \cdots & \sum_{j\neq Q} \omega_{Qj} \end{bmatrix} \tag{12}$$

for which

$$\boldsymbol{P\Omega P} = \boldsymbol{\Omega} \tag{13}$$

where $\boldsymbol{P}$ *is the orthogonal projection onto the linear span of the range histories, and*

$$\sum_{i=1}^{i=Q-1} \sum_{j=i+1}^{j=Q} \omega_{ij} \left(\rho_i(t) - \rho_j(t)\right)^2 - 1 = 0 \tag{14}$$

then $\boldsymbol{\Omega}$ *is the uniquely specified matrix,*

$$\boldsymbol{\Omega} = \boldsymbol{S}(\boldsymbol{S}'\boldsymbol{S})^{-2}\boldsymbol{S}'. \tag{15}$$

Proof of Uniqueness Theorem 2: See Appendix.

4. Computing Conventional Measurements of Size and Shape from the Invariants

4.1. Calculation of the Centered Inner Product Matrix

The uniqueness theorem has a significant corollary which shows us how to convert the invariants into a more common description of the size and shape of the configuration.

COROLLARY 1:

$$\boldsymbol{\Omega} = \boldsymbol{S}(\boldsymbol{S}'\boldsymbol{S})^{-2}\boldsymbol{S}' \tag{16}$$

is the Moore-Penrose generalized inverse of $\boldsymbol{SS}'$*, the centered inner product matrix of the configuration* $\boldsymbol{S}$.

Proof of Corollary 1: See Appendix.

4.2. *Construction of Specific Spatial Models*

The Uniqueness Theorem shows that the matrix of invariants, $\boldsymbol{\Omega}$, can be determined from range data (when the rank conditions are satisfied). The Corollary then shows us how to recover the matrix $\boldsymbol{SS}'$, by simply computing the Moore-Penrose generalized inverse of the matrix $\boldsymbol{\Omega}$. In this section, we show that $\boldsymbol{SS}'$ determines the Euclidean size and shape of the configuration matrix $\boldsymbol{S}$. This can be done by showing that $\boldsymbol{SS}'$ determines the point configuration $\boldsymbol{S}$ up to arbitrary isometries.

From the Q by Q matrix $\boldsymbol{SS}'$ alone, the Q by 3 centered configuration matrix, $\boldsymbol{S}$, cannot be uniquely recovered. This problem occurs because for any 3 by 3 orthonormal matrix, $\boldsymbol{O}$, $(\boldsymbol{SO})(\boldsymbol{SO})' = \boldsymbol{SOO}'\boldsymbol{S}' = \boldsymbol{SS}'$. But the matrix $\boldsymbol{SS}'$ can be factored, $\boldsymbol{SS}' = \boldsymbol{AA}'$, where $\boldsymbol{A}$ is a full rank Q by 3 matrix. This can be achieved using a Choleski factorization routine [14]. Alternatively, the factors can be obtained by computing the eigendecomposition of $\boldsymbol{SS}'$ and using the square roots of the three nonzero eigenvalues to scale their associated eigenvectors.

Each of the factorization methods mentioned in the previous paragraphs may lead to a different Q by 3 matrix $\boldsymbol{A}$ and neither of the methods may give you $\boldsymbol{S}$. The following theorem will show that each method gives you $\boldsymbol{SO}$ for some arbitrary 3 by 3 orthogonal matrix $\boldsymbol{O}$. This arbitrariness of the orthogonal matrix $\boldsymbol{O}$ corresponds to the arbitrary choice of the orthogonal coordinate system for three dimensional space. It occurs because the geometric invariants which we determine from the range histories are coordinate free quantities keeping the same values for any coordinate system.

THEOREM 3: (*The Factorization Theorem*). *Let* $\boldsymbol{A}$ *be a* Q *by* 3 *matrix such that* $\boldsymbol{AA}' = \boldsymbol{SS}'$ *and let* $\boldsymbol{B}$ *be a* Q *by* 3 *matrix such that* $\boldsymbol{BB}' = \boldsymbol{SS}'$. *Let* $\text{rank}(\boldsymbol{A}) = \text{rank}(\boldsymbol{B}) = \text{rank}(\boldsymbol{SS}') = 3$. *Then there exists an orthonormal matrix* $\boldsymbol{O}$ *such that* $\boldsymbol{AO} = \boldsymbol{B}$.

Proof of the Factorization Theorem 3: See Appendix.

A specific model of a scattering center configuration can be chosen, by choosing a specific factorization. This is often useful, providing a visualization of value for human interpretation. Some factorizations are more useful than others. A coordinate frame for which the axes correspond to what one would naturally call the length, width, and height of the object is often most useful, for objects for which those three concepts can unambiguously be defined.

Table 1. Exact and calculated three-dimensional point locations.

x	y	z	x	y	z
−0.40	0.94	0.00	−0.3999996693214	0.9399998930504	3.5035782221335e-07
0.38	−0.01	0.04	0.3800000092641	−0.0100000498730	0.0399999687885
0.59	0.51	1.06	0.5899995167458	0.5100001712217	1.0599996136383
−0.46	−0.29	−0.53	−0.4599997903727	−0.2900000505455	−0.5299998168783
−0.96	0.05	−0.33	−0.9599998513465	0.0500000417291	−0.3299997904918
0.21	−0.57	−0.56	0.2100001926660	−0.5700001243025	−0.5599999144483
0.64	−0.63	0.32	0.6399995923647	−0.6299998812802	0.3199995890338

However, for some potential automated applications, such as object matching, the introduction of any particular coordinate frame may unnecessarily complicate the process. Automated comparison of configurations might best done by direct comparison of the coordinate free geometric invariants. These could be the invariants $\{\omega_{ij}\}_{1\le i<j\le Q}$, or any other equivalent set of geometric invariants, such as the set of all pairwise distances between points of the object. In the next section we compare the three-dimensional point locations for the first, exact data example. After that, in two succeeding sections we compare a set of geometric invariants, the set of three-dimensional Euclidean distances between all pairs of the points.

5. A Numerical Example

To illustrate the theory, we have constructed a numerical example. Seven points were used, with the exact coordinates given in the first three columns of Table 1.

The points were chosen by truncating, to two decimal places, the output of a random number generator. But, that is not important. The only essential characteristic of the points is that they are not all co-planar.

Eight hundred illumination directions were used to simulate the time history of the sensor's angular positions with respect to the seven point configuration. Each of these illumination directions is a unit vector. Or, in other words, each one is the *xyz*-coordinates of a point on the unit sphere.

$$\begin{bmatrix} c_x(1) & c_y(1) & c_z(1) \\ c_x(2) & c_y(2) & c_z(2) \\ c_x(3) & c_y(3) & c_z(3) \\ \vdots & \vdots & \vdots \\ c_x(800) & c_y(800) & c_z(800) \end{bmatrix}.$$

The three components of these unit vectors are plotted as three time series in Figure 1.

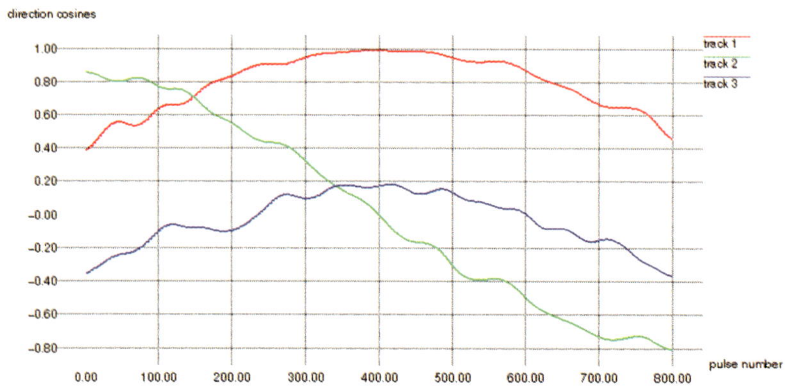

Figure 1. Exact illumination direction components; red = c_x, green = c_y, blue = c_z.

The essential characteristics of these unit vectors are that there be at least six of them; and that they do not all lie in the intersection of the unit sphere with some origin-centered elliptic cone. The geometric theory imposes no other requirements on the direction vectors. (These particular illumination directions were constructed by adding some smoothed random perturbations to an angular motion history calculated from a solution to the Euler equations [15] for a moving rigid body. This was convenient because we have been studying methods of recovering moments of inertia from the data. But the construction method does not really matter for the purely kinematic results presented here.)

From the seven point locations and the 800 illumination directions the seven far-field range histories were created. These are shown in Figure 2.

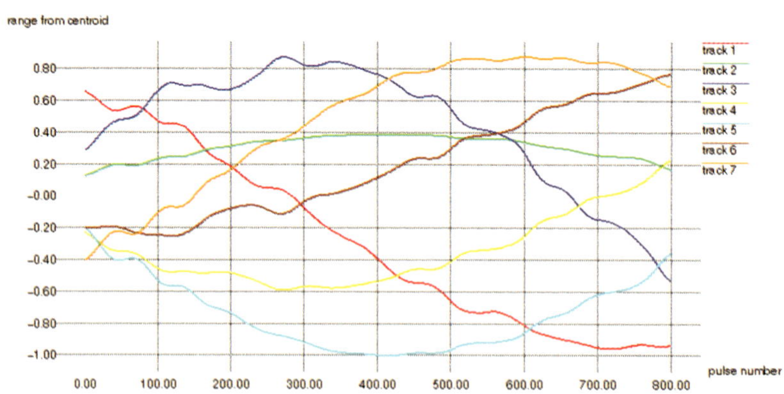

Figure 2. Exact far field range histories.

For the inverse problem we want to solve, we now assume that only these seven range histories are available to us. We now regard both the point configuration and the illumination directions as unknowns. Then using only the range histories, we determine the geometric invariants, which satisfy the conditions of the uniqueness theorem 2. From these invariants, the a specific model of the point locations is constructed, in accordance with the Moore-Penrose generalized inverse corollary 1 and the factorization theorem 3.

The calculated estimates of the locations of points are shown in last three columns of Table 1. These are within machine tolerance of the original point coordinates which were presented in the first three columns.

In addition to the point locations, we calculated the illumination directions from the range histories. Figure 3 contains both the original and calculated illumination directions. The calculated illumination directions completely overlay the originals. As expected, for noise-free data, we recover the exact details of the time varying angular relationship between the sensor and the object.

6. An Example with Noisy Ranges

In the noise free case, we can determine the invariant constants which exactly satisfy the conditions of the uniqueness theorem. But for real applications one must expect imperfect data. Hence, as in most inverse problems, the question of how to make accurate estimates of the parameters in the presence of perturbations in the data, becomes important.

There is much to say about the effects of perturbations or noise in the range history data, and about strategies for maintaining accurate answers in real world applications. The most detailed approach to such questions usually involves development of a statistical theory, which describes the characteristics of the estimated parameters that result from reasonable models for noise in the data. The theory is not simple, primarily because noise in the range measurements enters non-linearly into the invariant equations, but useful analytic results can still be derived.

Indeed, more simply than developing a statistical theory, one can just examine the conditions for continuous dependence of the invariant parameters on the range data. Careful consideration of the uniqueness theorem, reveals that the only conditions in which the invariant parameters fail to depend smoothly on the range data involve either the failure of the object points to be non-coplanar, or the restriction of the illumination directions to a fixed centered elliptic cone. These degeneracy conditions define only lower dimensional varieties in the space of possible shapes and motions. Hence the problem of determining the invariants from the data is generically well posed (in that the locus of degenerate conditions is a set of measure zero, first Baire category, etc).

The possible degeneracies are all manifested in the appearance of excess rank deficiencies of various matrices formed from the range data, so the approach to ill-conditioning can be measured by the size of certain eigenvalues. Lower bounds on the

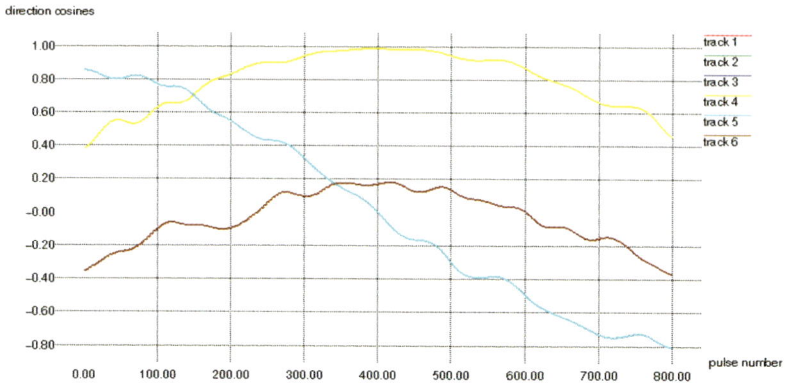

Figure 3. Exact and estimated illumination direction components; tracks 1-3 exact, tracks 4-6 estimated.

key eigenvalues, which physically represent bounds on the flatness of the points and the conicallity of the angular motions, determine bounds on the effects of perturbations in the range data upon estimates of the invariant parameters. Naturally, the same key eigenvalues enter into the statistical analysis, as factors causing the amplification or diminution of noise in the data.

But all the details of error analysis are for future papers. Herein, we present a simple simulation to illustrate just one typical example of the effect of errors in the range data. To show the behavior of the geometric calculations in the presence of noise, we added uncorrelated Gaussian pseudo-random noise to the range histories. Figure 4 shows the noisy range histories (and, for comparison, the original range histories).

From these noisy inputs, we calculated the invariant parameters. A method, based on least squares, was used to handle the noise. The naive least squares approach yields biased

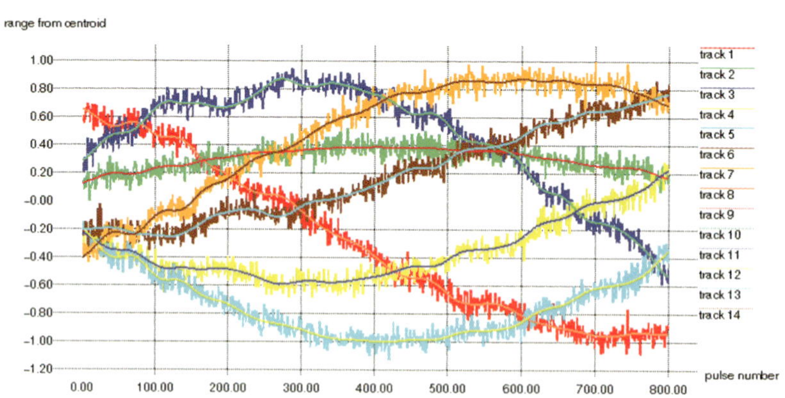

Figure 4. Exact and noisy range histories.

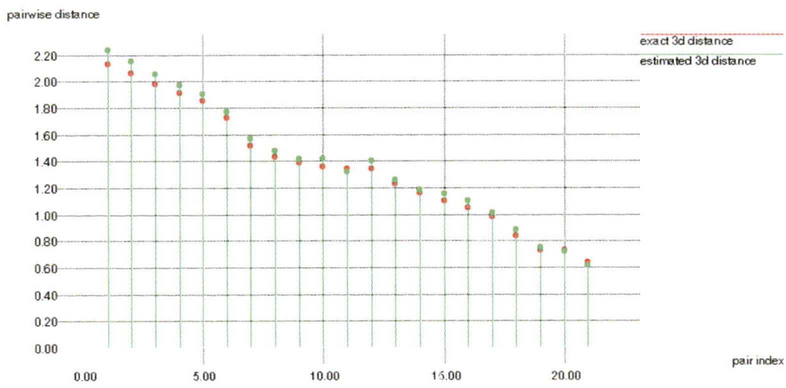

Figure 5. Exact and estimated three-dimensional Euclidean distances between pairs of points.

estimates because of the nonlinear entry of the noise into the equations. The statistical theory in the first author's thesis shows how the bias can be corrected, but the corrections were not fully implemented for this example.

To compare the estimated object geometry to the true geometry, we converted the invariants into the equivalent, invariant, set of pairwise distances between the points. We show these $\binom{7}{2} = 21$ distances in Figure 5. The red dots show exact distances while the estimated distances are shown in green. The presence of the noise in the range data has created errors in the distance measurements, but since the size of the errors is not a large percentage of the distance measurements, we conclude that the shape of the point configuration has not been severely distorted by the estimation. Most of the distances are slightly overestimated in this example, suggesting that the estimate has slightly

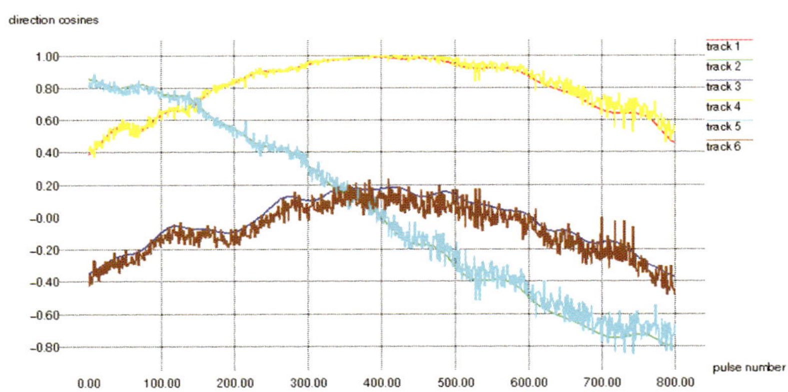

Figure 6. Exact and estimated illumination direction components; tracks 1-3 exact, tracks 4-6 estimated.

exaggerated the size of the object. Nevertheless, the maximum relative error for the distances in this set is only 5.6%.

Having estimated the invariants, we also estimated, from the noisy range data, the illumination directions. These are shown in Figure 6. The noise estimates are printed on top of the true illumination directions. The illumination direction estimates, while still containing noticeable noise, reflect the gross character of the original illumination conditions. A small bias is noticeable, particularly in the third direction cosine.

That the high frequency errors of the range data contribute significantly to the variance of the illumination directions is not unexpected for this problem. Here we had 800 time samples for 7 range histories. The natural averaging processes for these estimates cause the observation of more samples for each range function to contribute primarily toward the reduction of variance in the size and shape estimates, and not so much toward the diminution of variation in the estimates of the illumination directions. But, the observation of more range histories (from more points on the body) would directly reduce the noise in the illumination direction estimates.

7. Comments Regarding Application to Sensor Data

While there are sensors, such as radars, which collect signals that are functions of range, these sensors typically do not directly provide the ranges to distinct points of an object. To extract estimates of those range histories requires methods adapted to the particular sensor; these methods are not described in this paper. For the particular case of radar signals, with which we have experience, the main difficulties arise because of destructive interference among the signals reflected from scattering centers at approximately the same range from the radar. This makes it difficult to track a set of points through wide scanning angles. Second order difficulties due to the obscuration or angle dependent fading of scattered signals adds to the difficulty. Hence the geometric theory presented here offers only part of the solution to many ranging sensor problems of interest.

The level of difficulty depends on the arrangement of the scattering centers on the object and the bandwidth of the transmitted signals. For large objects with strong well resolved scattering mechanisms, sampled with high enough pulse repetition rate, the tracking of the scattering centers is not too difficult. But for small objects (relative to the resolution of the radar signals) advanced methods must be developed to estimate the range histories.

8. Conclusions

The theory developed herein shows that there exist motion-invariant functionals of the range measurements to points on a rigid body. These functionals can be generically determined from range-only data, with no prior knowledge of the size, shape, or motion of the body. Also the functionals completely characterize the Euclidean geometry of the configuration of points on the body.

The experiments show that the theory can used to construct a numerical algorithm for processing the range measurements in order to extract geometry and motion from range data. Such an algorithm is numerically stable and can be used to get useful geometric measurements even in the presence of modest perturbations of the range data.

Appendix A. Proofs

A.1. Proof of Lemma 1

$$(\boldsymbol{x}-\bar{x}\mathbf{1})'\boldsymbol{A}(\boldsymbol{x}-\bar{x}\mathbf{1}) = (\boldsymbol{x}-\bar{x}\mathbf{1})'\boldsymbol{C}'\boldsymbol{A}\boldsymbol{C}(\boldsymbol{x}-\bar{x}\mathbf{1})$$

$$= (\boldsymbol{x}-\bar{x}\mathbf{1})' \begin{bmatrix} a_{11}-a_{.1}-a_{1.}+a_{..} & a_{12}-a_{.2}-a_{1.}+a_{..} & \cdots & a_{1Q}-a_{.Q}-a_{1.}+a_{..} \\ a_{21}-a_{.1}-a_{2.}+a_{..} & a_{22}-a_{.2}-a_{2.}+a_{..} & \cdots & a_{2Q}-a_{.Q}-a_{2.}+a_{..} \\ \vdots & \vdots & & \vdots \\ a_{Q1}-a_{.1}-a_{Q.}+a_{..} & a_{Q2}-a_{.2}-a_{Q.}+a_{..} & \cdots & a_{QQ}-a_{.Q}-a_{Q.}+a_{..} \end{bmatrix}$$

$$\times\ (\boldsymbol{x}-\bar{x}\mathbf{1}) \tag{17}$$

But

$$\sum_{\substack{i=1\\ i\neq j}}^{i=Q} b_{ij} = \sum_{\substack{i=1\\ i\neq j}}^{i=Q} (-a_{ij}+a_{.j}+a_{i.}-a_{..})$$

$$= -(Qa_{.j}-a_{jj}) + (Q-1)a_{.j} + (Qa_{..}-a_{j.}) - (Q-1)a_{..}$$

$$= a_{jj}-a_{.j}-a_{j.}+a_{..}$$

So

$$(\boldsymbol{x}-\bar{x}\mathbf{1})'\boldsymbol{A}(\boldsymbol{x}-\bar{x}\mathbf{1}) = (\boldsymbol{x}-\bar{x}\mathbf{1})' \begin{bmatrix} \sum_{j\neq 1} b_{1j} & -b_{12} & \cdots & -b_{1Q} \\ -b_{21} & \sum_{j\neq 2} b_{2j} & \cdots & -b_{2Q} \\ \vdots & \vdots & & \vdots \\ -b_{Q1} & -b_{Q2} & \cdots & \sum_{j\neq Q} b_{Qj} \end{bmatrix} (\boldsymbol{x}-\bar{x}\mathbf{1}). \tag{19}$$

But expanding this quadratic form, we see that

$$(\boldsymbol{x} - \bar{x}\mathbf{1})'\boldsymbol{A}(\boldsymbol{x} - \bar{x}\mathbf{1}) = \sum_{i=1}^{i=Q}\left(\sum_{j\neq i} b_{ij}\right)(x_i - \bar{x})^2 - \sum_{i=1}^{i=Q-1}\sum_{j=i+1}^{j=Q}(b_{ij} + b_{ji})(x_i - \bar{x})(x_j - \bar{x})$$

$$= \sum_{i=1}^{i=Q-1}\sum_{j=i+1}^{j=Q} b_{ij}(x_i - \bar{x})^2 + \sum_{i=1}^{i=Q-1}\sum_{j=i+1}^{j=Q} b_{ji}(x_j - \bar{x})^2$$

$$- \sum_{i=1}^{i=Q-1}\sum_{j=i+1}^{j=Q} (b_{ij} + b_{ji})(x_i - \bar{x})(x_j - \bar{x})$$

$$= \sum_{i=1}^{i=Q-1}\sum_{j=i+1}^{j=Q} \left(b_{ij}(x_i - \bar{x})^2 + b_{ji}(x_j - \bar{x})^2 - (b_{ij} + b_{ji})(x_i - \bar{x})(x_j - \bar{x})\right)$$

$$= \sum_{i=1}^{i=Q-1}\sum_{j=i+1}^{j=Q} \left(b_{ij}(x_i - \bar{x})^2 + b_{ij}(x_j - \bar{x})^2 - 2b_{ij}(x_i - \bar{x})(x_j - \bar{x})\right)$$

$$= \sum_{i=1}^{i=Q-1}\sum_{j=i+1}^{j=Q} b_{ij}\left((x_i - \bar{x}) - (x_j - \bar{x})\right)^2 = \sum_{i=1}^{i=Q-1}\sum_{j=i+1}^{j=Q} b_{ij}(x_i - x_j)^2. \quad (20)$$

■

A.2. Proof of the Existence Theorem (Theorem 1)

If the rank of $\boldsymbol{S}$ is less than 3 then, for some non-zero $\boldsymbol{x} \in \mathbb{R}^3$, $\boldsymbol{Sx} = 0$. But such an $\boldsymbol{x}$ would be the normal vector to some plane containing all the points. This is not allowed by the hypothesis, so there must be no such vector $\boldsymbol{x}$. So the rank of $\boldsymbol{S}$ is 3.

Ignoring wavefront curvature on the scale of the rigid object, we can model the range history, $\rho_q(t)$, of a moving scattering center as

$$\rho_q(t) = \begin{bmatrix} c_1(t) & c_2(t) & c_3(t) \end{bmatrix} \begin{bmatrix} x_q \\ y_q \\ z_q \end{bmatrix} + b(t) \quad (21)$$

where $[c_1(t) \quad c_2(t) \quad c_3(t)]'$ is an unknown, time varying, unit vector which describes the illumination direction of the radar with respect to the reflecting object, $b(t)$ is an unknown, time varying translation, and $\begin{bmatrix} x_q & y_q & z_q \end{bmatrix}'$ are the coordinates of the scattering center in the inertial reference frame of the object. (See [8] for a more detailed discussion.)

In matrix form, we see that $\tilde{\rho}(t) = \boldsymbol{S}\boldsymbol{c}(t)\ \forall_t$, wherein $\tilde{\rho}_i(t) = \rho_i(t) - \frac{1}{Q}\sum_{q=1}^{q=Q} \rho_q(t)$. Hence, $\boldsymbol{S}'\tilde{\rho}(t) = \boldsymbol{S}'\boldsymbol{S}\boldsymbol{c}(t)\ \forall_t$. Then, since the rank of $\boldsymbol{S}$ is three, the rank of $\boldsymbol{S}'\boldsymbol{S}$ is three and $\boldsymbol{S}'\boldsymbol{S}$ is $\boldsymbol{S}'\boldsymbol{S}$ is hence invertible so

$$(\boldsymbol{S}'\boldsymbol{S})^{-1}\boldsymbol{S}'\tilde{\rho}(t) = \boldsymbol{c}(t) \quad \forall_t. \quad (22)$$

Hence

$$1 = \boldsymbol{c}'(t)\boldsymbol{c}(t) = \tilde{\rho}'(t)\boldsymbol{S}(\boldsymbol{S}'\boldsymbol{S})^{-1}(\boldsymbol{S}'\boldsymbol{S})^{-1}\boldsymbol{S}'\tilde{\rho}(t) = \tilde{\rho}'(t)\boldsymbol{S}(\boldsymbol{S}'\boldsymbol{S})^{-2}\boldsymbol{S}'\tilde{\rho}(t) \quad \forall_t. \tag{23}$$

But, by virtue of the lemma concerning centered quadratic forms, lemma 1,

$$\tilde{\rho}'(t)\boldsymbol{S}(\boldsymbol{S}'\boldsymbol{S})^{-2}\boldsymbol{S}'\tilde{\rho}(t) = \sum_{i=1}^{i=Q-1}\sum_{j=i+1}^{j=Q} \omega_{ij}\big(\rho_i(t) - \rho_j(t)\big)^2 \tag{24}$$

where

$$\omega_{ij} = -\Big(\boldsymbol{S}(\boldsymbol{S}'\boldsymbol{S})^{-2}\boldsymbol{S}'\Big)_{ij} + \Big(\frac{1}{Q}\mathbf{1}'_Q\boldsymbol{S}(\boldsymbol{S}'\boldsymbol{S})^{-2}\boldsymbol{S}'\Big)_j + \Big(\frac{1}{Q}\boldsymbol{S}(\boldsymbol{S}'\boldsymbol{S})^{-2}\boldsymbol{S}'\mathbf{1}_Q\Big)_i$$
$$-\frac{1}{Q^2}\mathbf{1}'_Q\boldsymbol{S}(\boldsymbol{S}'\boldsymbol{S})^{-2}\boldsymbol{S}'\mathbf{1}_Q. \tag{25}$$

But, because the column sums of $\boldsymbol{S}$ are null,

$$\mathbf{1}'_Q\boldsymbol{S} = \mathbf{0}, \quad \frac{1}{Q}\mathbf{1}'_Q\boldsymbol{S}(\boldsymbol{S}'\boldsymbol{S})^{-2}\boldsymbol{S}' = \mathbf{0},$$
$$\frac{1}{Q}\boldsymbol{S}(\boldsymbol{S}'\boldsymbol{S})^{-2}\boldsymbol{S}'\mathbf{1}_Q = \mathbf{0}, \quad \text{and} \quad \frac{1}{Q^2}\mathbf{1}'_Q\boldsymbol{S}(\boldsymbol{S}'\boldsymbol{S})^{-2}\boldsymbol{S}'\mathbf{1}_Q = 0. \tag{26}$$

So we have shown that

$$1 = \sum_{i=1}^{i=Q-1}\sum_{j=i+1}^{j=Q} \omega_{ij}\big(\rho_i(t) - \rho_j(t)\big)^2 \text{ for } \omega_{ij} = -\Big(\boldsymbol{S}(\boldsymbol{S}'\boldsymbol{S})^{-2}\boldsymbol{S}'\Big)_{ij} \quad 1 \le i < j \le Q. \tag{27}$$

■

A.3. Proof of Lemma 2

First we show that the q-rank of the set of illumination directions is 6. For any i and j,

$$\tilde{\rho}_i\tilde{\rho}_j = (\tilde{x}_i c_1 + \tilde{y}_i c_2 + \tilde{z}_i c_3)(\tilde{x}_j c_1 + \tilde{y}_j c_2 + \tilde{z}_j c_3)$$
$$= \tilde{x}_i\tilde{x}_j c_1^2 + \tilde{y}_i\tilde{y}_j c_2^2 + \tilde{z}_i\tilde{z}_j c_3^2 + (\tilde{x}_i\tilde{y}_j + \tilde{x}_j\tilde{y}_i)c_1c_2 + (\tilde{x}_i\tilde{z}_j + \tilde{x}_j\tilde{z}_i)c_1c_3$$
$$+(\tilde{y}_i\tilde{z}_j + \tilde{y}_j\tilde{z}_i)c_2c_3 \tag{28}$$

so the linear span of $\{\tilde{\rho}_i\tilde{\rho}_j\}_{1\le i\le Q, 1\le j\le Q}$ is contained in the linear span of $\{c_i\ c_j\}_{1\le i\le 3, 1\le j\le 3}$. Consequently, since the linear span of $\{\tilde{\rho}_i\tilde{\rho}_j\}_{1\le i\le Q, 1\le j\le Q}$ has dimension 6, the linear span of $\{c_i\ c_j\}_{1\le i\le 3, 1\le j\le 3}$ must have dimension at least 6. Hence, the q-rank of the illumination history is 6.

Next we need to show that $\kappa_l(\rho)$, the linear rank of the range data is 3. If, on the contrary, $\kappa_l(\rho) \leq 2$, then there exists ρ_{q1} and ρ_{q2} such that for each ρ_i, $\rho_i = l_{i1}\rho_{q1} + l_{i2}\rho_{q2}$. Consequently, for any product of a pair of ranges, $\rho_i\rho_j$,

$$\rho_i\rho_j = (l_{i1}\rho_{q_1} + l_{i2}\rho_{q_2})(l_{j1}\rho_{q_1} + l_{j2}\rho_{q_2}) = l_{i1}l_{j1}\rho^2_{q_1} + (l_{i1}l_{j2} + l_{i2}l_{j1})\rho_{q_1}\rho_{q_2} + l_{i2}l_{j2}\rho^2_{q_2} \tag{29}$$

and we have that $\rho_i\rho_j$ is in the linear span of the three functions, ρ^2_{q1}, $\rho_{q1}\rho_{q2}$, and ρ^2_{q2}. Since this argument can be made for each possible pair, we conclude that $\kappa_q(\rho) \leq 3$, contradicting the hypothesis of the lemma, namely that $\kappa_q(\rho) = 6$. So $\kappa_l(\rho) \geq 3$.

But the range histories are also in the span of the illumination directions, which is a set of three functions, so $\kappa_l(\rho) \leq 3$. Hence, $\kappa_l(\rho) = 3$.

Finally, to show that the linear rank of the illumination directions is 3, we note that the linear span of the centered ranges is contained in the linear span of the illumination directions; consequently, $\kappa_l(\rho) = 3$ implies $\kappa_l(c) \geq 3$. Hence, $\kappa_l(c) = 3$. ■

A.4. Proof of Lemma 3

Because the illumination vectors are unit vectors, $\boldsymbol{c}(t)'\boldsymbol{I}_3\boldsymbol{c}(t) = \mathbf{1}\ \forall_t$. Hence, $\boldsymbol{c}(t)'(\boldsymbol{A} - \boldsymbol{I}_3)\ \boldsymbol{c}(t) = 0\ \forall_t$. This implies, that for any six points, $t_1, t_2, \ldots, t_6$

$$\begin{bmatrix} c_1^2(t_1) & c_2^2(t_1) & c_3^2(t_1) & c_1(t_1)c_2(t_1) & c_1(t_1)c_3(t_1) & c_2(t_1)c_3(t_1) \\ c_1^2(t_2) & c_2^2(t_2) & c_3^2(t_2) & c_1(t_2)c_2(t_2) & c_1(t_2)c_3(t_2) & c_2(t_2)c_3(t_2) \\ \vdots & \vdots & \vdots & \vdots & \vdots & \vdots \\ c_1^2(t_6) & c_2^2(t_6) & c_3^2(t_6) & c_1(t_6)c_2(t_6) & c_1(t_6)c_3(t_6) & c_2(t_6)c_3(t_6) \end{bmatrix} \times \begin{bmatrix} 1 - a_{11} \\ 1 - a_{22} \\ 1 - a_{33} \\ 0 - 2a_{12} \\ 0 - 2a_{13} \\ 0 - 2a_{23} \end{bmatrix} = \begin{bmatrix} 0 \\ 0 \\ 0 \\ 0 \\ 0 \\ 0 \end{bmatrix}. \tag{30}$$

Since the q-rank of the illumination directions is 6, there must exist six points for which the rows of the coefficient matrix are linearly independent. Hence, $\boldsymbol{A} = \boldsymbol{I}_3$. ■

A.5. Proof of Lemma 4

For each t, $\tilde{\rho}(t)$ is in the range of $\boldsymbol{S}$, so the range of $\boldsymbol{P}$ is contained in the range of $\boldsymbol{S}$. Conversely, since $\kappa_l(c) = 3$, the three standard basis vectors for $\mathbb{R}^3$ are in the linear span of the set of illumination vectors, so all three of the columns of $\boldsymbol{S}$ are in the linear span of the set of centered range vectors.

Hence, the range of $\boldsymbol{S}$ is contained in the range of $\boldsymbol{P}$. Consequently, $\boldsymbol{P} = \boldsymbol{P_S}$.

Since $\boldsymbol{S}$ is full rank, the formula for $\boldsymbol{P_S}$, is $\boldsymbol{P_S} = \boldsymbol{S}(\boldsymbol{S}'\boldsymbol{S})^{-1}\boldsymbol{S}'$, which is given in many sources, such as [16]. ■

A.6. Proof of the Uniqueness Theorem (Theorem 2)

From the hypothesized invariant equation 14 and lemma 1, we know that

$$\sum_{i=1}^{i=Q-1} \sum_{j=i+1}^{j=Q} \omega_{ij}(\rho_i(t) - \rho_j(t))^2 = \tilde{\rho}'(t)\boldsymbol{\Omega}\,\tilde{\rho}(t) = 1 \tag{31}$$

and from the definition the illumination directions we know that

$$\tilde{\rho}'(t)\boldsymbol{\Omega}\,\tilde{\rho}(t) = \boldsymbol{c}'(t)\boldsymbol{S}'\boldsymbol{\Omega}\boldsymbol{S}\boldsymbol{c}(t) = 1 \quad \forall_t. \tag{32}$$

The matrix $\boldsymbol{S}'\boldsymbol{\Omega}\boldsymbol{S}$ is a three by three symmetric matrix. Therefore, by lemma 3, $\boldsymbol{S}'\boldsymbol{\Omega}\boldsymbol{S} = \boldsymbol{I}_3$. Hence,

$$\begin{aligned}
\boldsymbol{\Omega} &= \boldsymbol{P}\boldsymbol{\Omega}\boldsymbol{P} \\
&= \boldsymbol{S}(\boldsymbol{S}'\boldsymbol{S})^{-1}\boldsymbol{S}'\boldsymbol{\Omega}\boldsymbol{S}(\boldsymbol{S}'\boldsymbol{S})^{-1}\boldsymbol{S}' \\
&= \boldsymbol{S}(\boldsymbol{S}'\boldsymbol{S})^{-1}\boldsymbol{I}_3(\boldsymbol{S}'\boldsymbol{S})^{-1}\boldsymbol{S}' \\
&= \boldsymbol{S}(\boldsymbol{S}'\boldsymbol{S})^{-2}\boldsymbol{S}'.
\end{aligned} \tag{33}$$

■

A.7. Proof of Corollary 1

We verify by checking the Penrose conditions [17]:

1. $\boldsymbol{SS}'\boldsymbol{\Omega}\boldsymbol{SS}' = \boldsymbol{SS}'\boldsymbol{S}(\boldsymbol{S}'\boldsymbol{S})^{-2}\boldsymbol{S}'\boldsymbol{SS}' = \boldsymbol{SS}'$
2. $\boldsymbol{\Omega}\boldsymbol{SS}'\boldsymbol{\Omega} = \boldsymbol{S}(\boldsymbol{S}'\boldsymbol{S})^{-2}\,\boldsymbol{S}'\boldsymbol{SS}'\boldsymbol{S}(\boldsymbol{S}'\boldsymbol{S})^{-2}\boldsymbol{S}' = \boldsymbol{S}(\boldsymbol{S}'\boldsymbol{S})^{-2}\,\boldsymbol{S}' = \boldsymbol{\Omega}$
3. $(\boldsymbol{SS}'\boldsymbol{\Omega})' = (\boldsymbol{SS}'\boldsymbol{S}(\boldsymbol{S}'\boldsymbol{S})^{-2}\boldsymbol{S}')' = (\boldsymbol{S}(\boldsymbol{S}'\boldsymbol{S})^{-1}\boldsymbol{S}')' = \boldsymbol{S}(\boldsymbol{S}'\boldsymbol{S})^{-1}\boldsymbol{S}' = \boldsymbol{SS}'\boldsymbol{S}(\boldsymbol{S}'\boldsymbol{S})^{-2}\boldsymbol{S}' = \boldsymbol{SS}'\boldsymbol{\Omega}$
4. $(\boldsymbol{\Omega}\boldsymbol{SS}')' = (\boldsymbol{S}(\boldsymbol{S}'\boldsymbol{S})^{-2}\boldsymbol{S}'\boldsymbol{SS}')' = (\boldsymbol{S}(\boldsymbol{S}'\boldsymbol{S})^{-1}\boldsymbol{S}')' = \boldsymbol{S}(\boldsymbol{S}'\boldsymbol{S})^{-1}\boldsymbol{S}' = \boldsymbol{S}(\boldsymbol{S}'\boldsymbol{S})^{-2}\boldsymbol{S}'\boldsymbol{SS}' = \boldsymbol{\Omega}\boldsymbol{SS}'$.

■

A.8. Proof of Theorem 3

Let $\boldsymbol{O} = (\boldsymbol{A}'\boldsymbol{A})^{-1}\boldsymbol{A}'\boldsymbol{B}$, then

$$\boldsymbol{OO}' = (\boldsymbol{A}'\boldsymbol{A})^{-1}\boldsymbol{A}'\boldsymbol{BB}'\boldsymbol{A}(\boldsymbol{A}'\boldsymbol{A})^{-1} = (\boldsymbol{A}'\boldsymbol{A})^{-1}\boldsymbol{A}'\boldsymbol{AA}'\boldsymbol{A}(\boldsymbol{A}'\boldsymbol{A})^{-1} = \boldsymbol{I} \tag{34}$$

and

$$\begin{aligned}
\boldsymbol{A}\boldsymbol{O} &= \boldsymbol{A}\boldsymbol{O}(\boldsymbol{B}'\boldsymbol{B})(\boldsymbol{B}'\boldsymbol{B})^{-1}\\
&= \boldsymbol{A}[(\boldsymbol{A}'\boldsymbol{A})^{-1}\boldsymbol{A}'\boldsymbol{B}](\boldsymbol{B}'\boldsymbol{B})(\boldsymbol{B}'\boldsymbol{B})^{-1}\\
&= \boldsymbol{A}(\boldsymbol{A}'\boldsymbol{A})^{-1}\boldsymbol{A}'(\boldsymbol{B}\boldsymbol{B}')\boldsymbol{B}(\boldsymbol{B}'\boldsymbol{B})^{-1}\\
&= \boldsymbol{A}(\boldsymbol{A}'\boldsymbol{A})^{-1}\boldsymbol{A}'(\boldsymbol{A}\boldsymbol{A}')\boldsymbol{B}(\boldsymbol{B}'\boldsymbol{B})^{-1}\\
&= (\boldsymbol{A}\boldsymbol{A}')\boldsymbol{B}(\boldsymbol{B}'\boldsymbol{B})^{-1}\\
&= (\boldsymbol{B}\boldsymbol{B}')\boldsymbol{B}(\boldsymbol{B}'\boldsymbol{B})^{-1}\\
&= \boldsymbol{B}(\boldsymbol{B}'\boldsymbol{B})(\boldsymbol{B}'\boldsymbol{B})^{-1}\\
&= \boldsymbol{B}. \qquad (35)
\end{aligned}$$

■

Acknowledgments

This research has been partially supported by AFRL/SNAT under the project guidance of V. J. Velten, with advice from J. J. Westerkamp and G. Arnold.

References

1. E.T. Bell, *The Development of Mathematics*, New York, London: McGraw-Hill Book Company, Inc., 1945.
2. G.B. Gurevich, *Foundations of the Theory of Algebraic Invariants*, The Netherlands: P. Noordhoff, 1964.
3. H. Weyl, *The Classical Groups. Their Invariants and Representations*, Princeton, New Jersey: Princeton University Press, 1966.
4. P.J. Olver, *Classical Invariant Theory*, Cambridge University Press, 1999.
5. J.L. Mundy and A. Zisserman, *Geometric Invariance in Computer Vision*, Cambridge, Massachusetts, and London, England: The MIT press, 1992.
6. J.L. Mundy, A. Zisserman, and D. Forsyth, *Applications of Invariance in Computer Vision*, Berlin, New York: Springer-Verlag, 1993.
7. V.J. Velten, "Synthetic Aperture Radar (SAR) Geometric Invariants: Experimental Results," in *Proceedings SPIE*, vol. 3721, 1999, pp. 520–531.
8. M.A. Stuff, "Three-Dimensional Analysis of Moving Target Radar Signals: Methods and Implications for ATR and Feature Aided Tracking," in *Proceedings SPIE*, vol. 3721, 1999, pp. 485–496.
9. M.A. Stuff, "Three-Dimensional Invariants of Moving Target," in *Proceedings SPIE*, vol. 4053, 2000, pp. 526–537.
10. D.G. Arnold and K. Sturtz, "Geometric Object/Image Relations for RADAR," in *Proceedings SPIE*, vol. 4053, 2000, pp. 538–545.
11. V.J. Velten, "SAR Image Invariants from 3D Scattering Centers," in *Proceedings SPIE*, vol. 4382, 2001, pp. 367–378.

12. D.G. Arnold, R. Claypool, V.J. Velten, and K. Sturz, "Synthesizing Invariant 3-D Rigid Scattering Centers," in *Proceedings SPIE*, vol. 4382, 2001, pp. 395–403.
13. G.E. Shilov, *Introduction to the Theory of Linear Spaces*, Englewood Cliffs, New Jersey: Prentice Hall, 1961.
14. G.H. Golub and C.F. Van Loan, *Matrix Computations*, Baltimore, Maryland: The Johns Hopkins University Press, 1983, pp. 88–89.
15. A. Gray, *A Treatise on Gyrostatics and Rotational Motion: Theory and Applications*, New York: Dover, 1959.
16. G. Strang, *Linear Algebra and its Applications*, 2nd edition, Orlando, Florida: Academic Press, 1980.
17. A. Ben-Israel and T.N.E. Greville, *Generalized Inverses: Theory and Applications*, New York: John Wiley and Sons, 1974.

Multidimensional Systems and Signal Processing, 14, 183–205, 2003

A Wide-Band Approach to the Absolute Phase Retrieval in SAR Interferometry☆

N. VENEZIANI veneziani@iesi.ba.cnr.it
Istituto Elaborazione Segnali e Immagini - CNR, Bari, Italy

F. BOVENGA fabio.bovenga@ba.infn.it
Dipartimento Interateneo di Fisica, Bari, Italy

A. REFICE alberto.refice@ba.infn.it
Dipartimento Interateneo di Fisica, Bari, Italy

Received December 4, 2000; Revised May 2, 2002; Accepted May 10, 2002

Abstract. Because of possible multiple solutions allowed, the unwrapping of interferometric fringe patterns in the spatial domain is an ill-posed problem which needs some a priori knowledge of the ground morphology for the solution of ambiguities. This is especially true for interferometric SAR (Synthetic Aperture Radar) data. In this paper we propose a different approach to InSAR processing for retrieving the height of ground points independently from each other, unlike most conventional phase unwrapping procedures, which operate in the spatial domain. The basic idea is to repeat raw data focusing by using range sub-bands centered at different frequencies, in order to find a point history of the interferometric phase variation vs. frequency. We introduce the general framework of the method together with considerations on the theoretical limits of applicability, then we report results of our simulations related to a wide-band SAR system. We show that, under certain conditions, height values can be retrieved over a network of coherent and strong scatterers, even when enclosed into low-coherence areas.

Key Words: SAR interferometry, phase unwrapping, sub-band focusing

1. Introduction

The extraction of ground morphology from satellite data has been an active research field since halves of years '70. Different techniques have been proposed during these years, both in the optical and in the microwave region of e.m. spectrum: examples are photogrammetry, radargrammetry, shape from shading, and radar interferometry [1]. The exploitation of phase extracted from SAR (synthetic aperture radar) images is the key factor characterizing the interferometrical method among mentioned techniques. The reasons of the increased importance of SAR interferometry (InSAR) are: high spatial resolution (up to 5×10 m); height precision, with residual error generally less than 10 m; automated processing, and low cost per square kilometer. Moreover, differential interferograms [3] can be figured out by removing the signal related to ground morphology, which show the unique characteristic

☆ Work supported by the Italian Space Agency, under contract no. ASI 1999 - ARS-99-50.

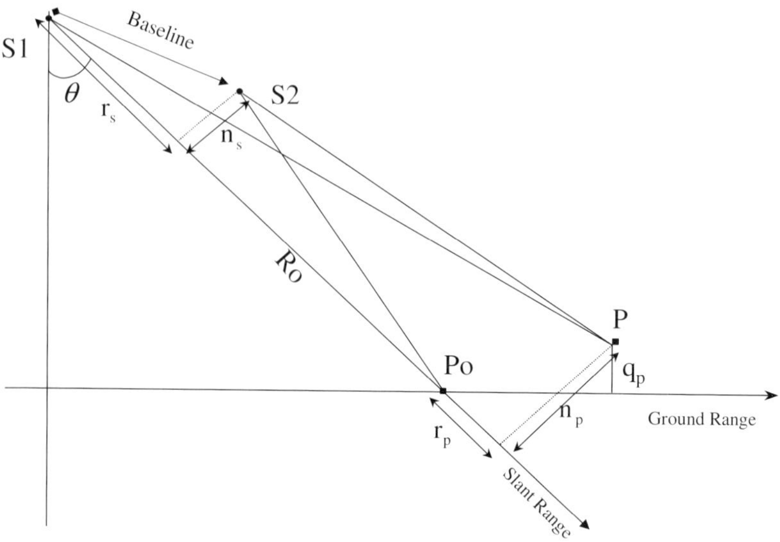

Figure 1. Geometry of acquisition for two satellite platforms located in S1 and S2.

to be sensitive to millimetric deformations of the soil as well as to atmospheric inhomogeneities. So, through the InSAR techniques, a two-dimensional distribution of integral water vapor in the troposphere can be achieved for example by meteorologists to a geometrical resolution higher than never in precedence [31], alternatively to geodynamic applications in the field of volcanology, subsidence, and tectonic, which are becoming possible after the filtering of the atmospheric effects on the phase [2].

In 1974 Graham [5] studied the possibility of using the phase information content of SAR coherent acquisitions for retrieving geometrical features of the ground. An interferometric phase field can be generated by a cross-multiplication of two microwave echo-signals of the same ground area, from two antennas separated by a distance known as *baseline*. The physical law is the same as that of the classical Young experiment: interference of coherent waves.

In a SAR image obtained by using a carrier frequency f, the phase of a point at the antenna-target distance R, can be generally written as $\varphi = \frac{4\pi R}{c} f$, where c is the speed of light. The interferometric phase depends on the difference between the antenna-target distances relative to two antenna positions: $\Delta R = R_2 - R_1$. If we consider the symbols in Figure 1, the relationships between the *interferometric* phase Φ_{int} and the height of a point with respect to the slant range plane n_p, or to the ground range q are [7]:

$$\Phi_{\text{int}}^{\text{slant}} = \frac{4\pi n_s f}{c R_0} n_p \ , \ \ \Phi_{\text{int}} = \frac{4\pi n_s f}{c R_0 \sin\theta} q \tag{1}$$

when referred to the origin in the near range point P_0. In eqs. (1), n_s is the interferometric normal baseline which determines both phase statistics and height sensitivity. From the same equations, the so-called *height of ambiguity*, $q_{2\pi}$, can be defined as the height that leads to a phase variation equal to 2π. The longer the baseline, the smaller the $q_{2\pi}$ is, and thus a better accuracy on height estimations can be achieved.

For an interferometric measurement, the echo-signals I_1, I_2 from S_1 and S_2 respectively have to be correlated, and this requires that their speckle patterns will be similar, or equivalently, that the random walk effect of the echoes from scatterers within a resolution cell will cancel out. Since this random walk effect (i.e. the result of the coherent superposition of radar returns from single point scatterers within a SAR resolution cell) depends, in first approximation, only on the acquisition geometry, if the baseline is greater then a certain critical value the images are completely decorrelated and the measure becomes impossible.

The interferometric coherence γ is defined as the normalized complex cross-correlation of signals I_1 and I_2:

$$\gamma = \frac{E[I_1 \cdot I_2^*]}{\sqrt{E[|I_1|^2]E[|I_2|^2]}} \tag{2}$$

where $E[\cdot]$ denotes the expectation value (usually realized in practice by means of an averaging window) and * is the complex conjugation operator. The absolute value of the interferometric coherence varies between 0 and 1 depending on all decorrelation causes between two SAR signals (geometrical, temporal, thermal [11]) and characterizes the interferometric phase from a statistical point of view [12]. The interferometric phase probability density function for different values of coherence is plotted in Figure 2: as the coherence increases the phase value distribution gets more and more peaked around the true value and the interferometric phase variance decreases.

If $|\hat{\gamma}|$ is the estimate of the true coherence value, n_l the number of independent looks used for the estimation and ϕ_o the noise-free phase value, the interferometric phase variance is, as in [10]:

$$\sigma_\phi^2 = E[(\phi - \phi_o)^2] = \frac{1 - |\hat{\gamma}|^2}{2n_l\,|\hat{\gamma}|^2}. \tag{3}$$

This relation, together with eq. (1), gives the height variance of the topographic map as a function of coherence.

Besides the decorrelation issues mentioned here, different kinds of systematic errors such as multipath, atmospheric effects, motion compensation (in airborne systems), processing inaccuracy and misregistration can locally reduce the correlation quality of an interferometric phase field with scene morphology. However, any type of decorrelation reflects itself in cumulative way in the coherence values of a complex interferogram, and this coherence can be estimated through eq. (2) by means of a small window centered on every pixel considered. So, in the following, a specific decorrelation mechanism will not be considered, and the interferometric data quality will be characterized simply by a global level of coherence. With respect to the matter that this paper concerns, a further cause of height error will be considered here, which occurs during the phase unwrapping step and phase-to-height conversion, primarily due to undersampling of the phase field.

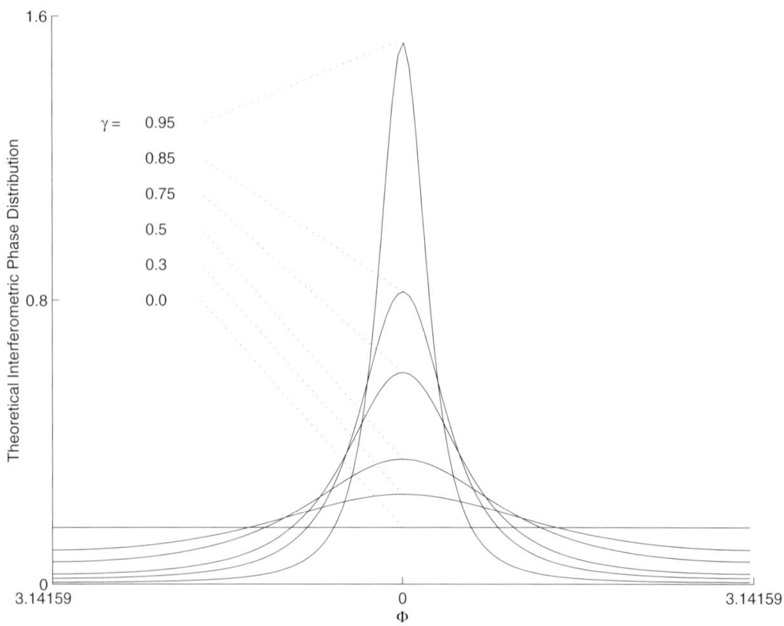

Figure 2. Interferometric Phase Probability Density Function for different values of Coherence.

In eq. (1) the relationships between terrain height and absolute phase are shown. Since we know only the interferometric phase within $(-\pi;\ \pi]$, before the phase-to-height conversion, a phase unwrapping step in the spatial domain is commonly required in a standard interferometric processing, together with the knowledge of the height of a reference point for offset subtraction. The unwrapping of fringe patterns is an ill-posed problem due to the fact that phase statistics and signal undersampling at specific values of local slope at ground cause phase ambiguities which make it impossible to find a univocal solution. Many phase unwrapping algorithms have been implemented to solve this mathematical problem [6]; in spite of different approaches, the reliability of the result is always strongly dependent on *a priori* knowledge of the scene. In order to retrieve the height of a point, we need a path within the interferometric phase field along which absolute phase differences between adjacent points are less than π. If aliasing takes place, a correct height reconstruction needs determination of the number of phase wrappings that have been lost. So, the knowledge of absolute height differences between isolated points could help to solve aliasing ambiguities during conventional unwrapping procedures, with a gain in robustness and efficiency.

In the following we describe a method to retrieve the height difference between couples of points on a coarse resolution field without the need to perform a preliminary spatial unwrapping procedure. This can be achieved by means of a multiple sub-band processing in range direction of the master and slave SAR images, which produces multiple interferometric phase fields at different frequencies with respect to the radar carrier.

A pixel-by-pixel analysis of phase vs. frequencies can then be performed with a linear fit proportional to the mean height of the ground resolution cell corresponding to the pixel analyzed. We present a formulation of the new technique and some preliminary considerations about limits and conditions of applicability, together with simulation results.

Before proceeding over, the reader should consider that some terms have been omitted in eq. (1), those establishing a relationship of the interferometric phase with variations in the atmospheric delay as well as with soil displacements occurred in the time between the detection of master image and slave image, how it happens in SAR interferometry at deferred passes. Implicitly it states that all considerations will be developed in the successive sections are referred to a bistatic system, based on a couple of radar antennas performing the survey simultaneously from their respective positions. Some considerations with respect to the multipass interferometry will be resumed only in the conclusions of the paper.

2. Chirp Spectral Properties and Focusing

The SAR signal is a pulse linearly modulated in frequency, know as *chirp*:

$$c(t) = e^{j2\pi(f_0 t + \frac{k}{2}t^2)} \cdot \mathrm{rect}(t, T) \tag{4}$$

where f_0 is the radar carrier frequency, T is the pulse duration, B is the chirp bandwidth and $k = \frac{B}{T}$ the frequency rate. The echo of a chirp pulse, $c(t)$, is received, base-band demodulated, sampled in phase and quadrature channels to generate a complex signal that represents the *raw data*. In first approximation the image focusing can be performed separately in range and azimuth direction because of the approximate independence of the azimuth focusing parameters with respect to those in the range direction. We consider now the process of SAR image generation along the slant range direction. The focusing step practically consists of deramping the frequency-time trend of the chirp with the correct rate, say K: this processing involves an optimum signal filtering, which improves the SNR [9], and increases the spatial resolution. It is performed by convolving the signal with an inverse chirp usually shaped for example by means of a suitable window in order to reduce sidelobes:

$$h(t) = w(t) \cdot c^*(-t) = w(t) \cdot e^{j2\pi\left(f_0 t - \frac{k}{2}t^2\right)}. \tag{5}$$

Before showing the result it is useful to consider the chirp spectral analysis. If we consider a chirp impulse of length T,

$$FT[c(t)] = C(f) = \begin{cases} \sqrt{\frac{T}{B}} \cdot e^{-j\pi\frac{T}{B}(f-f_0)^2} \cdot e^{j\frac{\pi}{4}} & \text{if } f_0 - \frac{B}{2} < f < f_0 + \frac{B}{2} \\ 0 & \text{otherwise} \end{cases} \tag{6}$$

i.e. the chirp spectrum has a constant amplitude inside the chirp-band (B centered in f_0) and a phase that varies with quadratic law. The higher the product BT, known as *compression rate*, the better the rectangular approximation of the chirp spectrum is.

Let us consider the result of focusing a chirp echo received from a target at distance R from the sensor. The time delay is $\tau = \frac{2R}{c}$; thanks to the Fourier transform properties we can calculate the spectrum of this echo after the base-band demodulation:

$$FT\left[c^{\text{demod}}(t-\tau)\right] = C(f) \cdot e^{-j2\pi f_0 \tau} \cdot e^{-j2\pi f \tau}. \tag{7}$$

Performing the convolution with the inverse chirp $h(t)$ in the frequency domain by using the approximate formulation in (6) and neglecting the weighting function in (5) results in:

$$\begin{aligned} s(t) &= FT^{-1}\left[e^{-j2\pi(f+f_0)\tau} C(f) * C^*(f)\right] = \frac{T}{B}\int_{-\frac{B}{2}}^{\frac{B}{2}} e^{-j2\pi(f+f_0)\tau} e^{j2\pi t f}\, df \\ &= \frac{T}{\pi} e^{-j2\pi f_0 \tau} \cdot \operatorname{sinc}[\pi B(t-\tau)]. \end{aligned} \tag{8}$$

The slant range resolution depends on the sinc main lobe width and, after a full band focusing, is $\frac{c}{2B}$. The phase under the main lobe depends on the product between the time delay τ and the radar carrier frequency f_0, and represents the scene geometrical information we need to retrieve. Unfortunately we only know its principal value and moreover, because of the high value of f_0 and R, the phase gradient modulus is high, so that phase aliasing takes place and unwrapping the signal does not give a unique solution. The interferometric technique reduces the absolute phase gradient amplitude by relating the phase to target height (instead of the sensor-target distance that is hundreds of thousands times larger). As mentioned before, the interferometric phase is also wrapped but, if there is no aliasing, a spatial unwrapping is able to reproduce the soil surface geometry except for a constant offset.

3. Multifrequency Approach

The last result of the previous section shows that the phase impulse response of a SAR system under the main amplitude lobe is only due to the radar carrier frequency, f_0, and the echo time delay (geometrical sensor-target distance), R. We highlighted already the constrains which allow to use with good approximation this result: $BT \gg 100$. Consequently, the interferometric phase depends linearly on the target height with respect to the slant range plane, and on the carrier frequency. Moreover, phase wrapping along the spatial field of the interferogram prevents a correct knowledge of the target height.

In 1992, Madsen and Zebker [13] investigated the idea of using two sub-band interferograms in order to extend the area over which the spatial averaging can be performed, thus overcoming the problems relating to estimation noise. The basic idea of the *split-spectrum* technique is to use an interferogram obtained by means of two "standard" interferograms of the same scene but referred to different frequencies within the chirp bandwidth. The phase depends on the frequency difference, which is hundreds of times less than f_0, and therefore varies with a small rate within the principal interval, without wrapping effects. The absolute phase can then be retrieved by means of a linear operator: $\Phi^{\text{absolute}} = \frac{f_0}{\Delta f} \cdot \Phi^{\text{differential}}$. Unfortunately the linear coefficient amplifies the noise too, leading to an unavoidable increase in the final height uncertainty. An averaging in spatial domain allows to retrieve the absolute phase over a coarse resolution grid. In the following, an innovative approach to height retrieval is proposed, which performs a differential analysis of the phase in the frequency domain by splitting the signal spectral bandwidth into multiple sub-bands. By using more than two sub-bands, target height can be determined by means of a phase-frequency linear fit with less uncertainty and more robustness against noise than in the spatial case.

Suppose to select a portion B_i of the chirp signal spectrum centered on the frequency f_i, and to use a focusing inverse chirp $h_i(t)$ covering only such a band:

$$h_i(t) = h(t) \cdot \text{rect}(t - t_i, T_i) \tag{9}$$

where $t_i = \frac{f_0 - f_i}{k}$ and $T_i = \frac{B_i}{B} \cdot T$.

This *partial chirp* is used to perform the signal range focusing, in order to process only a portion B_i of the full band available. Applying the Fourier transform properties to the signal demodulated and delayed of τ and (6) to calculate the spectrum of both the signal and the partial inverse chirp h_i, we obtain:

$$\begin{aligned} s_i(t) &= e^{-j2\pi f_0 \tau} c(t-\tau) * h_i(t) \\ &= \int_{f_0 - f_i - \frac{B_i}{2}}^{f_0 - f_i + \frac{B_i}{2}} \sqrt{\frac{T}{B}} e^{-j2\pi(f+f_0)\tau} e^{-j\pi \frac{T}{B} f^2} \cdot \sqrt{\frac{T_i}{B_i}} e^{-j\pi \frac{T_i}{B_i} f^2} e^{j2\pi t f}\, df \\ &= T_i \cdot e^{-j2\pi f_i \tau} \cdot e^{j2\pi(f_i - f_0)t} \cdot \text{sinc}[\pi B_i (t-\tau)]. \end{aligned} \tag{10}$$

Because the chirp rate K must be the same for both the chirp, (T, B), and the partial chirp, (T_i, B_i), we set to zero the quadratic phase term and to $\frac{T_i}{B_i}$ the square root product. This observation will be useful in the following. The signal in (10) exhibits the following characteristics:

- a slant range resolution $\Delta R = \frac{c}{2B_i}$, decreased of $\frac{B_i}{B}$ times with respect to the full-band one;
- a *modulating* phase term proportional to $\Delta f = f_i - f_0$;
- a constant phase term proportional to the geometrical information R and the frequency f_i which differs by Δf from the carrier of the original signal.

By shifting the band B_i within the signal band B, the frequency f_i shifts of the same quantity so that the geometrical phase term still depends on the distance R but with a different proportionality factor. If this processing is repeated with the same parameters for master and slave images, the generated interferometric phase, Φ_i^{int}, will reproduce, with slightly different wavelengths, the same topographic profile:

$$INT_i = s_i^1(t) \cdot [s_i^2(t)]^* \propto e^{j2\pi(\tau_2-\tau_1)f_i} \tag{11}$$

where indices 1 and 2 indicate the master and slave images respectively. This kind of processing allows to study the interferometric phase of a single point as a function of frequency with the following relationship:

$$\Phi^{\mathrm{int}} = \frac{4\pi n_s q}{cR_0 \sin\theta} f \tag{12}$$

where R_0 is the near range, n_s is the normal baseline and q the height of the point with respect to the ground plane. A set of *partial band interferograms* can be generated one for each sub-band central frequency f_i. By selecting a pixel, (x, y), its interferometric phase trend along frequencies can be studied. The value of $q(x, y)$ can be retrieved by means of a linear fit of phase values relative to this pixel and to different partial interferograms, $\Phi_i^{\mathrm{int}}(x, y)$;

$$\Phi_i^{\mathrm{int}}(x,y) = C_o + C_1(x,y) \cdot f_i. \tag{13}$$

Now the physical quantity of interest q can be calculated as a function of phase variation along frequencies instead of the interferometric phase directly:

$$q(x,y) = \frac{cR_0 \sin\theta}{4\pi n_s} \cdot \frac{\partial}{\partial f}\Phi^{\mathrm{int}}(x,y) = \frac{cR_0 \sin\theta}{4\pi n_s} \cdot C_1(x,y). \tag{14}$$

The height is proportional to the linear rate $C_1(x, y)$ and the number of interferograms N_f depends on the parameters B, B_i. When the sub-bands do not overlap, N_f represents the range multilook factor and it coincides with the term n_l in eq. (3).

The height of any pixel (with respect to the pixel at near range) is retrieved independently from the others and no phase unwrapping step is needed in the spatial domain. This allows to apply the method over areas of the interferometric phase field with low coherence, where only few pixels show a good stability.

The coarse resolution of the resulting interferogram depends on the spectral sub-band width, B_i, and can be set during the design step.

As in standard interferometric phase analysis in the spatial domain, the phase of a pixel can be wrapped along the frequencies, too. Hence before performing the linear fit with standard methods (such as a Least Mean Squares fit), a one-dimensional phase unwrapping step is needed; the constraint $|\Delta\Phi^{\mathrm{int}}| \leq \pi$ ensures to avoid aliasing along frequencies. A maximum height value without ambiguity, Q, represents the upper limit for correct height retrieving:

$$q \leq Q = \frac{cR_0 \sin\theta}{4 n_s \Delta f} \tag{15}$$

Δf is the frequency sampling spacing ($\Delta f = f_{i+1} - f_i, \forall i = 1 : N_f$), and can be adjusted to a certain extent to lead to Q below the maximum height magnitude foreseen over the area where the point analyzed is located. From (15) with $R_0 = 800$ km, $\theta = 23°$, $n_s = 400$ m, and $\Delta f = 20$ MHz results $Q \approx 3$ km.

3.1. Study of Applicability

The practical use of this technique strongly depends on the value of B which must be as wide as possible. The wider B, the better the resolution on partial interferometric phase field is, and more possible combinations of $(B_i, \Delta f)$ are allowed. Moreover, the effective band is usually shorter than the original due to the weighting for sidelobe reduction and spectral-shift phenomenon [8]. The problem is to establish a criterion to decide if the phase noise and the mission radar parameters allow a height estimation with error below a certain value. To this purpose a height error formulation as function of coherence, B, B_i and Δf is presented in the following. The difference in coherence between the partial interferograms and the principal one, due to the change in bandwidth and frequency, will cause a change in geometrical and thermal decorrelation terms [11] so that:

$$\gamma_i^{\text{tot}} = \gamma^{\text{tot}} \cdot \frac{\gamma_{\text{Geo}}(B_i, f_i)}{\gamma_{\text{Geo}}(B, f)} \cdot \frac{\gamma_{\text{Th}}(B_i)}{\gamma_{\text{Th}}(B)} = \gamma^{\text{tot}} \cdot \mathcal{F}(B, f_0, B_i, f_i). \tag{16}$$

Since the system is designed to work with full band signal, the thermal decorrelation (in the hypothesis of white band noise) will not increase for the sub-band signals. The main cause of coherence loss is the geometrical decorrelation due to a loss in resolution:

$$\mathcal{F}(B, f_0, B_i, f_i) \approx \frac{1 - \frac{n_s f_i}{R_0 B_i \tan\theta}}{1 - \frac{n_s f_0}{R_0 B \tan\theta}}. \tag{17}$$

This relationship, together with (3) allows to evaluate the phase variance for the sub-band interferogram as a function of the full-band interferometric coherence:

$$\sigma_{\phi_i} = \sqrt{\frac{1 - (|\gamma| \mathcal{F})^2}{2n_l (|\gamma| \mathcal{F})^2}}. \tag{18}$$

The uncertainty on ϕ_i yields an error in the fit coefficients evaluation. Performing the linear fit with a Least Mean Square method weighed by $\sigma_{\phi i(x,y)}$, the error on the rate $C_1(x, y)$ is [15]:

$$\sigma_{C_1} = \sqrt{\frac{\sum_{i=1}^{N_f} \frac{1}{\sigma_{\phi_i}^2}}{\sum_{i=1}^{N_f} \frac{1}{\sigma_{\phi_i}^2} \cdot \sum_{i=1}^{N_f} \left(\frac{f_i}{\sigma_{\phi_i}}\right)^2 - \left(\sum_{i=1}^{N_f} \frac{f_i}{\sigma_{\phi_i}^2}\right)^2}}. \tag{19}$$

Line	B	B_i	Δf	N_f
a	100	40	20	3
b	200	100	20	5
c	200	40	20	9
d	400	100	20	15
e	400	40	20	19
f	400	20	10	39

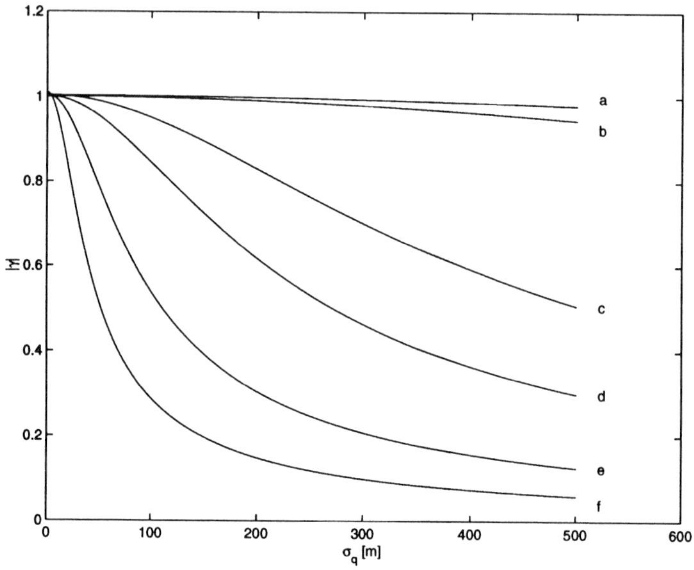

Figure 3. Parametric $|\gamma|$ vs. σ_q trends for B, B_i, Δf, N_f reported in the table.

With good approximation we can assume $\mathcal{F}$ to be independent from f_i:

$$\gamma_i(x, y, B_i, f_i) = \gamma_i(x, y, B_i) \quad \forall i, (x, y) \tag{20}$$

so that, given B_i, σ_{ϕ_i} is the same for all partial interferograms. Hence the height error σ_q as a function of $|\gamma|$, B, B_i, f_0 (from (18), (14) and (19)) is:

$$\sigma_q = \frac{cR_0 \sin\theta}{4\pi n_s} \cdot \sigma_{C_1} = \frac{cR_0 \sin\theta}{4\pi n_s} \sqrt{\frac{1 - (\mathcal{F}\,|\gamma|)^2}{2n_l(\mathcal{F}\,|\gamma|)^2 \left[\sum_{i=1}^{N_f}(f_i)^2 - \frac{1}{N_f}\left(\sum_{i=1}^{N_f} f_i\right)^2\right]}}. \tag{21}$$

This relationship allows to verify whether, for a given value of full band coherence, a possible configuration of processing parameters exists that leads to an height error

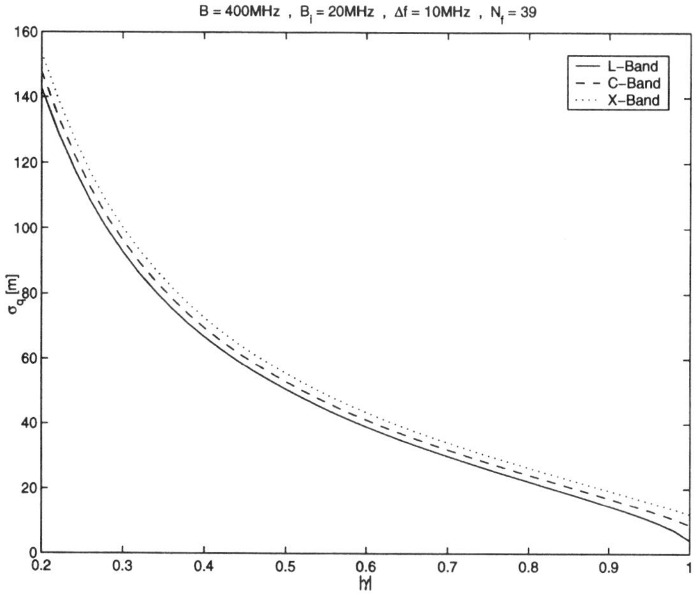

Figure 4. Parametric σ_q vs. $|\gamma|$ trends with B, B_i, Δf, N_f as in the case **f** of the table in the previous figure, and for L, C and X band: σ_q increases with the carrier frequency.

sufficiently small. Alternatively, the inverse relationship gives, for a configuration of processing parameters, the minimum value of the full-band coherence that a point must have for retrieving its height with an acceptable error:

$$|\gamma|_{min} = \frac{\mathcal{F}^{-1}(B, f_0, B_i)}{\sqrt{1 + 2n_l\left(\frac{4\pi n_s}{cR_0 \sin\theta}\right)^2 \left[\sum_{i=1}^{N_f}(f_i)^2 - \frac{1}{N_f}\left(\sum_{i=1}^{N_f} f_i\right)^2\right]\sigma_q^2}}. \tag{22}$$

In order to calculate the number of 2π cycles between two points, independently from spatial aliasing problems, we need a precision of half a $q_{2\pi}$ corresponding to an error $\sigma_q^{\max} = \frac{cR_0 \sin\theta}{4n_s f_0}$. This value allows to determine the lower limit of the coherence for successful application of the technique. If this limit is exceeded, the best B_i and f_i are those which minimize σ_q under the constraint given by (15). Figure 3 shows $|\gamma|$ vs. σ_q trends, for $f_0 = 1.2$ GHz and different values of B, B_i f_i, N_f reported in the table at the top: as N_f increases the height error for a given coherence value decreases. The parameter n_l in the last two equations takes into accounts the sub-bands overlapping: it is equal to N_f until there is no overlapping, then its value has to be corrected by a factor proportional to $\delta f/Bi$.

Instead of performing a WLMS linear fit, the problem of height retrieving can be solved also as a parameter estimation of a complex sinusoid from a finite number of noisy samples by calculating the maximum of the signal spectrum [14]. This strategy leads to an height error trend w.r.t. coherence and band splitting parameters analogous to the previous equations, but it shows more robustness w.r.t. noise fluctuation, overcoming the need of the phase unwrapping step along the frequencies.

The previous theoretical result shows (see Figure 4) that, for B equal to 400 MHz, a configuration of B_i, Δf exists so that, for coherence values better than 0.5, the height of a pixel can be retrieved with uncertainty less than 60 m. Therefore, by designing the InSAR system with baseline and radar wavelength leading to a height ambiguity $q_{2\pi}$ greater than 120 m, it is possible to solve phase unwrapping ambiguity problems due to aliasing or noise in the fringe pattern over areas with coherence better than 0.5. The analysis of phase values in the partial-band interferograms as a function of frequency can be performed for pixels with coherence above the threshold value. A grid of reference points of absolute height estimated can be finally built, superimposed on a "coarse" interferogram with resolution depending on the sub-band width. On the other hand, when the coherence value is near 0.9, as in SRTM (Shuttle Radar Topographic Mission) experiment, then an height accuracy of about 10 m is expected.

The statistical model considered in our formulation assumes as mutually independent the random contributions of the scatterers within a given resolution cell, as a function of the sub-band central frequency. It should therefore be considered as a worst-case evaluation (w.r.t. for example a single scatterer configuration); consequently, the σ_q values inferred from the previous figures may be too conservative with respect to real cases.

Different and more favorable situations can be considered. As will be shown in the next section, targets can be simulated for which the technique of multiple sub-band interferograms is able to retrieve the height value with accuracy better than the previous theoretical estimates, even for small B and, consequently, N_f values. In particular, in order to test the technique in a simpler and more controlled situation, the processing chain from sub-band focusing to the interferometric phase fit, has been applied to special simulated targets, consisting of a random array of scattering elements, one of which has a radar echo dominant with respect to the others. In this configuration the SAR system response results polarized by this strong scatterer, ensuring high coherence values and a uniform spectral response. This is particularly suitable e.g. for interferograms with large incoherent area including few single coherent and pointlike strong scatterers, which can be thus used as control points in the subsequent spatial unwrapping step.

In the following the simulation results are reported.

4. Simulation Results

Validation of theoretical results requires either raw data availability or a sufficiently realistic simulation of the SAR signal generation and of the acquisition process. In this section we describe schematically the *raw data simulation* procedure and the following processing steps. The relevant parameters are reported for the simulated data-set together with the results of our multifrequency analysis.

4.1. Scene Description

Many simulation studies of raw SAR signals are available in the literature, e.g. [16]. In order to simulate statistical and deterministic SAR imaging properties in amplitude and phase, care must be taken in designing both soil electromagnetic and geometric features and scattering model. Since we are not interested in retrieving the soil dielectric features but only topography (signal phase), we designed an *ad hoc* simplified scattering model.

As in the previous theoretical description we simulate the phenomenon only in the slant-range direction referring to the plain geometry in Figure 1. The scene is discretized in a range array whose elements represent natural scatterers. The n-th scatterer is defined by a height with respect to the slant direction, q^{slant}, and a backscattering complex coefficient $A(n)$:

$$
\begin{aligned}
q^{\text{slant}}(n) &= (n-1)\cdot\frac{\Delta x}{\tan(\theta-\alpha)} + dq_n \\
A(n) &= a_n \cdot e^{j\phi_n}
\end{aligned}
\tag{23}
$$

where Δx is the slant range spacing between simulated scatterers, α the terrain slope, ϕ_n a gaussian random phase term and dq_n a uniformly distributed height variation factor. We choose a particular height profile defined by means of an analytical linear function. This height definition allows to investigate the interferometric phase variation for different slopes and at the same time simplifies the following processing steps. The stochastic component of the simulated phenomenon is simulated by randomizing the phase and the height terms ϕ_n and dq_n. Due to such randomized heights, scatterers within a resolution cell are distributed in a volume that has a deep of the order of the wavelength. Therefore the simulation takes account also for the noise component due to volumetric scatterring.

A resolution cell, whose size depends on the bandwidth processed, contains a number of pointy scatterers depending on the slant range spacing Δx. Statistical considerations (related to the stochastic parameters) suggest us to set this number equal to 10. The real part of back-scattering coefficient, a_n, can be arbitrary set for every scatterer. During the experimentation, special resolution cells have been simulated where a single scatterer has a dominant role: this simulation is performed by setting the back-scattering coefficient of the scatterer to a value higher than the background. We denote by K_{ref} the ratio between the strongest scatterer amplitude and the (constant) amplitude of all the other targets in the line. Depending on K_{ref}, two different cell structures can be simulated: for K_{ref} greater than 1 the SAR system response results polarized by a single scatterer, ensuring high coherence values and a uniform spectral response. Such target behavior should be similar to that of Permanent Scatterers described in [4]. For K_{ref} equal to 1 the amplitude values have a wider distribution. However, since the heights randomization term depends only on the carrier frequency, the resolution cell shows yet a volumetric distribution stable w.r.t. all the frequencies within the signal bandwidth.

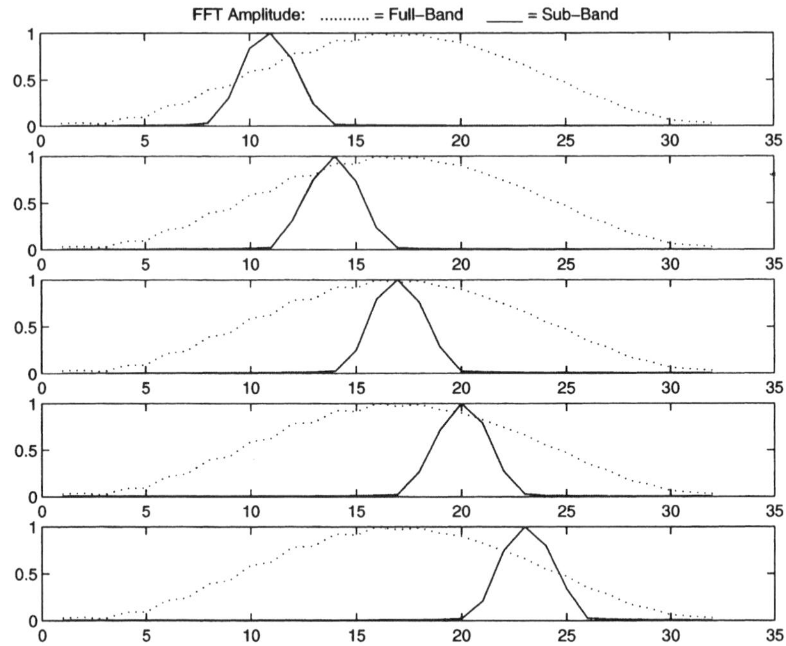

Figure 5. Sub-Bands (B_i, f_i) selected inside the Full-Band: $B = 200$ MHz, $f_0 = 1.2$ GHz, $B_i = 40$ MHz, $f_i = f_0 - (\{2, 1, 0, -1, -2\} \cdot \Delta f)$ with $\Delta f = 20$ MHz.

This configuration leads to a behavior of scatterers that is similar for all the sub-bands. A more general situation would produce a more noisy behavior in the frequency domain.

4.2. SAR Raw Data and Focusing

The output of SAR simulator is a complex raw signal demodulated in the base-band. For this, a complex chirp pulse is generated with spectral bandwidth equal to B, central frequency equal to zero and sampling frequency $f_s > B$. The output signal is computed as the superposition of elementary echoes scattered by a simulated soil. Every echo is shifted by a quantity proportional to the scatterer height and to the radar carrier frequency f_0. Its amplitude is defined by $a_n = |A(n)|$, while its phase depends on the deterministic and stochastic height values as well as on the backscatterer coefficient phase:

$$s(i) = \sum_{n=1}^{N} a_n \cdot e^{j\phi_n} \cdot e^{-j2\pi f_0 \tau_n} \cdot e^{j2\pi \frac{k}{2}\left(\frac{i}{f_s} - \tau_n\right)^2} \cdot \mathrm{rect}\left(\frac{i}{f_s} - \tau_n, T\right). \tag{24}$$

The focusing step is performed with a numerical convolution between the signal and the inverse chirp.

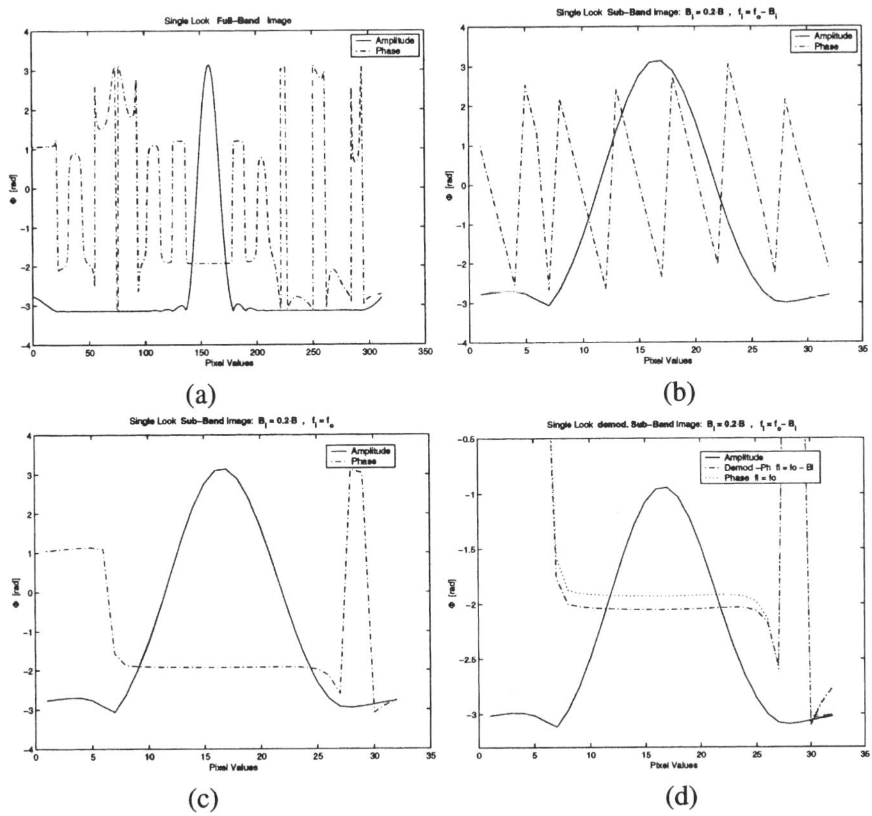

Figure 6. (a) Amplitude and Phase of Full-Band master image: inside the sinc main lobe the phase is constant as predicted in (8). (b) Amplitude and phase of Sub-Band focused master image with $B = 200$ MHz, $f_0 = 1.2$ GHz, $B_i = 40$ MHz, $f_i = f_0 - 2 \cdot \Delta f$. Since $f_i \neq f_0$ a linear phase term (wrapped) occurs inside the sinc lobe as predicted in (10). (c) Amplitude and phase of Sub-Band focused master image with $B = 200$ MHz, $f_0 = 1.2$ GHz, $B_i = 40$ MHz, $f_i = f_0$: the sinc lobe is wider and the phase is constant because the modulating term $\propto f_i - f_0$ is zero. (d) After demodulation step the phase is constant and has a value different from the central Sub-Band focused image $f_i = f_0$.

4.3. Multifrequency Approach

In order to find the interferometric phase trend vs. frequency, the focusing step is repeated for master and slave images by using time-windowed inverse chirps (see Figure 5). The windows have a Hamming shape and are centered at instants: $t_i = \frac{1}{f_s} \mathrm{int}\left(f_i \cdot \frac{Tf_s}{B}\right)$ with a frequency uncertainty equal to $\frac{B}{Tf_s} \approx 3$ kHz. During numerical inverse chirp definition we took care of setting the same value of chirp rate K. In Figure 6 the outputs of these steps are shown. As predicted by (10) the resolution decreases, the phase inside the sinc lobe is constant for full-band image and for sub-band image with $f_i = f_0$, whereas when $f_i \neq f_0$ a linear phase trend occurs (in (b) its wrapped trend is plotted). In order to generate sub-band

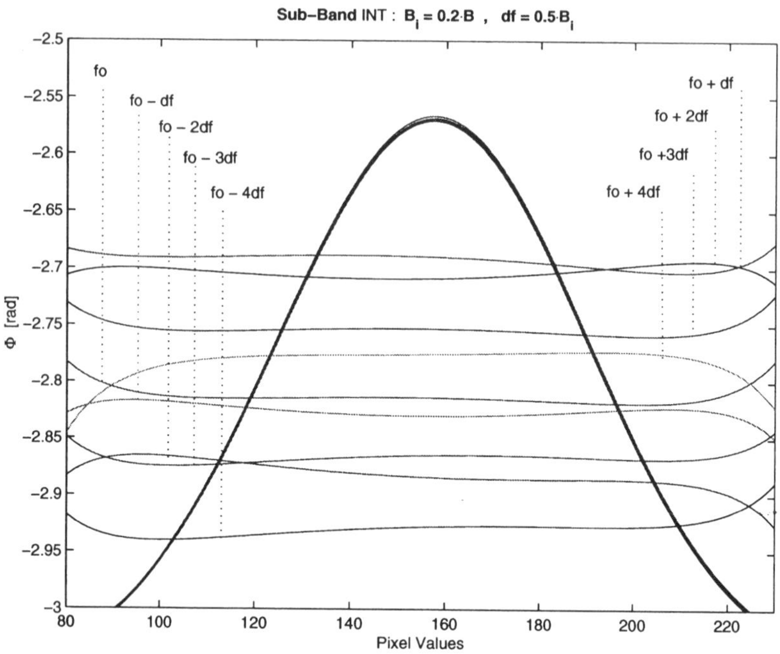

Figure 7. Sub-Band Interferometric Phase for $B = 200$ MHz, $f_0 = 1.2$ GHz, $B_i = 40$ MHz, $f_i \in [1.12, 1.28]$ GHz with $df \equiv \Delta f = 20$ MHz.

interferograms, the master and slave images are interpolated and coregistered for every sub-band. Before interpolation, the linear trend described in eq. (10) is subtracted by demodulating the sub-band images. Sub-band interferometric phase values are shown in Figure 7, where different phase values correspond to different sub-bands.

4.4. *Result Analysis*

Three sets of input parameters are given to the SAR simulator related to the radar specifications (f_0, f_s, B, f_i, B_i, K), to the geometry of acquisitions (R_1, $B_\perp$, $B_\parallel$, θ, α) and to the ground radiometric properties (Δx, $A(n)$). Simulations were performed ensuring a number of reflectors inside a resolution cell at least equal to 10.

The outputs of simulation process are the full-band interferogram and the sub-band interferograms reported, after the flattening step [7], to the ground range reference. The overall process takes into account for geometrical decorrelation, processing errors (focusing and interpolation) and volumetric scattering within the resolution cell. On the other hand, since we are interested only in the phase content of data, SAR image effects such as layover, shadowing, and foreshortening are not reproduced on the amplitude images.

In Figures 8, 9, 10 we present the main results of the processing described. In Figure 8, phase fits for $f_0 = 5.3$ GHz, $B = 40$ MHz, $B_i = 20$ MHz are plotted. The shorter normal

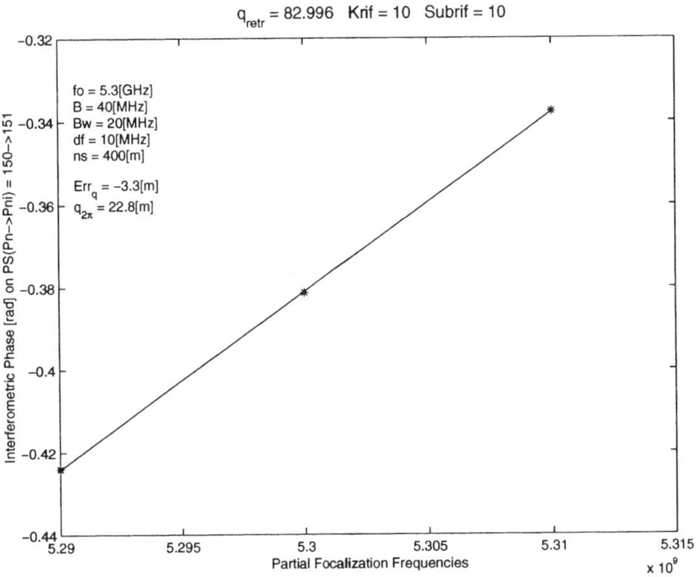

(a)

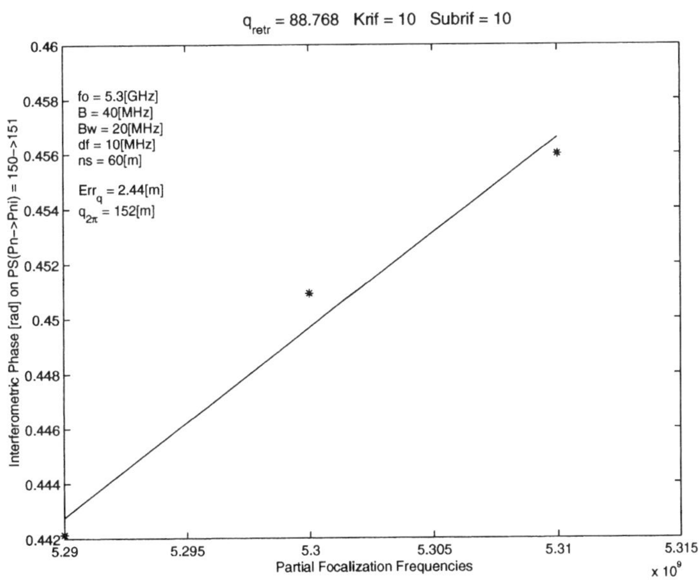

(b)

Figure 8. Linear Fit for $f_0 = 5.3$ GHz, $B = 40$ MHz, $B_w \equiv B_i = 20$ MHz, $df \equiv \Delta f = 10$ MHz, $K_{\text{ref}} = 10$ and $n_s = 400$ m (a), $n_s = 60$ m (b). As geometric decorrelation decreases ($|\gamma| = 0.85$ (a), 0.98 (b)), height error decreases too: 3.3 m (a), 2-h m (b).

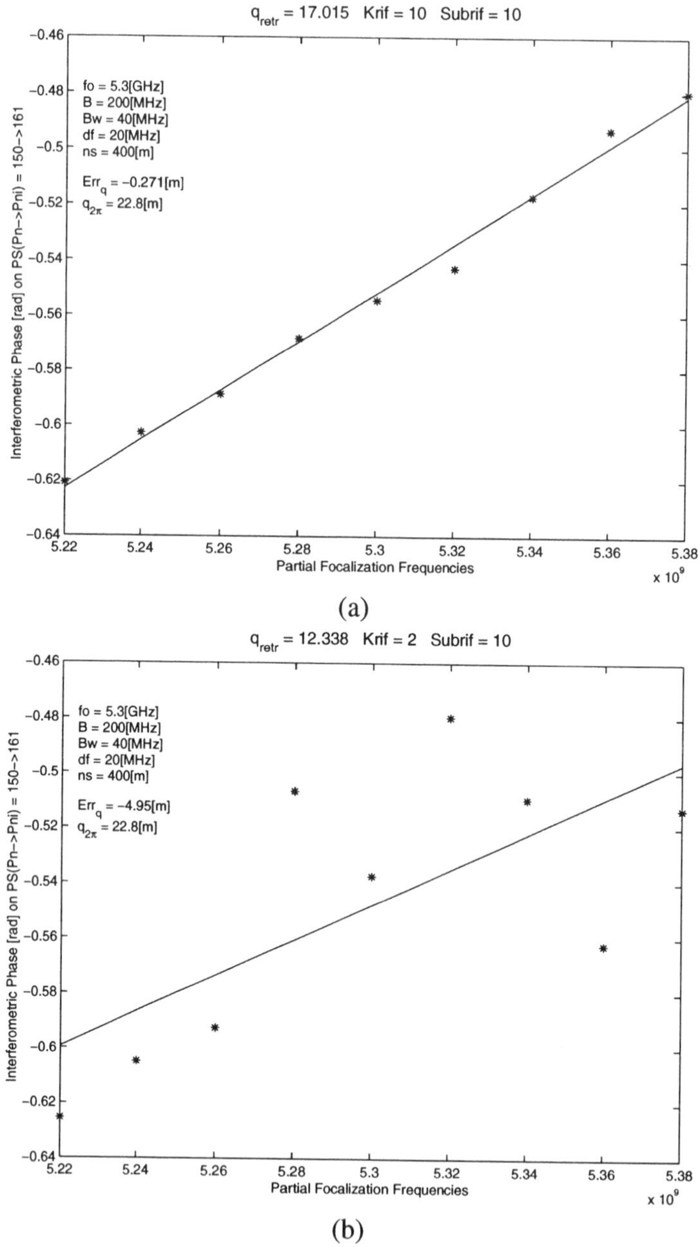

Figure 9. Linear Fit for $f_0 = 5.3$ GHz, $B = 200$ MHz, $B_w \equiv B_i = 40$ MHz, $df \equiv \Delta f = 20$ MHz, $n_s = 400$ m, $|\gamma| = 0.97$: as the difference between the stronger scatterer and the others reflectors inside the resolution cell decreases ($K_{ref} = 10$ (a) vs. $K_{ref} = 2$ (b)), the height error increases (0.27 m (a) vs. 4.95 m (b)).

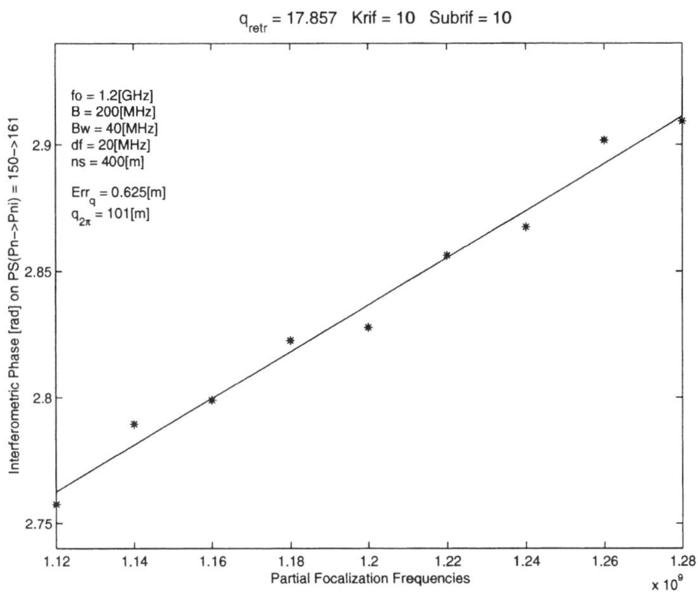

(a)

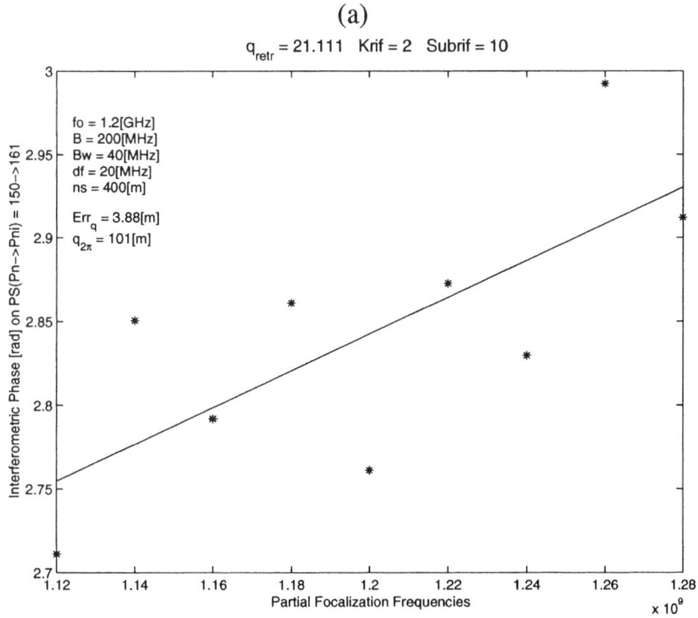

(b)

Figure 10. Linear Fit for $f_0 = 1.2$ GHz, $B = 200$ MHz, $B_w \equiv B_i = 40$ MHz, $df \equiv \Delta f = 20$ MHz, $n_s = 400$ m, $|\gamma| = 0.99$ and $K_{ref} = 10$ (a), $K_{ref} = 2$ (b): as the difference between the stronger scatterer and the others decreases the height error increases (0.625 m (a) vs. 3.88 m (b)).

baseline in (b) with respect to (a) leads to better height error because of a decreased geometrical decorrelation. In Figure 9 ($f_0 = 5.3$ GHz, $B = 200$ MHz, $B_i = 40$ MHz, $\Delta f = 20$ MHz) the difference in the height errors between (a) and (b) is due to K_{ref}, which is equal to 2 in the last plot, making less deterministic the backscattered response. Moreover, by comparing Figure 9.a with Figure 8.a it can be noticed how the wider bandwidth (200 MHz vs. 40 MHz) allows to generate more sub-band interferograms (9 vs. 3) without great loss in spatial resolution (40 MHz vs. 20 MHz): height error from 3.3 m decreases to 0.267 m. In Figure 10 the radar parameters are the same as in Figure 9 except for $f_0 = 1.2$ GHz. The figures show comparable values of height errors in the two band. However, if we use the multifrequency analysis for the solution of aliasing ambiguity in a standard procedure for phase unwrapping, then the L-band results the better choice. In fact, in this case we are interested to optimize the ratio between the height error and the height sensitivity, $\text{Err}_q^r = \frac{\text{Err}_q}{q_{2\pi}}$, which means that we need errors less than the height corresponding to one phase cycle. For the previous simulations it results:

- L-Band: $\text{Err}_q^r = \frac{\text{Err}_q}{101\text{ m}} = 0.006$
- C-Band: $\text{Err}_q^r = \frac{\text{Err}_q}{22.8\text{m}} = 0.012$

with a relative error that is better in L-band than in C-band.

5. Conclusions

In this work a new approach to the processing of data from a SAR-interferometer has been introduced. The technique is based on a differential analysis of interferometric phases in the frequency domain, produced by the modulation band of a SAR transmitter. It is applicable to wide band sensors which allow to produce multiple interferograms at partial–band, so that each of these can be referred to a different frequency in the transmitter bandwidth.

An application has been also introduced, related to scatterers whose radar echo is dominant with respect to the background elements inside the resolution cell. In this case, the height of a scatterer has been evaluated (with respect to the reference point at near range) without performing any phase unwrapping procedure in the spatial domain, but only analyzing the variation of interferometric phase with frequencies at the single pixel considered. This result, in particular, is independent from the presence of aliased fringes in the spatial domain, and could be used to set fiducial markers useful in conditioning standard phase unwrapping procedures or in order to verify a posteriori the correctness of the output DEM.

The following aspects stand out:

- the SAR signal bandwidth B is crucial for successful applications. In fact the wider the bandwidth, the greater the number of sub-bands can be; moreover, the single sub-band can be wider leading to a better resolution and to a lower height error;

– the accuracy of results depends on the quality of target's backscattering; we verified the applicability of our technique only with respect to strongly-scattering *stable targets* whose response is marginally influenced by background signal;
– a long wavelength is better than a short one because of lower decorrelation and higher $q_{2\pi}$ values.

What about the effects of the atmosphere on the interferometric phase? And what about the technique we are introducing, when a bistatic survey is not allowed? It is known that the atmospheric phase screen (APS) is not avoidable when the multiple-pass interferometry is in demand, even with the aim to observe dynamic phenomena of the ground or the same atmosphere, as mentioned in section 1. Assuming the absence of deformation between the satellite observations, two phase terms should be added to the time delay of signals, accounting to the optical path variations due to the effect of Ionosphere such as to the effect of water vapor in the Troposphere. Simple expressions for these phase terms are, from [3]:

$$\Phi_{Iono} = \frac{4\pi}{\lambda_i} \cdot \left(-\frac{40.28}{f_i^2 \cos(\theta)} \cdot \Delta TEC\right) \tag{25}$$

$$\Phi_{Tropo} = \frac{4\pi}{\lambda_i} \cdot \Delta R_{atm} = \frac{1.45}{\cos(\theta)} \cdot \Delta W \tag{26}$$

where TEC is the *total (free) electron content* and holds about 0 during the night, $20 \cdot 10^{16}\ m^2$ at the minimum of solar cycle and $100 \cdot 10^{16}\ m^2$ at its maximum; W is the *liquid water content of clouds* $[g/m^3]$; ΔTEC and ΔW are the differences of these quantities between the two acquisitions.

So, if we consider a single scatterer in the scene, in order to obtain the right elevation of it both terms need to be compensated because of they alter the estimation of the true distance between the target and the two antenna positions S1 (master) and S2 (slave). This could be accomplished for Φ_{Iono}, thanks to its dependence from $1/f^2$, by sharing the same technique used by GPS receivers. In fact these receivers exploit at this aim a microwave signal with two frequencies, L1 = 1.58 GHz and L2 = 1.23 GHz. The phase Φ_{Tropo}, on the contrary, is not separable at all from the morphological phase looking at the global Φ_{int} as a function of the frequency set f_i. However, in spite of these difficulties the situation is not different from the standard interferometry where multiple interferograms are considered at different times in order to filter out the random phase terms due to the atmosphere. By using a single interferogram, height variations can be estimated yet, by comparing phase terms between pixels at relative distance within the correlation length of the atmospheric phenomena.

Our early results by using C-band data from ERS and SIR-C proved that available spaceborne systems are really not suitable for the multifrequency analysis here introduced. Their bandwidth in fact, of 15.5 MHz and 40 MHz respectively, resulted too narrow in comparison to our requirements. So, suitable SAR data should be analyzed, in order to assess the true limits of this technique and to determine the optimal combination of involved parameters.

Finally new results from different authors are corroborating our analysis. In the meantime of the paper revision two works appeared confirming the results here reported.

The first one from Fezzetti et al. [4], shows on real SAR data the great potentiality of Permanent Scatterers (PS), which should have the same nature of those simulated in this paper when K_{ref} is set to 10. By using a large time series of SAR images, these authors show as PS can be detected and used in order to remove APS from interferograms and to reduce the height errors (on PS's) below 0.5 m. The strength of this approach derives from the use of PS, and from the analysis of interferometric phases in the multi-baseline context provided by the multi-pass interferometry. Let us underline that the sensibility of morphological phase Φ_{int} to baseline variations is greater than to the f_i excursion we are supposing along the transmitter's modulation band. On the contrary, the atmospheric terms (eq. (25) and eq. (26)) don't depend at all on the baseline value, so explaining the different performance provided by this method w.r.t. the filtering of APS.

The second paper [17], introduces an approach similar to ours, in the sense that a large number of sub-band interferograms are used. But in this case the frequency information is exploited in the framework of a classification scheme for phase unwrapping, based on a maximum likelihood estimation technique. A simulated data set has been used in order to verify the overall performance of the method. Differently from our test, an interferometric pair of *single look complex* images is simulated there, relatively to a 2D scene and referring to generic targets like those we consider in section 4.1. The experimental results of this paper are congruent with the case f in Figure 3. In conclusion, from reference [17] a useful suggestion can be derived, which allows to improve the statistical independence of partial band interferograms. In fact, overlapped sub-bands along the range-spectrum can be made independent by exploiting the split-spectrum technique on the azimuth bandwidth.

References

1. T. Toutin and L. Gray, "State–of–the–art of Elevation Extraction from Satellite SAR Data," *ISPRS Journal of Photogrammetry and Remote Sensing*, vol. 55, 2000, pp. 13–33.
2. D. Massonet and K.L. Feigel, "Radar Interferometry and its Application to Changes in the Earth's Surface," *Reviews of Geophysics*, vol. 36, November 1998, pp. 441–494.
3. R.F. Hanssen, *Radar Interferometry: Data Interpretation and Error Analysis*, Dordrecht: Kluwer Academic Publishers, 2001.
4. A. Fezzetti, C. Prati, and F. Rocca, "Permanent Scatterers in SAR Interferometry," *IEEE Trans. On Geosc. And Remote Sensing*, vol. 39, no. 1, Jan. 2001, pp. 8–20.
5. L.C. Graham, "Synthetic Interferometer Radar for Topographic Mapping," *Proc. IEEE*, vol. 62, 1974, pp. 763–768.
6. D.C. Ghiglia and M.D. Pritt, *Two Dimensional Phase Unwrapping: Theory, Algorithms, and Software*, New York: Wiley, 1998.
7. C. Prati, F. Rocca, A. Monti Guarnieri, and P. Pasquali, *Report on ERS-1 SAR Interferometric Techniques and Applications*, ESA report n.3-7439/92/HGE-I, Dip. Di Elettronica - Politecnico di Milano, June 1994.
8. F. Gatelli, A. Monti Guarnieri, F. Parizzi, P. Pasquali, C. Prati, and F. Rocca, "The Wavenumber Shift in

SAR Interferometry," *IEEE Transactions on Geoscience and Remote Sensing*, vol. 32, no. 4, July 1994, pp. 885–864.

9. J.C. Curlander and R.N. Mc Dounough, *Syntetic Aperture Radar: Systems and Signal Processing*, Wiley Interscience, 1991.
10. M.S. Seymour and I.G. Cumming, "Maximum Likelihood Estimation for SAR Interferometry," *Proceedings of IGARSS'94*, Pasadena, CA.
11. H.A. Zebker and J. Villasenor, "Decorrelation in Interferometric Radar Echoes," *IEEE Transactions on Geoscience and Remote Sensing*, vol. 30, no. 5, Sept. 1992, pp. 885–864.
12. R. Touzi and A. Lopes, "Statistics of the Stokes Parameters and the Complex Coherence Parameters in One-Look and Multilook Speckle Fields," *IEEE Transactions on Geoscience and Remote Sensing*, vol. 34, no. 2, March 1996, pp. 519–531.
13. S.N. Madsen and H.A. Zebker, "Automated Absolute Phase Retrieval in Across-Track Interferometry," *Proceedings of IGARSS'92*, vol. II, 1992, pp. 1582–1584.
14. D.C. Rife and R.R. Boorstyn, "Single-Tone Estimation from Discrete-Time Observations," *IEEE Transactions on Information Theory*, vol. 20, Sept. 1974, pp. 591–598.
15. W.H. Press, S.A. Tenkolsky, W.T. Vetterling, and B.P. Flannery, "Numerical Recipes in FORTRAN 77: The Art of Scientific Computing," Volume I, Cambridge University Press, 1986–1992.
16. G. Franceschetti, M. Migliaccio, D. Riccio, and G. Schirinzi, "SARAS: A Synthetic Aperture Radar (SAR) Raw Signal Simulator," *IEEE Transactions on Geoscience and Remote Sensing*, vol. 30, no. 1, January 1992, pp. 110–122.
17. V. Pascazio and G. Schirinzi, "Estimation of Terrain Elevation by Multifrequency Interferometric Wide Band SAR Data," *IEEE Signal Processing Letters*, vol. 8, 2001, pp. 7–9.

Multidimensional Systems and Signal Processing, 14, 207–222, 2003

Scattering-Based Tomography for HRR and SAR Prediction

B. S. DENNEY[1] bdenney@neuralsystems.com
Neural Computing Systems, 2081 Business Center Drive, Suite 206, Irvine, California, USA 92612

R. J. P. DE FIGUEIREDO
Neural Computing Systems, 2081 Business Center Drive, Suite 206, Irvine, California, USA 92612

Received December 8, 2000; Revised May 20, 2002; Accepted May 21, 2002

Abstract. This paper introduces a method for predicting HRR radar signatures and SAR images by creating a parametric three-dimensional scattering model from existing measured or model-based HRR signatures and/or SAR images. The method identifies potential three-dimensional *persistent scatterers* and estimates their scattering patterns. The results are parametric HRR signature and SAR image reconstruction functions of range, azimuth, and elevation.

The modeling is accomplished through a scattering-based tomography technique. This technique localizes potential scatterers by using a filtered back-projection algorithm for the inverse radon transform. Once found, potential scatterers may then have their two-dimensional (azimuth and elevation) scattering patterns parameterized through the use of a truncated spherical harmonic series.

Results using the reconstructions from HRR data are presented. A M109 model is reconstructed based on HRR signatures. The model allows us to predict what the vehicle would look like from any arbitrary orientation using SAR. Finally an M548 vehicle is modeled using 26 measured HRR signatures. The model is shown to be better than the synthetic model data. Additionally we show that the new model results can be combined with the synthetic data to provide a better target model for signature matching.

Key Words: ATR, HRR, SAR, tomography, radar prediction

1. Introduction

This paper describes a new parametric technique for modeling scattering objects based on observation data collected from high range resolution (HRR) radar and/or synthetic aperture radar (SAR). The three-dimensional models from this technique may be used to rapidly predict SAR and HRR observations from any azimuth and depression angle.

The modeling technique is essentially a two-part process. The first part localizes potential three-dimensional scatters. The second part estimates the scattering pattern associated with these scatterers. The scatterer localization is accomplished through scattering-based tomography. And the scattering patterns are estimated with two-dimensional polar function approximators.

1.1. Motivation

The work described in this paper was spawned by attempts to improved HRR radar signature model prediction based on available measured data. Early attempts included a method that *tweaked* the model reflectance parameters to account for observed data [1]. This was accomplished and refined through sophisticated data structures which captured ray-trace paths from multi-faceted objects [2].

Later attempts to interpolate HRR radar signatures were made by using splines to morph the signatures over azimuth [3]. The interpolation idea was extended by interpolating along the projection tracks (over azimuth) of potential scatterers [4]. This most recent approach recognized the potential of the application of inverse Radon transform for scatterer localization.

This paper will extend the scatterer-track interpolation concept.

1.2. Background

The relationship between radar imaging and tomography dates back many years. Munson et al. [5] pointed out the equivalence between spotlight mode SAR image formation and tomographic reconstruction. In their analysis, they point out that spotlight mode SAR is essentially a narrow-band version of tomography. Dai et al. [6] showed that radar returns from sine modulated exponential pulse (SMEP) processed with 2-D tomographic techniques can produce edge enhanced images of air targets.

In 1995 Allen et al. [7] showed that aspect (Azimuth) dependent reflectance characteristics of objects in radar could be useful in distinguishing between different objects such as trees and man made objects.

The realization that aspect dependent scattering characteristics could be parameterized was studied by several authors. Trintinalia et al. [8] introduced a scattering center parametrization technique in 1997. In this approach the authors used an adaptive Gaussian technique to represent unimodal scattering patterns with respect to aspect (azimuth) angle for two-dimensional planar targets. In 1999, Gerry et al. [9] introduced a different parametric scattering model. In their 2-D model, the scatterer parameters could be varied to represent different scattering geometries. The model also includes a parameter that gives the model frequency dependence.

Later Dai et al. [10] introduced an approach to limited-view reconstruction that tracks the positions of dominant scattering centers in target sinograms. The 2-D scatterer model extrapolates the radar observations to improve target imaging for limited aspect views.

In this paper we take an approach that combines some of the ideas of Dai et al. [10] and Trintinalia et al. [8] extended to three dimensions that creates a parametric scattering model given limited or sparse aspect views of a target. In addition we show that through the principles of tomography the approach is applicable to both SAR and HRR radar data and predictions.

One glaring difference between our work and previous work is our attempt to approach the problem from a non-coherent processing perspective. There are several reasons for this approach and they will be explained in the next section.

1.3. Non-Coherent Prediction Model

The approach of non-coherent processing taken by the authors in the model generation is motivated by several factors.

First, the method is intended to be more of a parametric function approximation approach than a physical model approach. While the parametric form used is in many respects physically motivated, it neglects coherent wave interaction. This simplifies the approach from a complex to a real function approximation technique.

Secondly, radar returns contain coherent sums from multiple scatterers. The interaction of these scatterers creates constructive and destructive interference. The interference may be difficult to model when not all the scatterers on the target are included in the model. This is the case in our approach where we choose to only model so-called *persistent scatterers*. Transient scatterers in aspect are not captured by the model and hence their interference effects are unknown. Therefore we argue that the process of modeling interference properly might be very difficult.

Finally, we argue that the non-coherent approach may be more robust to slight changes in sensor frequencies due to drift or the use of multiple sensors and multiple sensor modes such as HRR and SAR.

2. SAR and HRR Projections of Isotropic Scattering Points

To begin with we will consider objects with isotropic scattering points. Isotropic scatterers are scattering points which reflect radiation in the same relative manner regardless of the direction of radiation incidence.

Furthermore since we are interested in energy reflected back to the radiative source (transmitter), we may only concern ourselves with the reflective properties at the angle of incidence. This means that for all practical purposes we can represent an isotropic scattering point with a single reflectivity value.

So now we represent the reflectivity of point (x, y, z) as $f(x, y, z)$. And for a matter of convenience we let $\hat{f}_{\theta,\phi}(r, s, t)$ be a rotated version [11] of f such that $\hat{f}$ has an azimuth of θ and a elevation of ϕ.

$$\begin{aligned}\hat{f}_{\theta,\phi}(r,s,t) = f(&r\cos\theta\cos\phi - t\cos\theta\sin\phi - s\sin\theta,\\ &r\sin\theta\cos\phi - t\sin\theta\sin\phi + s\cos\theta,\\ &r\sin\phi + t\cos\phi)\end{aligned} \tag{1}$$

where r, s, and t are the coordinates in the rotated coordinate system $\{\hat{r}, \hat{s}, \hat{t}\}$. $\hat{r}$ is *range* axis and represents the direction of look from the sensor to the target. $\hat{s}$ is the *cross-range* axis: it is orthogonal to the range axis and parallel to the ground. $\hat{t}$ is the third axis forming a right handed coordinate system: $\hat{t} = \hat{r} \times \hat{s}$.

2.1. HRR Projections

From the radar transceiver, reflections returning at a fixed time, t_1, are reflections of a fixed path length, $2r_1$. If we ignore all multi-bounce paths then we may assume that the return signal is from the combination of all scatterers at range r_1.

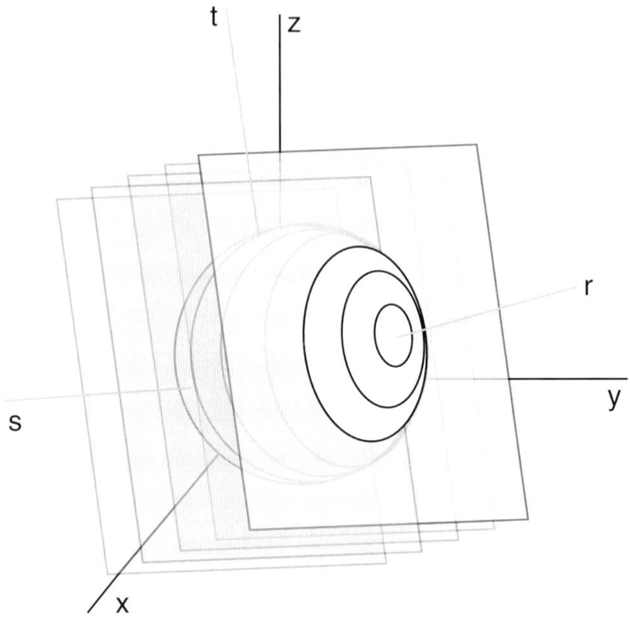

Figure 1. HRR signatures can be represented as integrals of planes of scatterers at various ranges along the look vector.

If we naively assume that the combination of scatterer returns is the sum of the return magnitudes (this is not the case since the radar signal is coherent), we may say that the return HRR signal amplitude is the summation of all scatterers at range r_1.

Since the size of the object of interest is much smaller than the range from the radar source, the set of all points on the object which are at a distance of r_1 lie approximately on a plane.

Now, let $h_{\theta,\phi}(r)$ be the projection of all scatterer returns at a range of r for azimuth θ and elevation ϕ. This projection is a plane integral of $\hat{f}$:

$$h_{\theta,\phi}(r) = \int_{-\infty}^{\infty}\int_{-\infty}^{\infty} \hat{f}_{\theta,\phi}(r,s,t)\, ds\, dt. \tag{2}$$

Figure 1 shows the integration planes of the HRR projection. The planes are orthogonal to the look vector, and the range is measured along the look vector, r.

2.2. SAR Projections

In SAR imagery, the scatterers are projected from three dimensions to two. The range and cross-range directions are preserved due to range and Doppler measurements [12].

This gives us the projection of $\hat{f}$ as a line integral across the third dimension.

$$g_{\theta,\phi}(r,s) = \int_{-\infty}^{\infty} \hat{f}_{\theta,\phi}(r,s,t)\, dt. \tag{3}$$

Here we assume the SAR image represents a projection of scatterers at the given aspect angles. In reality, the SAR image is the magnitude of a coherent sum across an aspect window of one-dimensional projections [13]. We argue that SAR image magnitudes indicate, through intensity, the likely presence of persistent scattering features. Thus, this representation is sufficient for our purposes.

3. Scattering Point Filtered-Back-projection

The problem of reconstructing three-dimensional volumes of emitting or attenuating mediums from projections is common in medical imaging. These techniques rely on the Fourier Slice Theorem which relates the Fourier Transform of the projection to the Fourier Transform of the medium attenuation/emission [14].

In this section we will explain the filtered-back-projection technique as it applies to the HRR Projections described above. We will also present the filter-back-projection method for the SAR Projections.

3.1. *Filtered-Back-projection for 1-D Projections*

Let $H_{\theta,\phi}(w)$ be the Fourier transform of $h_{\theta,\phi}(r)$.

$$\begin{aligned}
H_{\theta,\phi}(w) &= \int_{-\infty}^{\infty} h_{\theta,\phi}(r) e^{-j2\pi wr}\, dr && (4)\\
&= \int_{-\infty}^{\infty}\int_{-\infty}^{\infty}\int_{-\infty}^{\infty} \hat{f}_{\theta,\phi}(r,s,t)\, ds\, dt\, e^{-j2\pi wr}\, dr\\
&= \int_{-\infty}^{\infty}\int_{-\infty}^{\infty}\int_{-\infty}^{\infty} f(x,y,z) e^{-j2\pi w(x\cos\theta\cos\phi + y\sin\theta\cos\phi + z\sin\phi)}\, dx\, dy\, dz\\
&= F(a,b,c) && (5)\\
&= \tilde{F}(w,\theta,\phi) && (6)
\end{aligned}$$

where F is the Fourier transform of f, $\tilde{F}(w, \theta, \phi)$ is the polar coordinate version of the Fourier Transform of f, and

$$a = w\cos\theta\cos\phi \tag{7}$$

$$b = w\sin\theta\cos\phi \tag{8}$$

$$c = w\sin\phi. \tag{9}$$

Then we may write $f(x, y, z)$ as the inverse Fourier transform in polar form:

$$f(x,y,z) = \int_{-\infty}^{\infty}\int_{0}^{2\pi}\int_{0}^{\pi/2} \tilde{F}(w,\theta,\phi)e^{-j2\pi w(x\cos\theta\cos\phi+y\sin\theta\cos\phi+z\sin\phi)} \times w^2 \cos\phi \, d\phi \, d\theta \, d\phi \tag{10}$$

and from equation (6)

$$f(x,y,z) = \int_{0}^{2\pi}\int_{0}^{\pi/2}\int_{-\infty}^{\infty} H_{\theta,\phi}(w)w^2 e^{j2\pi wr} \, dw \cos\phi \, d\phi \, d\theta \tag{11}$$

$$= -\int_{0}^{2\pi}\int_{0}^{\pi/2} h''_{\theta,\phi}(r) \cos\phi \, d\phi \, d\theta \tag{12}$$

where r is

$$r = x\cos\theta\cos\phi + y\sin\theta\cos\phi + z\sin\phi. \tag{13}$$

Equations (11) and (12) tell us that $f(x, y, z)$ can be recovered from an infinite set of projections covering all viewing angles of the medium. The reconstruction consists of integrating a range-filtered version of the projections over all azimuth and elevation angles. The integration is weighted to give less weight to high elevation angle coverages since there is little variability in these projections over change in azimuth.

When we don't have infinite coverage of the object, we may approximate the scatterer reconstruction by using a set of N available projections:

$$f(x,y,z) \approx \frac{1}{N}\sum_{i=1}^{N} \cos\phi_i \, \mathcal{H}_{\theta_i,\phi_i}(x\cos\theta_i\cos\phi_i + y\sin\theta_i\cos\phi_i + z\sin\phi_i) \tag{14}$$

where $\mathcal{H}$ is the filtered version of h:

$$\mathcal{H}_{\theta_i,\phi_i}(r) = h_{\theta_i,\phi_i}(r) * \mathcal{F}^{-1}\{w^2\}. \tag{15}$$

When dealing with real data that includes some measurement noise, the w^2 (2nd derivative) filter can enhance the noise more than the signal. Therefore, it is often necessary to use a filtered version of h which filters noise and also aids in the image reconstruction. One common approach is to add a Hamming window to the filtering stage shown in equation (15).

3.2. *Filtered-Back-projection for 2-D Projections*

The filtered-back-projection technique for two-dimensional projections of three-dimensional targets follows along similar lines as the 1-D case. The projection filter is now a two-dimensional circularly symmetric version of the Ram-Lak [15].

Let $G_{\theta,\phi}(w, v)$ be the two-dimensional Fourier transform of $g_{\theta,\phi}(r, s)$ with range r and cross-range s.

$$f(x,y,z) = \int_0^{2\pi}\int_0^{\pi/2}\int_{-\infty}^{\infty} G_{\theta,\phi}(w,v)(w^2+v^2)^{1/2} e^{j2\pi(wr+vs)}\, dw\, dv \cos\phi\, d\phi\, d\theta \quad (16)$$

$$= \int_0^{2\pi}\int_0^{\pi/2} \mathcal{G}_{\theta,\phi}(r,s)\cos\phi\, d\phi\, d\theta \quad (17)$$

where r and s are

$$r = x\cos\theta\cos\phi + y\sin\theta\cos\phi + z\sin\phi \quad (18)$$

$$s = -x\sin\theta + y\cos\theta. \quad (19)$$

Again, in the two-dimensional projection case, $\mathcal{G}$, the filtered version of g, may include a two-dimensional windowing function to reduce observation noise.

The resulting reconstruction for M SAR observations then becomes

$$f(x,y,z) \approx \frac{1}{M}\sum_{i=1}^{M} \cos\phi_i\, \mathcal{G}_{\theta_i,\phi_i}(r,s). \quad (20)$$

3.3. *Fusion of 1-D and 2-D Data*

We may simply combine the scatterer reconstructions functions for the one-dimensional and two-dimensional cases given in equation (14) and (20). The two equations are fused based on a relative weighting factor α which typically would assign more weight to the SAR-based two-dimensional projections due to the fact that the SAR data contains more information than HRR projections. Thus, the fusion algorithm is

$$f(x,y,z) \approx \frac{(1-\alpha)}{N}\sum_{i=1}^{N} \cos\phi_i\, \mathcal{H}_{\theta_i,\phi_i}(r) + \frac{\alpha}{M}\sum_{i=N+1}^{M+N} \cos\phi_i\, \mathcal{G}_{\theta_i,\phi_i}(r,s). \quad (21)$$

4. Dealing with Anisotropic Scatterers

In the previous section we made the assumption that all scatterers were isotropic. In reality, however, scatterers have non-zero extent and have returns which vary greatly with the viewing angle of the scatterer. The variations can be due to the extended scatterer geometry, reflecting properties, and occlusions of the scatterer from various viewing angles.

The scattering function reconstruction techniques described in the previous section do not capture angular viewing dependency of the scatterers. The reconstruction instead captures regions where global scattering features might exist.

4.1. *Anisotropic Scattering Point Localization*

The reconstruction technique presented in Section 3 is equivalent to taking a line integral along the (range filtered) projection path of every potential scatterer over changing viewing angles. This means that persistent scatterers in the projections will create regions where $f(x, y, z)$ is large.

Previous studies of SAR data from ground vehicles demonstrate the existence of persistent scatters [16]. These scatterers are scatters that are viewable over an extended range of azimuth and elevation angles.

The Inverse Radon transform (back-projection method described in previous section) can be used to detect possible persistent scatter locations. This is analogous to the way the Hough transform [17] is used in image processing to detect possible global line features in images.

In fact, in the anisotropic scatterer case, we consider $f(x, y, z)$ to be the *potential scatterer* function. This function indicates the three-dimensional locations that are potentially related to a physical scatterer. The peaks of f represent scatterer tracks with significant average signal intensity. We essentially project these potential scatterer locations back to the observation space (i.e. HRR or SAR) and verify whether the scatterer candidate has persistent scattering characteristics in the observations.

By applying a low-pass filter along the candidate scatterer's track in azimuth we can reject many false scatterers as well as transient scatterers.

Previous techniques used for the localization of persistent scatterer models relied on two-dimensional models from the super-position of SAR images at a fixed depression/elevation angle and a clean algorithm for localization [18]. Or they used unimodal two-dimensional non-cyclic functions to approximate the scattering functions as a function of azimuth and frequency [8]. Other three-dimensional techniques were model-based approaches from physical optics [19], [20] which required a good object model.

The scatterer localization method we have explained here enables three-dimensional localization without having a good object model *a priori*.

4.2. *Estimating the Scattering Response*

Once we have used the back-projection technique to localize potential scatterers, we estimate the response of each scatterer. The response, $\rho_i(\theta, \phi)$, of scatterer i over azimuth and elevation needs to be estimated for each i such that we may reconstruct any SAR image or HRR signature from any view.

The approximated HRR and SAR models are naively assumed to be the summation of the scatterer reflected energies approximated by the response patterns, $\rho_i(\theta,\phi)$. This gives us our HRR signature model:

$$\hat{h}_{\theta,\phi}(r) = \sum_{\forall i: R_i(\theta,\phi)=r} \rho_i(\theta, \phi). \tag{22}$$

And the SAR image model is:

$$\hat{g}_{\theta,\phi}(r,s) = \sum_{\substack{\forall i: R_i(\theta,\phi)=r, \\ \text{and } S_i(\theta,\phi)=s}} \rho_i(\theta,\phi) \tag{23}$$

where $R_i(\theta, \phi)$ and $S_i(\theta, \phi)$ are the tracks of scatterer i in range and cross-range respectively:

$$R_i(\theta,\phi) = x_i \cos\theta \cos\phi + y_i \sin\theta \cos\phi + z_i \sin\phi \tag{24}$$

$$S_i(\theta,\phi) = -x_i \sin\theta + y_i \cos\theta. \tag{25}$$

To obtain $\rho_i(\theta, \phi)$ one must approximate the observed responses with a two-dimensional cyclic function in θ and ϕ. These functions may be approximated with the summation of spherical-harmonics. In fact any function of two angular variables may be expanded in terms of spherical harmonics [21].

$$\rho(\theta,\phi) = \sum_{n=0}^{\infty} \sum_{m=-n}^{n} a_{n,m} Y_n^m(\theta,\phi) \tag{26}$$

where $Y_n^m(\theta, \phi)$ is the spherical harmonic written in terms of associated Legendre functions, $P_n^m(\alpha)$

$$Y_n^m(\theta,\phi) = (-1)^m \left[\frac{2n+1}{4\pi} \frac{(n-m)!}{(n+m)!}\right]^{1/2} P_n^m(\cos\theta) e^{jm\phi} \tag{27}$$

$$P_n^m(\alpha) = (-1)^m (1-\alpha^2)^{m/2} \frac{d^m}{d\alpha^m} P_n(\alpha) \tag{28}$$

where $P_n(\alpha)$ is a Legendre polynomial:

$$P_n(\alpha) = \frac{(-1)^n}{2^n n!} \frac{d^n}{d\alpha^n} (1-\alpha^2)^n. \tag{29}$$

The harmonic coefficients, $a_{n,m}$ may be obtained from

$$a_{n,m} = \int_0^{2\pi} \int_{-1}^{1} [Y_n^m(\theta,\phi)]^* \, \rho(\theta,\phi) \, d(\cos\theta) \, d\phi \tag{30}$$

where $\rho(\theta, \phi)$ is taken from the SAR or HRR projections along the track of the scatterer of interest.

In our two dimensional scattering response reconstruction we can use a truncated form of the series given by equation (26). Only the first few terms of the first summation are needed since our technique focuses on the reconstruction of *persistent* scatterers. In other words, the higher order harmonics are not needed since they represent smaller features.

If the elevation/depression angle is fixed or is within a restricted range for our observations and reconstructions, then we may simplify the estimation of ρ_i by a truncated Fourier Series in θ (azimuth):

$$\rho_i(\theta, \phi)|_{\phi=\phi_0} = a_{0,i} + \sum_{n=1}^{N} \left(a_{n,i} \cos n\theta + b_{n,i} \sin n\theta\right) \tag{31}$$

where the coefficients may be obtained from:

$$a_{n,i} = \frac{(1-\alpha)}{N} \sum_{i=1}^{N} \cos n\theta_i \, h^{-}_{\theta_i,\phi_i}(r_i) + \frac{\alpha}{M} \sum_{i=N+1}^{M+N} \cos n\theta_i \, g^{-}_{\theta_i,\phi_i}(r_i, s_i) \tag{32}$$

$$b_{n,i} = \frac{(1-\alpha)}{N} \sum_{i=1}^{N} \sin n\theta_i \, h^{-}_{\theta_i,\phi_i}(r_i) + \frac{\alpha}{M} \sum_{i=N+1}^{M+N} \sin n\theta_i \, g^{-}_{\theta_i,\phi_i}(r_i, s_i). \tag{33}$$

In the above expression the function h^- and g^- are the residual functions from the observations and the reconstruction based on previously detected scatterers. With these expressions one can build a scattering center model by first detecting potential scatters one by one from the back-projection algorithm, then estimating and adding the scatter contributions to the model, and using the residual model to detect and estimate more scatters.

5. Alignment Considerations

For the back-projection techniques described thus far, we required that all projections be aligned such that there exists a three-dimensional fixed point which projects to the same range (and cross-range) in the projections, regardless of the viewing angles, θ and ϕ. This requirement can be somewhat problematic because real HRR and SAR observations are often self aligned in arbitrary ways which do not preserve a three-dimensional fixed point.

Therefore the problem of alignment is critical to our scattering center model reconstruction technique.

One common, and sometimes practical method for alignment of observations of an object (or target) is to use low fidelity synthetic target models for predicting HRR signatures and SAR images. The prediction code typically preserves a three-dimensional fixed rotation point. The low fidelity templates may then be used as alignment guides for incoming observations.

When the target in question is a ground vehicle, alignment may be accomplished without a synthetic model. Techniques for unknown vehicle alignment are currently under development by the authors. The results are based on some recent signature observations [22] and hull alignment techniques [23].

6. Results

To demonstrate the techniques described in this paper, we present two examples.

The first experiment covers model reconstruction for a complex target: a ground vehicle from synthetic HRR observation data. This experiment will demonstrate the ability of the model to predict SAR images from HRR signatures.

Next we will use the scattering-based tomography method to generate a model from 26 measured HRR signatures of a ground vehicle. This experiment will compare the reconstructed model to previously available synthetic signatures.

In these experiments we used a five coefficient Fourier expansion for the scattering pattern: $\{a_0, a_1, b_1, a_2, b_2\}$. We also chose the number of scattering centers to detect to be much larger than the number of actual scattering centers. By over selecting the number of

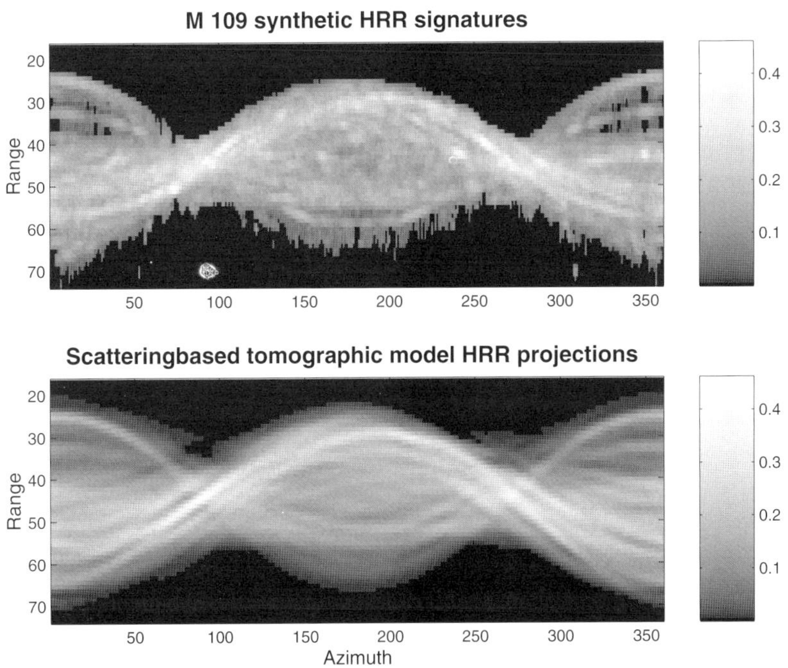

Figure 2. The synthetic HRR signatures for a M109 tank (top). The scattering-based tomographic model projections from the synthetic signatures (bottom). Note that the model projections capture the large scale scattering features in azimuth.

scattering centers we make sure that some are not missed. If a false scattering center is included in the model, its affect can be negligible because the scattering coefficients can be close to zero (i.e. the scatter does not reflect).

6.1. Complex Ground Vehicle Reconstruction

In this experiment we use HRR data generated from synthetic electro-magnetic models. The HRR data used is simulated X-band 1 foot resolution. First we used 360 HRR observations (every one degree) to build the scattering-based tomographic model. This model consists of 1000 scattering centers.

Figure 2 shows the synthetic HRR data and the scattering-based tomographic model projections. We observe that the model appears to capture the large scale (*persistent* over azimuth) scattering features of the vehicle.

Using the detected model we may also predict SAR images for the vehicle. Figure 3 shows the predicted SAR images at various azimuths for the M109 at a 15 degree

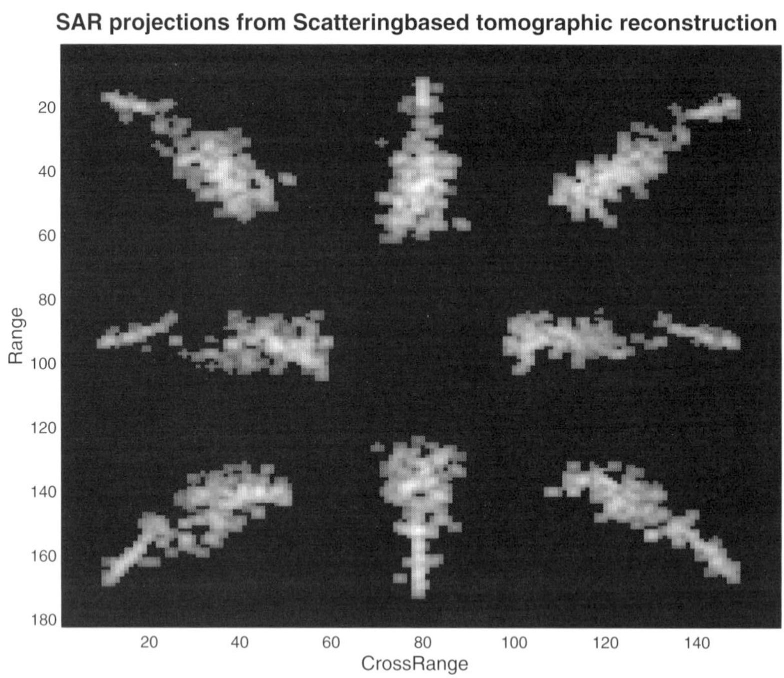

Figure 3. The SAR images predicted from scattering-based tomographic model projections generated from synthetic HRR signatures.

depression angle. Note that all of the HRR observations were from a 31 degree depression angle.

6.2. *Measured Ground Vehicle Experiment*

26 simulated measured HRR observations of a M548 were taken over 360 degrees of azimuth and at a fixed depression angle. This data was generated from MSTAR SAR data (X-band, spotlight mode, 1 foot resolution). The observations were aligned to signatures generated from a M548 synthetic model. The aligned observations were then used to build a scattering-based tomographic model (described in this paper). The model consisted of 1000 scattering centers.

Next we took the remaining measured signatures available at the same depression angle for the vehicle (there were 234 remaining) and found the minimum mean squared errors between them the following data sets:

- The nearest neighbor from the set of 26 measured (sparse measured).
- The best match in a plus or minus five degree window of the synthetic data set (synthetic).
- The best match in a plus or minus five degree window of the reconstructed measured signatures from the scattering-based tomographic model (Reconstructed sparse measured).
- The best match in a plus or minus five degree window of the reconstructed measured signatures or the synthetic data (Model hybrid).
- The best match in a plus or minus five degree window of the 270 measured signatures excluding the signature being tested (measured on measured).

Figure 4 shows the results from the reconstructed measured data experiment. From this experiment we see that reconstructed sparse measured data has smaller errors than the full synthetic data set. Additionally, the synthetic and reconstructed data sets seem to have different regions of effectiveness. This was demonstrated previously by the authors in [4]. In that paper the authors showed the feasibility of combining sparse measured data with synthetic data. In the previous experiment with five targets, the four methods, *sparse measured*, *synthetic*, *reconstructed sparse measured*, and *hybrid* achieved classifications rates of 42.5%, 73.2%, 80.2%, and 82.8% respectively. The *measured on measured* classification rate was 89.6%.

6.3. *Computational Time*

The computation required to build the scattering models varies linearly with the number of observations. In our example, of 10 HRR observations every 36°, a model of 1000 scatterers was generated in 3 minutes and 48 seconds using un-optimized Matlab code on an 600MHz. Intel Pentium III processor. The computation to generate an HRR signatures from this model is about 5 milliseconds per signature using un-optimized Matlab code.

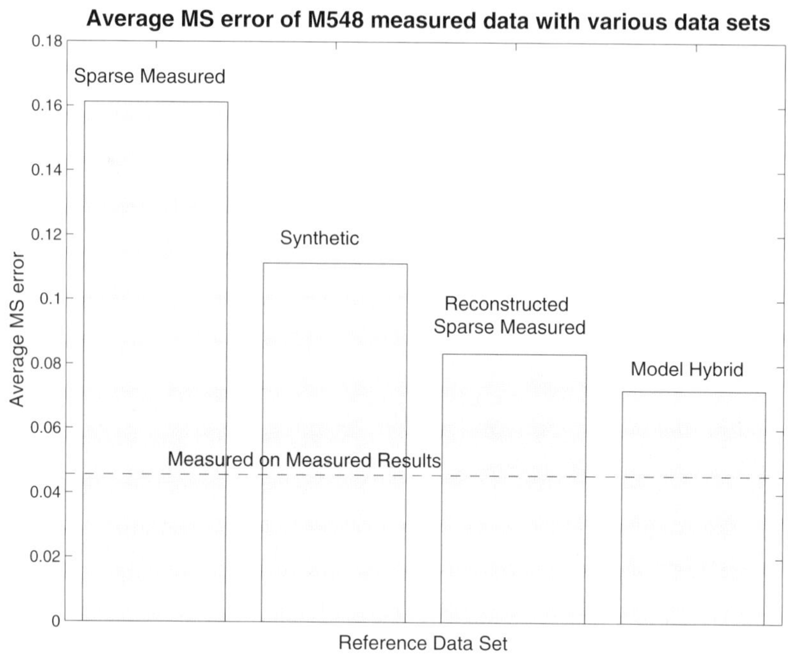

Figure 4. The average mean squared error between measured data and various data sets. Note that the signatures reconstructed from the sparse measured data exhibit levels of errors better than the synthetic model data. The hybrid result shows us that the two models (reconstructed model from sparse and synthetic model) perform better in different regions. It has been previously shown that this hybrid approach improves target identification. The errors presented are compared with the average error rate of using measured data (about every 1 to 2 degrees) for signature matching. In previous experiments with five targets, the four methods, *sparse measured*, *synthetic*, *reconstructed sparse measured*, and *hybrid* achieved classifications rates of 42.5%, 73.2%, 80.2%, and 82.8% respectively. The *measured on measured* classification rate was 89.6%.

7. Conclusions & Future Work

In this paper we have presented a technique for modeling three-dimensional scattering objects with a point scatterer model from SAR and HRR data. The approach is unique because it can find a three-dimensional model and predict HRR signatures and SAR images from any orientation. The model generation effectively allows for sensor (modality) fusion between HRR and SAR data. Also, it does not require a good synthetic model to infer any model assumptions.

Future work in this subject will study an adaptive auto-alignment technique that would completely eliminate the need for a pre-determined synthetic model. This would allow tracking and signature prediction of unknown vehicles. Also we will study the number of HRR and/or SAR observations required to form a reliable target model. We also plan to study the robustness of this technique to pose errors. Also, prediction experiments using SAR observations will be tested and the accuracy of SAR predictions will be quantified.

Note

1. Further author information: (Send correspondence to B. S. Denney).

References

1. J.D. Leonard, B.S. Denney, G.C. Lai, R.J.P. de Figueiredo, and A.R. Nolan, "Combining Measured Radar Signatures with Computer Generated Signatures for Aircraft Identification," in *Sixth Automatic Target Recognizer Systems and Technology Conference*, 1997.
2. A.R. Nolan and J.A. Hughes, "Synthetic Signature Analysis Using Ray Trace Decomposition," in *SPIE Conference on Algorithms for Synthetic Aperture Radar Imagery VI*, April 1999.
3. B.S. Denney, S.N. Richman, and R.J.P. de Figueiredo, "Model Based HRR Radar ATR: Combining Synthetic Models with Limited Measured Observations," in *SPIE Conference on Algorithms for Synthetic Aperture Radar Imagery VI*, April 1999, pp. 437–447.
4. B.S. Denney, R.J.P. de Figueiredo, and R. Williams, "A New Approach to Parametric HRR Signature Modeling for Improved 1-D ATR," in *SPIE Conference on Algorithms for Synthetic Aperture Radar Imagery VII; Proceedings of SPIE*, April 2000, pp. 372–383.
5. D.C. Munson, J.D. O'Brien, and W.K. Jenkins, "A Tomographic Formulation of Spotlight-Mode Synthetic Aperture Radar," *Proceedings of the IEEE*, vol. 71, no. 8, August 1983, pp. 917–925.
6. Y. Dia, E.J. Rothwell, K.M. Chen, and D.P. Nyquist, "Time-Domain Imaging of Radar Targets Using Algorithms for Reconstruction from Projections," *IEEE Transactions on Antennas and Propagation*, vol. 45, no. 8, August 1997, pp. 1227–1235.
7. M.R. Allen, J.M. Jauregui, and L.E. Hoff, "Fopen-Sar Detection by Direct Use of Simple Scattering Physics," *Proc. SPIE, Algorithms for Synthetic Aperture Radar Imagery II*, vol. 2487, 1995, pp. 45–55.
8. L.C. Trintinalia, R. Bhalla, and H. Ling, "Scattering Center Parameterization of Wide-Angle Backscattered Data Using Adaptive Gaussian Representation," *IEEE Transactions on Antennas and Propagation*, vol. 45, no. 11, November 1997, pp. 1664–1668.
9. M.J. Gerry, L.C. Potter, I.J. Gupta, and A. van der Merwe, "A Parameteric Model for Synthetix Aperture Radar Measurements," *IEEE Transactions on Antennas and Propagation*, vol. 47, no. 7, July 1999, pp. 1179–1188.
10. Y. Dia, E.J. Rothwell, K.M. Chen, and D.P. Nyquist, "Time-Domain Imaging of Radar Targets Using Sinogram Restoration for Limited-View Reconstruction," *IEEE Transactions on Antennas and Propagation*, vol. 47, no. 8, August 1999, pp. 1323–1329.
11. B.S. Denney, "*Rotation Clustering for Robot Vision*," PhD thesis, University of California, Irvine, Irvine, California, 1996.
12. M. Skolnik (Ed.), *Radar Handbook*, New York: McGraw-Hill, 1990.
13. W. Smith, T. Irons, J. Riordan, and S. Sayre, "Peak Stability Defived from Phase History in Synthetic Aperture Radar," *Proc. SPIE, Algorithms for Synthetic Aperture Radar Imagery VI*, vol. 3721, 1999, pp. 450–461.
14. A. Rosenfeld and A.C. Kak, *Digital Picture Processing*, volume 1, 2nd ed., New York: Academic Press, 1982.
15. Z.H. Cho, J.P. Jones, and M. Singh, *Foundations of Medical Imaging*, New York: Wiley-Interscience, 1993.
16. J. Wissinger, R. Ristroph, J. Diemunsch, W. Severson, and E. Freudenthal, "Mstar's Extensible Search

Engine and Model-Based Inferencing Toolkit," *SPIE Conference on Algorithms for Synthetic Aperture Radar Imagery VI*, April 1999, pp. 554–570.

17. R. Gonzalez and R. Woods, *Digital Image Processing*, New York: Addison-Wesley, 1992.
18. K. Segalovitz and B.R. Frieden, "A "Clean"-Type Deconvolution Algorithm," *Astron. Astrophys.*, vol. 70, 1978, pp. 335–343.
19. R. Bhalla and H. Ling, "Three-Dimensional Scattering Center Extraction Using the Shooting and Bouncing Ray Technique," *IEEE Transactions on Antennas and Propagation*, vol. 44, no. 11, 1996, pp. 1445–1453.
20. R. Bhalla, J. Moore, and H. Ling, "A Global Scattering Center Representation of Complex Targets Using the Shooting and Bouncing Ray Technique," *IEEE Transactions on Antennas and Propagation*, vol. 45, no. 12, 1997, pp. 1850–1854.
21. P. Dennery and A. Krzywicki, *Mathematics for Physicists*, New York: Harper & Row, 1967.
22. J. Schmitz and R. Williams, "Ground Target Range Extent Analysis Using 1-D HRR Profiles," *SPIE Conference on Algorithms for Synthetic Aperture Radar Imagery VII; Proceedings of SPIE*, April 2000, pp. 384–391.
23. D. Gross, J. Schmitz, and R. Williams, "Statistical Analysis of 1-D HRR Target Features," *SPIE Conference on Algorithms for Synthetic Aperture Radar Imagery VII; Proceedings of SPIE*, April 2000, pp. 623–630.

Multidimensional Systems and Signal Processing, 14, 223–240, 2003

An Algorithm to Detect the Presence of 3D Target Motion from ISAR Data

JUNFEI LI fei@panam.edu
Department of Electrical Engineering, The University of Texas-Pan American, Edinburg, TX 78539

HAO LING
Department of Electrical and Computer Engineering, The University of Texas at Austin, Austin, TX 78712-1084

VICTOR CHEN
Airborne Branch, Naval Research Laboratory, Washington, DC 20375

Received December 1, 2000; Revised December 1, 2000; Accepted August 7, 2001

Abstract. We present an algorithm to detect the presence of 3D target motion from ISAR data. Based on the 3D point scatterer model, we first examine the effect of 3D motion on ISAR imaging. It is shown that existing motion compensation algorithms cannot properly focus targets exhibiting 3D motion during the imaging interval. An algorithm is then derived to blindly detect the degree of 3D target motion from raw radar data. It is based on measuring the linearity of phases between two or more point scatterers on the target. The phase estimation is implemented using the adaptive joint time-frequency technique. Examples are provided to demonstrate the effectiveness of the 3D motion detection algorithm with both simulation and real ISAR data. The detection results are corroborated with the truth motion data from on-board motion sensors and correlated with the resulting ISAR images.

Key Words: radar imaging, motion compensation, 3D motion detection

1. Introduction

High-resolution inverse synthetic aperture radar (ISAR) imaging is regarded as an effective tool in automatic target recognition [1], [2]. Ideally, the desired target motion is uniform rotation without translational motion, under which a two-dimensional (2D) Fourier transform brings the radar data in the (frequency)-(dwell time) domain into the (range)-(Doppler frequency) domain. Otherwise, motion compensation is needed as an intermediate step to form a focused ISAR image.

Since target motion can always be decomposed into translational motion and rotational motion, a typical motion compensation algorithm consists of two steps. First, a point on the target is focused through translational motion compensation. When there is non-uniform rotational motion, other points on the target are not necessarily focused. Rotational motion compensation is then applied to focus these other points. Existing motion compensation algorithms usually assume that the rotational motion of a target is confined to a 2D plane during the dwell duration [1], [2], [3], [4], [5], [6], [7]. We shall use the term *2D motion* to refer to target rotational motion of this type. Under the 2D motion assumption, rotational compensation of a second point on the target will focus the whole target. When there is *3D motion*, i.e., when the rotational motion is not confined to a 2D plane, rotational compensation of a second point cannot focus the whole target.

Recently, several independent research groups have reported that, for aircraft undergoing fast maneuvers or ships on rough seas, the motion of a target may be highly chaotic and does not always obey the 2D motion model [8], [9], [10]. As a result, the image formed using the standard motion compensation algorithms is blurred. In [8] and [9], the effect of 3D motion on ISAR imaging is discussed. However, target motions are assumed to be known from other auxiliary sensor data that are usually not accessible in real operational environment. In [10], the imaging interval is adaptively chosen based on the resolved target feature in the radar image to overcome the 3D motion issue. It requires sound knowledge of the target under consideration, which is often not known to the end users of ISAR data.

The objective of this paper is to develop an algorithm to detect the presence of 3D motions during the imaging interval from ISAR data. Based on the 3D point scatterer model, we first examine the effect of 3D motion on existing imaging algorithms. We then develop an algorithm to blindly detect the existence of 3D motion. For this purpose, only the estimation of phases of several prominent point scatterers is needed. It can be accomplished by the joint time-frequency analysis [6]. With the detection algorithm, we have the ability to distinguish the time intervals when the target undergoes smooth 2D motion from those containing more chaotic 3D motion. As a result, the good imaging intervals where focused images are more easily formed can be automatically determined.

The paper is organized as follows. First, the ISAR imaging problem is formulated in terms of a point scatterer model in Section 2. In Section 3, the 2D motion assumption in existing motion compensation algorithms is analyzed. We show the reason why 3D motion is a problem for ISAR imaging. Section 4 discusses the 3D motion detection algorithm in detail. Examples from both simulation and measurement data are presented in Section 5. The conclusions are given in the last section.

2. 2D and 3D Motion Models

The standard model used in ISAR processing is the point scatterer model given as

$$E(f, t_D) = \sum_{i=1}^{Ns} \sigma_i(x_i, y_i) \exp\left\{ -j\frac{4\pi f}{c}[r(t_D) + x_i + y_i\varphi(t_D)]\right\} \quad (1)$$

where f is the radar frequency and t_D is the dwell time. The radar echo data $E(f, t_D)$ is in the (frequency)-(dwell time) domain. x and y represent the target range and cross-range positions, respectively. The target consists of N_s point scatterers, with the ith point scatterer depicted by position (x_i, y_i) and strength $\sigma(x_i, y_i)$. The target motion includes both the translational motion described by $r(t_D)$ and the rotational motion

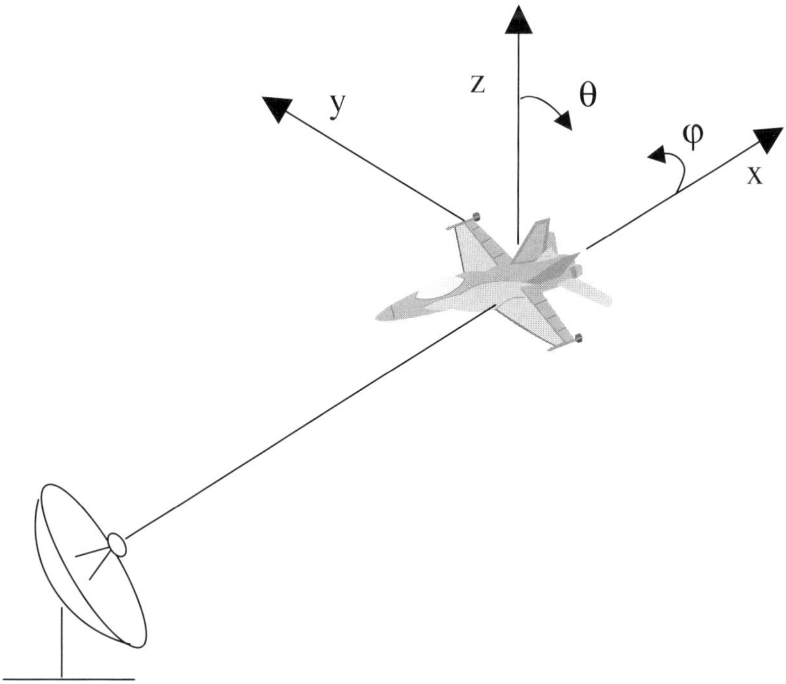

Figure 1. Geometry of an ISAR problem.

described by $\varphi(t_D)$. When there is no translational motion and the rotational motion is uniform, it is seen that a 2D Fourier transform brings the radar data $E(f, t_D)$ into a radar image $\sigma(x, y)$. Otherwise, motion compensation is a critical step in ISAR imaging.

The above model is what we call a 2D problem since the target rotational motion is confined to a 2D plane and describable in terms of only one angular parameter φ. When there is 3D motion of the target, a more general 3D model is required:

$$E(f, t_D) = \sum_{i=1}^{Ns} \sigma_i(x_i, y_i, z_i) \exp\left\{ -j\frac{4\pi f}{c}[r(t_D) + x_i + y_i\varphi(t_D) + z_i\theta(t_D)] \right\}. \quad (2)$$

In the above expression, a third coordinate z of the target is included to represent the 3D target and another independent angular motion parameter θ is introduced to describe the 3D rotational motion (see Figure 1). It is possible to perform 3D target imaging if the target motion is known exactly [11], [12]. In practical ISAR scenarios, however, we have no access to the target motion. Our objectives here are to examine the effect of 3D motion on ISAR imaging and devise an algorithm to detect the

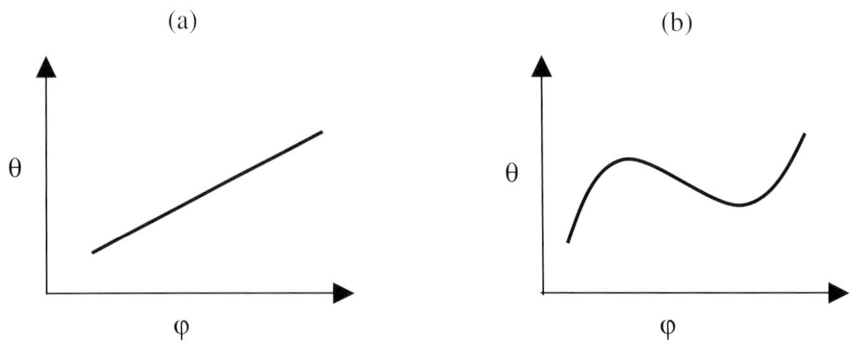

Figure 2. Illustration of 2D motion vs. 3D motion.

presence of 3D motion from the radar data itself, i.e., without any additional knowledge of the target motion.

3. Problem of Existing Motion Compensation Algorithms with 3D Target Motion

First, we show that the more general 3D model degenerates into the 2D model under two conditions. The first case is when there is a linear relationship between φ and θ (Figure 2), i.e.,

$$\theta(t_D) = b\varphi(t_D). \tag{3}$$

This allows us to cast equation (2) into the form

$$E(f, t_D) = \sum_{i=1}^{Ns} \sigma_i(x_i, y_i, z_i)\exp\left\{ -j\frac{4\pi f}{c}[r(t_D) + x_i + (y_i + bz_i)\varphi(t_D)]\right\}. \tag{4}$$

Comparing (4) with (1), we see that if we define $y_i' = y_i + bz_i$, then the rotational motion is in fact a two-dimensional one and the resulting 2D image $\sigma(x_i, y_i')$ is the projection from the 3D target $\sigma(x_i, y_i, z_i)$ onto the 2D motion plane.

The second case is when the z dimension of the target is so small that the third phase term in (2) can be neglected. For example, suppose a radar operates at a frequency of 10 GHz and the θ variation is limited to 0.5 degree. If the target thickness in the z-dimension is less than 0.2 m, then the third phase term is less than $\pi/4$ and the 2D model is adequate.

From the above consideration, we see that the 2D model is applicable if either the rotational motion is 2D or the target is of 2D in extent. When there exists 3D motion on a full 3D target, any motion compensation algorithms based on the 2D model is not expected to focus the target well. We will now examine this issue in more detail. Since the

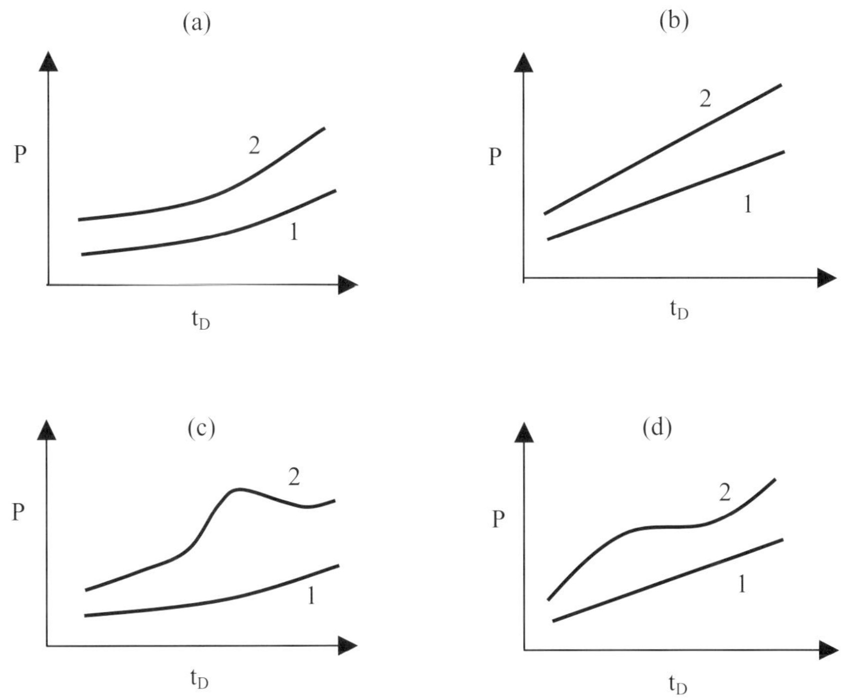

Figure 3. Phase linearization achieved by rotational motion compensation. (a) Phases of two point scatterer with 2D rotational motion. (b) Both phases are linearized with rotational motion compensation. (c) Phases of two point scatterers with 3D motion. (d) Only one phase is linearized with rotational motion compensation.

translational motion compensation is independent of the models in (1) and (2), only the rotational motion compensation needs to be investigated. With 2D rotational motion present, the phase of a point scatterer i due to the rotational motion is

$$P_i(t_D) = y_i\varphi(t_D). \tag{5}$$

Here, the constant $4\pi f/c$ has been suppressed for simplicity. As we can see from (5), the phases of all the point scatterers are linearly related to each other (through the ratio of their cross range positions). If we make one of the phases a linear function of time, then all the phases are linearized simultaneously, and the whole target can be focused after the Fourier transform. This is the basis of most 2D rotational motion compensation algorithms based on the point scatterer model [3], [4], [5], [6]. This concept is illustrated in Figure 3. Figure 3(a) shows the phase functions of two point scatterers under 2D rotational motion. Figure 3(b) shows that both points can be made linear functions of time after we force one of them to be a linear function.

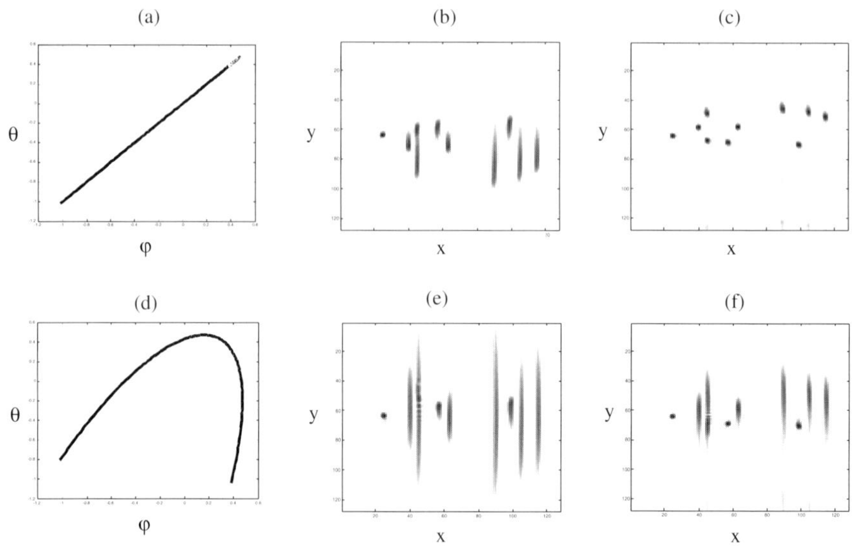

Figure 4. Problem with a typical motion compensation algorithm. (a) Target undergoes a 2D motion. (b) Image after translational motion compensation. (c) Image after rotational motion compensation. (d) Target undergoes a 3D motion. (e) Image after translational motion compensation. (f) Image after rotational compensation.

With 3D motion, the phase of a point scatterer due to the rotational motion is

$$P_i(t_D) = y_i\varphi(t_D) + z_i\theta(t_D). \tag{6}$$

In this case, the phases of the point scatterers are no longer linearly related. If we make one of the phases a linear function of time, the phases of the other point scatterers are not automatically made linear functions of time, as was the case of 2D motion. Figure 3(c) shows the phase functions of two point scatterers with 3D motion. As we can see from Figure 3(d), after one point is forced to be of linear phase, the phase of the other point remains nonlinear.

Figure 4 illustrates some simulation results of the effect of the rotational motion compensation based on the model in (1) on the final images under 2D and 3D target motion. The adaptive joint time-frequency (AJTF) algorithm reported in [6] is used for motion compensation. Ten points in 3D space are used to simulate the radar data. Figure 4(a) shows an assumed 2D rotational motion. Figure 4(b) shows the image after the translational motion compensation. The image shows one point being focused in range cell 25 while other points are unfocused due to the rotational motion. Figure 4(c) shows the image after the 2D rotational motion compensation in which a point scatterer in range cell 57 is selected for focusing. All the point scatterers are focused in the image. The situation with an assumed 3D target motion is shown in Figures 4(d)–4(f).

Figure 4(d) shows the assumed 3D motion. Figure 4(e) shows the image after translational motion compensation. Figure 4(f) shows the final image after the 2D rotational motion compensation. The two points in range cells 25 and 57 are focused, as expected. Another point scatterer in range cell 99 is also focused as it happens to be in the same 2D motion plane as the point scatterer in range cell 57. However, it is not possible to focus all the points simultaneously with an existing algorithm based on the 2D motion model.

4. 3D Motion Detection Algorithm

Since existing motion compensation algorithms cannot handle 3D target motion, it is desirable to develop a general compensation algorithm that can accommodate 3D motion. However, this is a difficult task (see [13] for background on this problem) and outside the scope of this work. Our goal here is to develop an algorithm to detect the presence of 3D motion from radar data. If we can reliably detect those time intervals where 2D target motions are predominant, we can use the existing 2D motion compensation algorithms to form well-focused ISAR images.

As discussed in the last section, 2D motion can be represented by a linear relationship between θ and φ. Therefore, we set out to detect the existence of a nonlinear relationship between θ and φ in our 3D motion detection algorithm. First, we write the relationship between θ and φ into a linear and a nonlinear part as follows:

$$\theta(t_D) = b\varphi(t_D) + m(t_D) \tag{7}$$

where b is the linear constant and $m(t_D)$ is the nonlinear part which indicates deviation from 2D target motion, or the degree of 3D motion. Next we try to gather target motion information by analyzing the phases of two point scatterers on the target. Let us write the relationship between the phase functions P_1 and P_2 of two point scatterers as:

$$P_2(t_D) = aP_1(t_D) + n(t_D). \tag{8}$$

The relationship is again decomposed into the linear part, where a is the linear constant, and the nonlinear part $n(t_D)$. Our goal is to derive a relationship between $m(t_D)$ and $n(t_D)$ so that the presence of m can be detected by observing n.

After the standard translational motion compensation, the time-varying phase of a point scatter is in the form of

$$P_i(t_D) = \Delta y_i\varphi(t_D) + \Delta z_i\theta(t_D) \tag{9}$$

where Δy_i and Δz_i are differential positions of point scatterer i relative to the reference point chosen during translational motion compensation. Substituting (7) into (9) and then evaluating (9) at point scatterers 1 and 2, we have

$$P_1(t_D) = (\Delta y_1 + b\Delta z_1)\varphi(t_D) + \Delta z_1 m(t_D) \tag{10a}$$

$$P_2(t_D) = (\Delta y_2 + b\Delta z_2)\varphi(t_D) + \Delta z_2 m(t_D). \tag{10b}$$

We next substitute (10) into (8), which leads to

$$a[(\Delta y_1 + b\Delta z_1)\varphi(t_D) + \Delta z_1 m(t_D)] + n(t_D) = (\Delta y_2 + b\Delta z_2)\varphi(t_D) + \Delta z_2 m(t_D) \tag{11}$$

Notice that if there is only 2D motion, then the phases of the two point scatterers must be linear. This means if $m = 0$, then $n = 0$. By using this fact and equating the coefficients of $\varphi(t_D)$ in (11), the constant a can be derived:

$$a = \frac{\Delta y_2 + b\Delta z_2}{\Delta y_1 + b\Delta z_1}. \tag{12}$$

By substituting (12) into (11), we finally arrive at

$$m(t_D) = \frac{\Delta y_1 + b\Delta z_1}{\Delta z_2\Delta y_1 - \Delta y_2\Delta z_1} n(t_D). \tag{13}$$

Equation (13) states that once the nonlinear phase term n is known, it is proportional to nonlinear motion m. Therefore, the steps to determine the degree of 3D target motion are as follows. First, we extract the phases of two point scatterers from the radar data. Next we find the nonlinear phase function n using a minimum least squares fit of equation (8). Once n is known, we use equation (13) to decide on the degree of 3D motion. The remaining issues are: (i) how to determine the phase functions of the point scatterers, (ii) how to define the degree of nonlinearity and the degree of 3D motion once n is known, and (iii) how to compare the degree of 3D motion from one imaging interval to another. These three issues are discussed in the following subsections.

4.1. Phase Estimation Using Adaptive Joint Time-Frequency Projection

After the translational motion compensation, the radar signal contains only rotational motion. To estimate the phase of a prominent point scatterer, we utilize the adaptive joint time-frequency (AJTF) projection technique discussed in [6]. We begin with the radar data in the (range)-(dwell time) domain. Within a fixed range cell, the data can be written as

$$E_s(t_D) = \sum_{i=1}^{Nr} \sigma_i \exp\Big(-j\frac{4\pi f_c}{c}(y_i\varphi(t_D) + z_i\theta(t_D))\Big) \tag{14}$$

where f_c is the center frequency. Among the N_r point scatterers within the range cell,

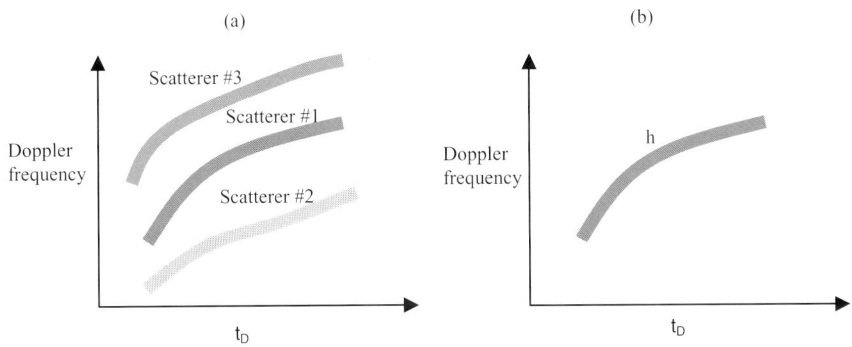

Figure 5. (a) (Dwell time)-(Doppler frequency representation of radar signal in a range cell with three point scatterers. (b) The basis function that is best matched to the dominant point scatterer is found by the AJTF project method.

we express the phase behavior of the strongest one as a polynomial function:

$$\phi_M(t_D) = (f_1 t + f_2 t^2 + f_3 t^3 + ..) \tag{15a}$$

and consider

$$h(t) = \exp[-j\varphi_M(t_D)] \tag{15b}$$

as a basis for the radar signal. The phase parameters are then found by searching for the maximum projection from the radar signal onto the basis function:

$$\langle f_1, f_2, f_3, \ldots \rangle = \arg\max \left| \int E(t_D) h^*(t_D) dt_D \right|. \tag{16}$$

Equation (16) means that the phase function parameters are estimated to give a maximum projection from the radar data onto the basis function for that prominent point scatterer. In the search procedure, the first term f_1 can be obtained by using the fast Fourier transform, while all other higher order terms $f_2, f_3, \ldots$ are obtained using exhaustive search. Figure 5 illustrates the process of AJTF phase estimation. Figure 5(a) shows the radar signal in one range cell with three point scatterers in the joint (dwell time)-(Doppler frequency) plane. The tilted trajectory of the prominent point scatterer 1 implies there exist higher-order terms in the phase function. Figure 5(b) shows the basis function $h(t_D)$. During the search, we change the position (f_1), tilting (f_2) and curvature ($f_3, \ldots$) of h until the projection of h onto the radar signal is maximized.

4.2. *Measure of Nonlinearity Between Two Phase Functions*

We notice that in (8), the two phase functions are formulated with a linear relationship plus a nonlinear residual part. After the two phase functions are estimated using the AJTF technique, a least-squares fitting can be performed to generate the best-fit linear part. The actual

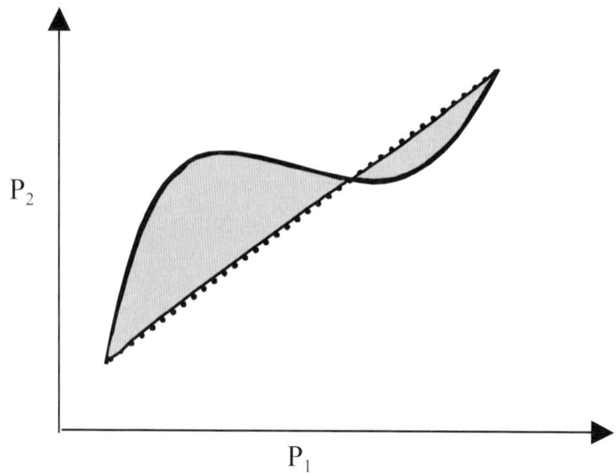

Figure 6. Measure of the nonlinearity of two phase functions.

phases deviates from this linear relationship. The deviation n is integrated over the dwell time to represent the degree of phase nonlinearity over the imaging interval as follows:

$$N_{12} = \int |n_{12}(t_D)| \, dt_D. \tag{17}$$

The process is illustrated in Figure 6. The solid line is the actual relationship between the two phase functions P_1 and P_2. The dotted line is the linear approximation of the relationship. The area of the shadowed region is N_{12}.

In a similar fashion, we define the degree of 3D motion as the deviation from a linear relationship between θ and φ over the dwell interval as follows:

$$M = \int |m(t_D)| \, dt_D. \tag{18}$$

Based on (13), we see that M and N_{12} are directly related:

$$N_{12} = \beta_{12} M \tag{19a}$$

where

$$\beta_{12} = \left| \frac{\Delta z_1 \Delta y_2 - \Delta y_1 \Delta z_2}{\Delta y_1 + b \Delta z_1} \right|. \tag{19b}$$

Thus by finding the observable N_{12}, we can obtain the degree of 3D motion M to within a proportionality constant.

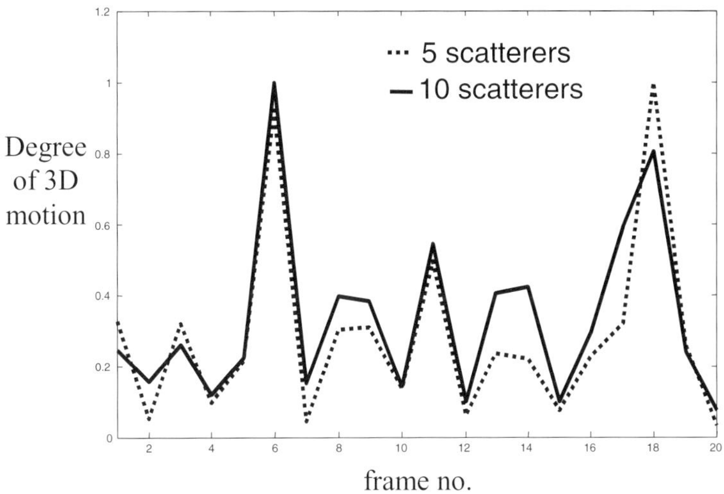

Figure 7. Effect of the number of point scatterers used on 3D motion detection result.

4.3. 3D Motion Comparison among Different Imaging Intervals

As indicated by (19), the phase nonlinearity of two point scatterers N is proportional to the degree of 3D motion M, so we can use the detected phase nonlinearity as a measure of 3D motion. However, we notice that the constant of proportionality is dependent on the point scatterer positions. A problem arises when we need to compare the detection result from one imaging interval to that from another imaging interval. Since we cannot guarantee that we track the same set of points from frame to frame, the proportionality constant can change from frame to frame, and we cannot reliably observe M from N across frames. To overcome this difficulty, we track more than two point scatterers within each frame and compute N_{ij} for each pairing of scatterers i and j ($i \neq j$). Then we generate an average value $<N_{ij}>$ from all the possible phase relationships. From (19), we have

$$\langle N_{ij} \rangle = \langle \beta_{ij} \rangle M. \tag{20}$$

We postulate that, from a statistical point of view, $<\beta_{ij}>$ approaches a constant that is independent of frames if we average over a sufficient number of point scatterers. If this is true, $<N_{ij}>$ should become a good indicator of M.

We test the effectiveness of this approach on the detection result by simulation. We input a set of motion parameters and generate the phase functions based on the 3D motion model. 20 point scatterers from an airplane model is used. We then randomly choose a number of point scatterers and use their phase functions to compute $<N_{ij}>$. We examine how the results vary as different number of point scatterers is used. We find that the results begin to converge after about 5 scatterers. Figure 7 shows a plot of $<N_{ij}>$ versus the frame number if

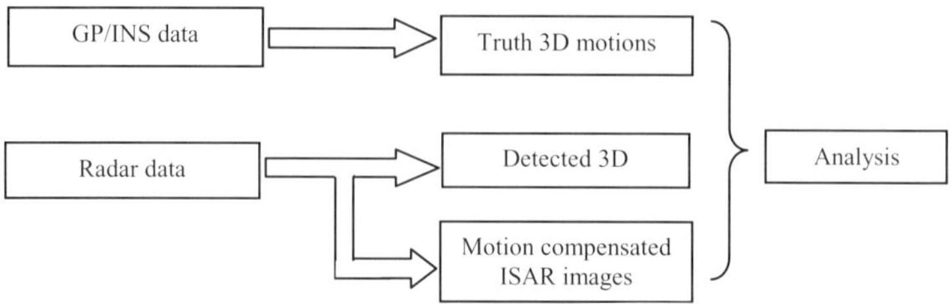

Figure 8. Data processing flow chart.

we use 5 point scatterers (10 phase pairs). If we increase the number of point scatterers to 10 (45 phase pairs), there is only minor change in the detection output. Therefore, $<N_{ij}>$ can be used to indicate the degree of 3D motion given a sufficient number of point scatterers.

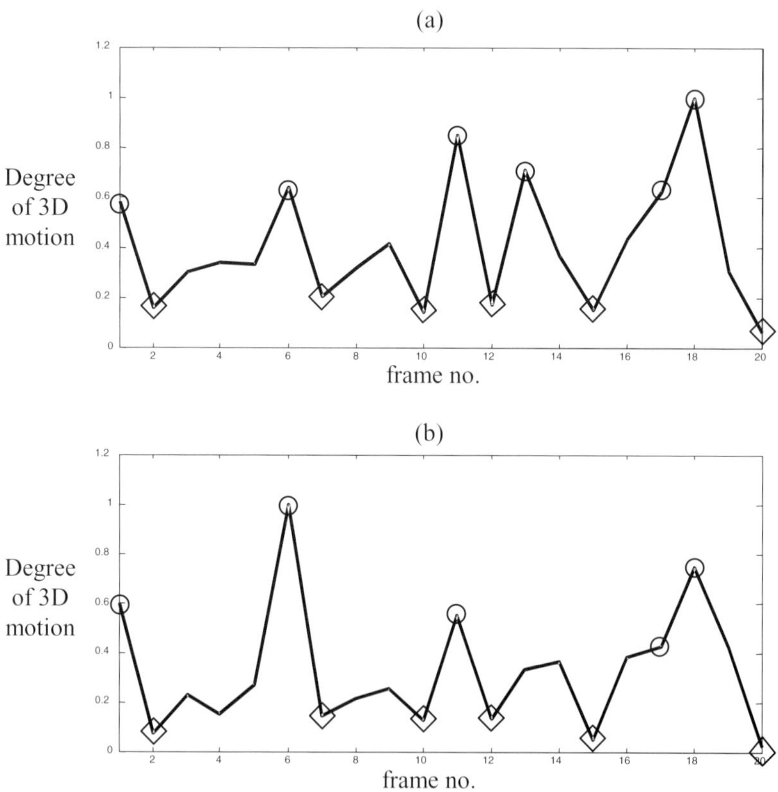

Figure 9. (a) Detected 3D motion from simulated radar data. (b) Degree of 3D motion from truth motion data.

5. Results

To demonstrate the effectiveness of the 3D motion detection algorithm, we test our algorithm on radar data from two targets. The first target is an aircraft, which flew in a large clockwise circle during a 9-minute interval. We also have access to the target motion data through the GPS (global positioning system) and INS (inertial navigation system) sensors carried on-board the aircraft [14]. Figure 8 shows our processing flow chart. The GPS/INS data is used to establish the truth target motion. The raw radar data is used as input to the 3D motion detection algorithm. We can also generate the ISAR images using our AJTF motion compensation algorithm [6]. We are therefore able to both compare the detection result with the truth motion, and observe the effect of the 3D motion on the ISAR image quality.

We first test the 3D motion detection algorithm on simulated radar data. To generate the simulation data, we use the actual motion data from the GPS/INS sensors in conjunction with a point scatterer model. From the aircraft model, 60 point scatterers are selected to simulate the radar data based on the actual motion data and equation (2). Five range cells are then chosen for phase analysis in the detection procedure. Figure 9(a) shows the detected degree of 3D motion for 20 image frames from the simulated radar data. For comparison, Figure 9(b) shows the degree of 3D motion obtained based on the truth motion data. The frames with significant 3D motion are highlighted with circles and the frames with 2D motion are highlighted with diamonds. It is seen that the two results agree fairly well.

Next, we test the detection algorithm using the actual radar measurement data. Figure 10(a) shows the detected 3D motion from the radar data over 20 frames. The corresponding imaging interval for each frame is 2.3 seconds while the total flight duration is 5 minutes. The four frames with the most significant 3D motion based on our detection algorithm are labeled as circles. They are frames 6, 14, 17 and 18. Figure 10(b) shows the degree of 3D motion obtained based on the truth motion data. We observe that the truth motion data indeed contains a high degree of 3D motion at those four frames detected by our algorithm.

To further examine the quality of the ISAR images when 3D motion is present, we generate images using our motion compensation algorithm in Figures 11 to 14. Figure 11(a) shows the plot of θ vs. ϕ derived from the truth motion data for frame 18, which is a frame found to contain substantial 3D motion. The actual motion is shown in the solid curve and the dashed line is the best-fit 2D motion approximation. It is clear that the curve deviates significantly from the dashed line and the actual motion cannot be well approximated with 2D motion. Figure 11(b) shows the resulting image obtained after the motion compensation, and is blurred in the Doppler dimension (vertical axis). As expected, the 2D motion compensation algorithm cannot focus all the points due to the 3D target motion. Figures 12(a) and 12(b) show the same conclusion for frame 14, which is another frame identified as having significant 3D motion. In Figure 13, we show the results for frame 2, which has very little 3D motion. As we can see from Figure 13(a), the actual motion can be well approximated by a line in the θ-φ plot. The image shown Figure 13(b) is well focused. In particular, the point scatterers on the target show nearly equal range and Doppler extent, contrary to the previous two images. The aircraft body line is clearly recognizable.

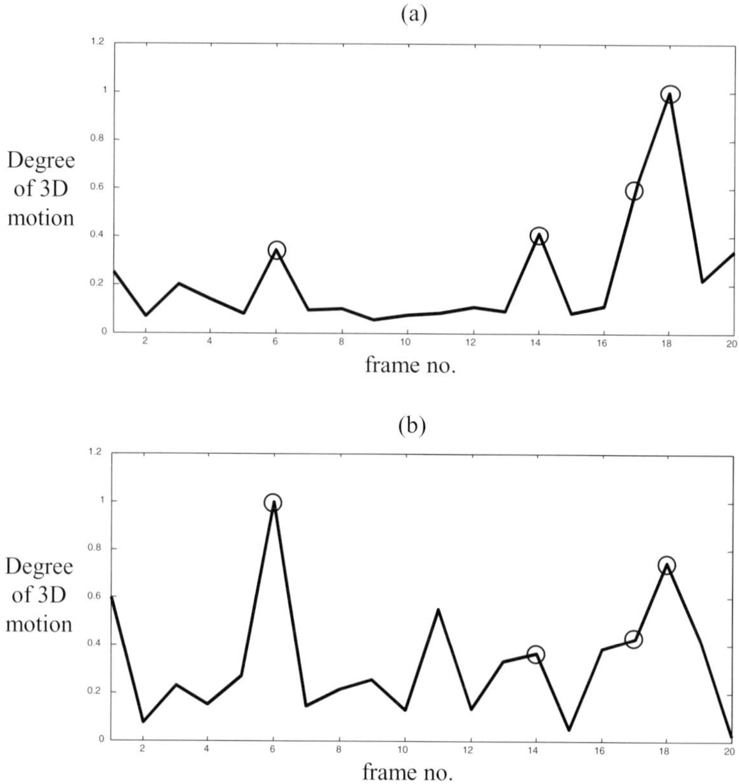

Figure 10. (a) Detected 3D motion from aircraft radar data. (b) Degree of 3D motion from truth motion data.

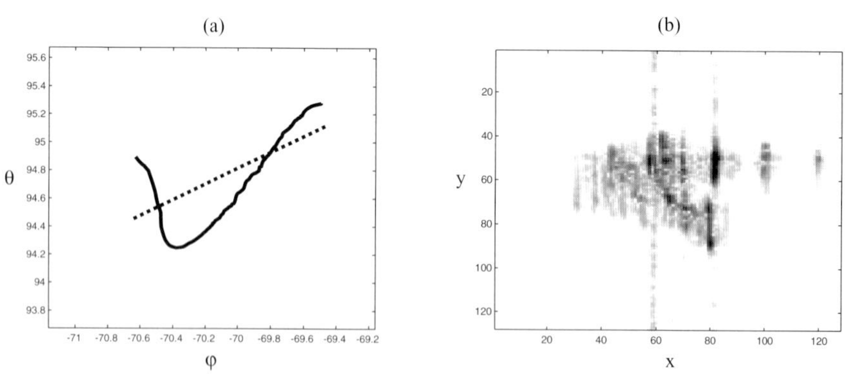

Figure 11. 3D motion and resulting ISAR image (frame 18).

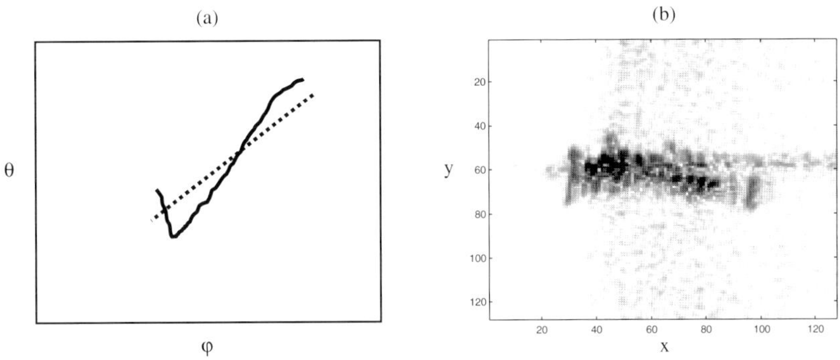

Figure 12. 3D motion and resulting ISAR image (frame 14).

From Figure 10, we notice that there exists a discrepancy in frame 11, where the detection result does not indicate any 3D motion while the truth motion data shows a significant amount of 3D motion. The truth motion is shown in Figure 14(a), confirming the presence of 3D motion. One explanation is that those prominent points used by the detection algorithm lie nearly on a 2D plane so that they still can be focused. As we have discussed in Section 3, the 2D model is applicable if either the motion is 2D or the target is of 2D in extent. It is likely that the latter condition is met for this frame. This is confirmed by the image shown in Figure 14(b). We see that the image quality is actually not so bad. Therefore, our detection algorithm objectively reflects the quality of the images generated by the 2D motion compensation.

A second data set is used to test our 3D motion detection algorithm. This data set consists of the ISAR data collected from a small ship on the ocean. Because of the surface movement of the sea, the target is believed to have considerable 3D motion during the imaging intervals. The 3D motion detection result is shown in Figure 15(a) with the peaks

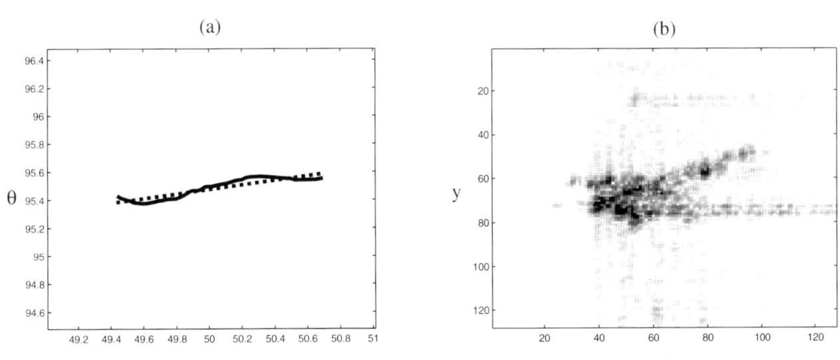

Figure 13. 2D motion and resulting ISAR image (frame 2).

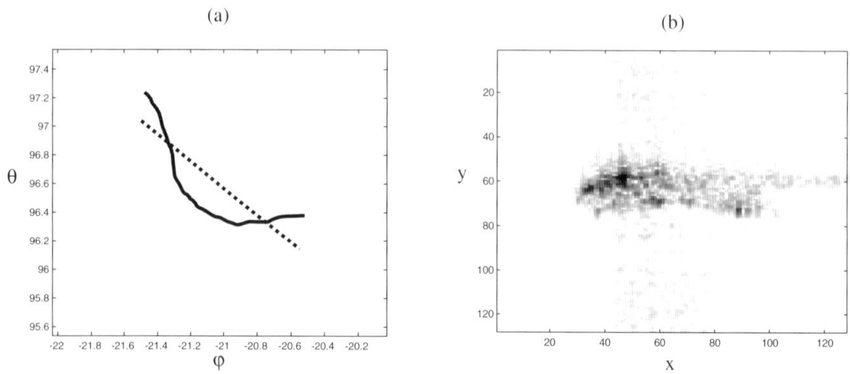

Figure 14. Frame no.11 showing focused ISAR image with 3D motion.

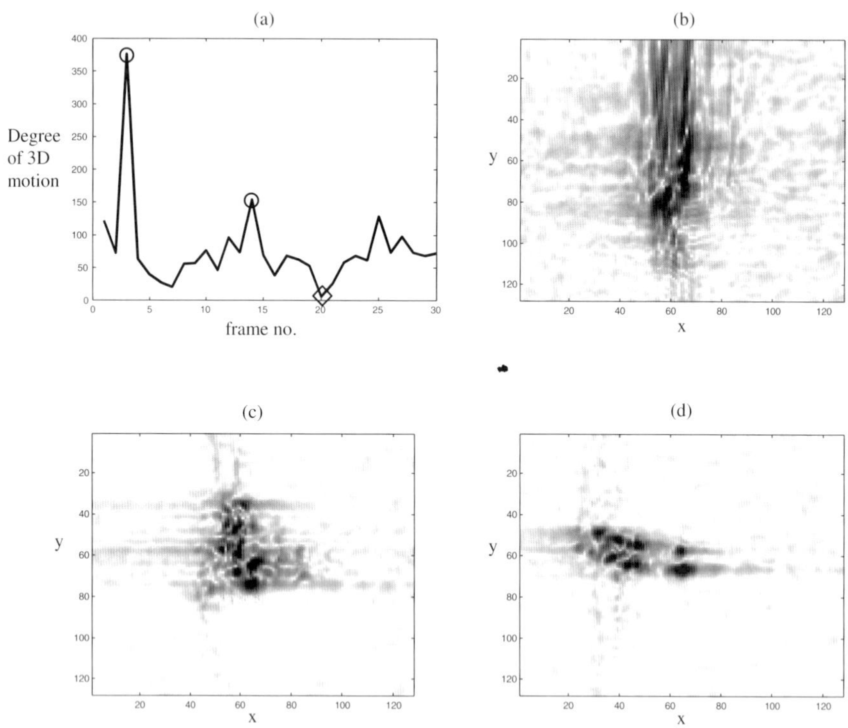

Figure 15. 3D motion detection result from radar data of a ship. (a) Detected 3D motion. (b) ISAR image from frame 3. (c) ISAR image from frame 14. (d) ISAR image from frame 20.

corresponding to regions with 3D motion. The total data duration is 20 seconds and the imaging dwell time is 0.64 second per frame. For this data set, reliable truth target motion is not available. Instead, we generate the motion compensated images shown in Figures 15(b) to 15(d) to demonstrate the effect of 3D motion on ISAR image quality. The image frame with the largest detected 3D motion, frames 3, is shown in Figure 15(b). It is poorly focused. Figure 15(c) shows the image from frame 14, which is the frame with the second highest detected 3D motion. The frame with the smallest 3D motion based on our algorithm, frame 20, is shown in Figure 15(d). It shows a well-focused ISAR image. This test confirms the effectiveness of our algorithm in detecting good imaging intervals from those imaging intervals containing large 3D motion.

6. Conclusions

In this paper, we set out to develop an algorithm to detect the presence of 3D target motion from ISAR data. Based on the 3D point scatterer model, we first examined the effect of 3D motion on ISAR imaging. It was shown that the existing motion compensation algorithms could not properly focus targets exhibiting 3D motion during the imaging interval. We then derived an algorithm to blindly detect the degree of 3D target motion from raw radar data. It is based on measuring the linearity of phases between two or more point scatterers on the target. The phase estimation was implemented using the adaptive joint time-frequency technique. Examples were provided to demonstrate the effectiveness of the 3D motion detection algorithm with both simulation and real ISAR data. The detection results were corroborated with the truth motion data from on-board motion sensors and correlated with the resulting ISAR images. With the detection algorithm, we have the ability to distinguish the time intervals when the target undergoes smooth 2D motion from those containing more chaotic 3D motion. As a result, the good imaging intervals where focused images are more easily formed can be automatically selected.

Acknowledgments

This work is supported by the Office of Naval Research under contract No. N00014-98-1-0615.

References

1. C.C. Chen and H.C. Andrews, "Target Motion Induced Radar Imaging," *IEEE Trans. Aerospace Electron. Syst.*, vol. 16, no. 1, Jan. 1980, pp. 2–14.
2. A. Ausherman, A. Kozma, J.L. Waker, H.M. Jones, and E.C. Poggio, "Developments in Radar Imaging," *IEEE Trans. Aerospace Electron. Syst.*, vol. 20, no. 4, July 1984, pp. 363–400.
3. W.G. Carrara, R.S. Goodman, and R.M. Majewski, *Spotlight Synthetic Aperture Radar - Signal Processing and Algorithms*, Boston, MA: Artech House, 1995.

4. S. Werness, W. Carrara, L. Joyce, and D. Franczak, "Moving Target Imaging Algorithms for SAR Data," *IEEE Trans. Aerospace Electron. Syst.*, vol. 26, no. 1, Jan. 1990, pp. 57–67.
5. D.E. Wahl, P.H. Eichel, D.C. Ghiglia, and C.V. Jakowatz, "Phase Gradient Autofocus- A Robust Tool for High Resolution SAR Phase Correction," *IEEE Trans. Aerospace Electron. Syst.*, vol. 30, no. 3, July 1994, pp. 827–835.
6. Y. Wang, H. Ling, and V.C. Chen, "ISAR Motion Compensation via Adaptive Joint Time-Frequency Techniques," *IEEE Trans. Aerospace Electron. Syst.*, vol. 34, no. 2, Apr. 1998, pp. 670–677.
7. Z. Liu, R. Wu, and J. Li, "Complex ISAR Imaging of Maneuvering Targets via the Capon Estimator," *IEEE Trans. Signal Processing*, vol. 47, no. 5, May 1999, pp. 1262–1271.
8. V.C. Chen and W.J. Miceli, "Effect of Roll, Pitch and Yaw Motions on ISAR Imaging," in *SPIE Proc. Radar Processing, Technology and Applications VI*, vol. 3810, Denver, CO, July 1999, pp. 149–158.
9. J. Li, Y. Wang, R. Bhalla, H. Ling, and V.C. Chen, "Comparison of High-Resolution ISAR Imagery from Measured Data and Synthetic Signatures," *SPIE Proc. Radar Processing, Technology and Applications VI*, vol. 3810, Denver, CO, July 1999, pp. 170–179.
10. A.W. Rihaczek and S.J. Hershkowitz, "Choosing Imaging Intervals for Identification of Small Ships," *SPIE Proc. Radar Processing, Technology and Applications VI*, vol. 3810, Denver, CO, July 1999, pp. 139–148.
11. S. Xiao and D.C. Munson, "Spotlight-Mode SAR Imaging of a Three-Dimensional Scene Using Spectral Estimation Techniques," in *Proc. IEEE Inter. Geosci. and Remote Sensing Symp.*, vol. 2, Seattle, WA, July 1998, pp. 642–644.
12. J.T. Mayhan, M.L. Burrows, K.M. Cuomo, and J.E. Piou, "High Resolution 3D 'Snapshot' ISAR Imaging and Feature Extraction," *IEEE Trans. Aerospace Electron. Syst.*, vol. 37, no. 2, Apr. 2001, pp. 630–641.
13. M.A. Stuff, "Three-Dimensional Analysis of Moving Target Radar Signals: Methods and Implications for ATR and Feature-Aided Tracking," *SPIE Proc. Algorithms for Synthetic Aperture Radar Imagery VI*, vol. 3721, Aug. 1999, pp. 485–496.
14. H. Ling, Y. Wang, J. Li, and R. Bhalla, "ISAR Image Formation of TIRA Data Using Adaptive Joint Time-Frequency Processing," in *NATAO Sensors and Electronics Technology Symposium on High Resolution Radar Techniques*, Granada, Spain, Mar. 1999, pp. 61.1–61.7.

Multidimensional Systems and Signal Processing, 14, 241–263, 2003

Experimental Evaluation of Adaptive Beamforming Methods and Interference Models for High Frequency Over-the-Horizon Radar Systems

G. A. FABRIZIO[1] joe.fabrizio@dsto.defence.gov.au
Intelligence, Surveillance and Reconnaissance Division, Defense Science and Technology Organisation, PO Box 1500, Edinburgh, South Australia, 5111, Australia
Cooperative Research Centre for Sensor Signal and Information Processing, SPRI Building, Warrendi Rd, Mawson Lakes, South Australia, 5095, Australia
The University of Adelaide, Department of Electrical and Electronic Engineering, South Australia, 5005, Australia

D. A. GRAY dgray@cssip.edu.au
Cooperative Research Centre for Sensor Signal and Information Processing, SPRI Building, Warrendi Rd, Mawson Lakes, South Australia, 5095, Australia
The University of Adelaide, Department of Electrical and Electronic Engineering, South Australia, 5005, Australia

M. D. TURLEY mike.turley@dsto.defence.gov.au
Intelligence, Surveillance and Reconnaissance Division, Defense Science and Technology Organisation, PO Box 1500, Edinburgh, South Australia, 5111, Australia

Received December 1, 2000; Revised October 10, 2001; Accepted October 24, 2001

Abstract. This paper experimentally evaluates the interference cancellation performance of different adaptive beamforming schemes applicable to high frequency (HF) over-the-horizon (OTH) radar systems. Such systems are known to receive multipath and diffusely scattered radio frequency interference produced as a result of reflection from the stratified, dynamic and spatially inhomogeneous ionospheric propagation medium. Apart from quantifying the effectiveness of operational adaptive beamformers in the HF (3–30 MHz) environment, realistic interference models are described and experimentally evaluated in terms of their ability to predict the observed interference cancellation performance which is not well represented by traditional models. Adaptive beamforming algorithms with robustness against "jammer motion" are also described and their effectiveness is experimentally demonstrated using interference data collected by 32 narrowband receivers of the very wide aperture (2.8 km) Jindalee OTH radar uniform linear array located near Alice Springs in central Australia.

Key Words: adaptive beamforming, interference models, over-the-horizon radar

1. Introduction

OTH radar antenna arrays are often required to mitigate sources of radio frequency interference (RFI) propagated long-distances by the ionosphere. Conventional or data-independent beamformers are sub-optimal when strong directional interference is present and for this reason it is often sought to improve the output signal-to-interference plus noise ratio (SINR) by means of adaptive beamforming.

Due to the stratified nature of the ionosphere, a HF (3–30 MHz) interference source is usually received as number of signal components or propagation modes which are reflected from different layers in the ionosphere. In addition to the multipath propagation, the dynamic and spatially inhomogeneous reflection characteristics of individual ionospheric layers is responsible for the production of diffusely scattered modes which exhibit some degree of Doppler spread and angular spread [1], [2].

Although adaptive beamforming has been applied in many diverse practical applications, the uniqueness of the HF environment requires a specific study to evaluate the performance benefits or improvements in output SINR which can be realised in practice. There is currently a paucity of experimental results to indicate the performance of adaptive beamformers in OTH radar systems or to determine the validity of currently available interference models. Such a study would demonstrate the extent to which theoretical and/or computer simulation studies based on existing interference models can be relied upon to quantify the performance benefits encountered in practice, as well as to provide valuable information for the purpose of justifying the implementation of adaptive beamformers in operational OTH radar systems.

Specific experimental research on adaptive beamforming for OTH radar was conducted in [3] where interference rejection improvements of 20 dB with respect to the conventional beamformer were quoted. However, important quantities such as the mean improvement in output SINR and the distribution of its variability over different data sets, coherent processing intervals and beam steer directions were not reported. Moreover, separate training samples containing only interference were not available so the adaptive beamformer had to be tuned on data which additionally contained clutter and desired signals. Another study [4] demonstrated that the adaptive beamforming weights fluctuated over time scales of 1 second or less in order to achieve interference and noise levels 20 dB lower than the conventional beamformer but the cause of this variation was not experimentally confirmed.

A more recent paper [5] used the MUSIC algorithm to analyse the spatial structure of two HF interference sources propagated by the ionosphere and experimentally confirmed the nonstationary (or more precisely the time-varying) nature of the spatial structure or wavefront shape of individual propagation modes over time scales in the order of seconds but models for the observed spatial dynamics of the interference were not discussed. A model for ionospherically-propagated HF interference was proposed in [6] but a parameter estimation technique was not supplied and the model itself was not specifically assessed in terms of its ability to predict the observed performance of adaptive beamformers in HF antenna arrays.

This paper is concerned with the unification and extension of previous research in three areas; 1) the statistical quantification of adaptive beamformer performance in the HF environment for the purpose of indicating the most promising methods for OTH radar and the reliability of analysis based on traditional but commonly used (fixed plane wave) interference models, 2) the description of a more realistic multipath and space-time distributed HF interference model and its experimental evaluation to guide the development and testing of more advanced adaptive beamforming algorithms and 3) the application of recently proposed robust adaptive beamforming algorithms which exploit the information contained in this model to enhance the performance of OTH radar systems.

Section 2 of this paper describes the experimental facility and the data collection procedure while the multi-sensor HF interference model and the issue of parameter estimation from real data are discussed in section 3. Section 4 describes various adaptive beamforming schemes which may be applied in operational OTH radar systems while section 5 presents the experimental results which statistically quantify their performance. This section also demonstrates the accuracy with which the multi-sensor HF interference model can forecast the experimentally measured performance.

2. Experimental Facility and Data Collection

Two separate experiments were conducted to collect the data required for this study. Data collected in the first experiment was used to determine the propagation characteristics of a 1265 km ground distance mid-latitude ionospheric path and to estimate the parameters of a HF interference model. The second experiment, which involved the propagation of HF interference over the same ionospheric path, was performed to evaluate the experimental performance of different adaptive beamforming methods and to test the validity of the estimated HF interference model.

2.1. Experiment 1: Space-Time Channel Scattering Function

In this experiment a 20 kHz bandwidth linear sawtooth sweep-frequency modulated continuous waveform (FMCW) was transmitted at a centre frequency of 16.110 MHz from Darwin and propagated southwards over a 1265 km mid-latitude ionospheric path to the Jindalee OTH radar receiver array located near Alice Springs in central Australia. The Jindalee OTH radar receiver array consists of 462 dual-fan antenna elements which span a linear aperture of 2.8 km, the array boresight is oriented north-west at an angle of 22 degrees relative to Darwin. The dual-fan outputs are grouped to form 32 uniformly spaced subarrays and hardware beamformed by a delay-line steering method to produce 32 subarray outputs. The subarray outputs are then demodulated by a common local oscillator generating a linear sawtooth sweep-FMCW (a synchronised replica of transmitted waveform), analog filtered and sampled at the Nyquist rate by 32 calibrated digital receivers.

In this experiment, the data were recorded as a sequence of coherent processing intervals (CPI) or radar dwells, where each CPI is composed of P phase coherent pulse repetition intervals (PRI) often referred to as FMCW sweeps. Samples resulting from digital spectral analysis within each PRI (range processing) are referred to as "range" samples; range processing is applied to separate the propagation modes reflected from different ionospheric layers on the basis of time-of-arrival. The temporal sequence of a particular range sample from one PRI to another within a certain receiver are termed *slow-time* samples and provide information regarding the Doppler characteristics of individual propagation modes. Similarly, the sequence of samples recorded by the different receivers across the array at a particular range cell is commonly known as an *array snapshot* which provides information regarding the spatial characteristics of individual propagation modes.

In this experiment, each CPI is of approximately 4 seconds duration and contains a total of $S_{tot} = N \times K \times P$ complex digital samples, where $N = 32$ receivers, $K = 42$ range samples and $P = 256$ PRI or FMCW sweeps. A sequence of 47 dwells were recorded in this manner on 1 April 1998 between 06:17–06:21 UT to measure the multipath and space-time (angle-Doppler) distribution of the prevailing HF ionospheric reflection channel.

2.2. Experiment 2: Co-Channel Radio Frequency Interference

In the second experiment the Jindalee OTH radar array was tuned to receive a 3 kHz bandwidth interference source also situated in the Darwin region. This data were recorded between 06:22 and 06:32 UT on 1 April 1998 at a carrier frequency of 16.050 MHz. Note that the space-time channel scattering function data described in experiment 1 were collected for the same mid-latitude ionospheric path immediately prior to this experiment (06:17–06:21 UT) on a nearby frequency channel (16.110 MHz). Hence, the ionospheric channel parameters estimated from experiment 1 may be used to construct a model for the interference received during experiment 2.

To cover a range of potentially usable OTH radar CPI lengths, the interference was received as a sequence of longer CPI with a duration of approximately 50 seconds. This was achieved by maintaining the parameters $N = 32$ receivers, $K = 42$ range samples and $P = 256$ PRI used in the first experiment and decreasing the pulse repetition frequency (PRF) from 60 Hz to 5 Hz. In this experiment, the radar bandwidth was matched to the interference bandwith of 3 kHz to avoid the reception of signals from other users. The resulting "white noise" interference signal was seen to spread across the entire range-Doppler search space in a significant number of beam steer directions after conventional (FFT-based) radar signal processing was performed. The interference source was switched off momentarily during the recording period to sample background noise only which was 50–60 dB lower in power on the same frequency channel.

To summarise, data from experiment 1 contained a reference signal used to characterise the space-time ionospheric reflection channel scattering function, while data from experiment 2 contained realistic interference propagated over the same channel in order to quantify the gain in interference cancellation provided by the adaptive beamforming implementations considered and to compare such gains with those predicted from the estimated interference model.

3. Multi-Sensor Signal Model and Parameter Estimation

The multi-sensor signal model adopted for this study [6] is a space-time generalisation of the single receiver model considered in [7], in terms of its spatial second order statistics the adopted model is similar to those discussed in [8], [9]. The essential features of this model are described and analysed below to introduce notation and to explain its relationship with the previously described experiments. This section also discusses the issue of parameter estimation and explains how the ionospheric reflection channel parameters were extracted from the channel scattering function data recorded in experiment 1.

3.1. *Multi-Sensor Signal Model*

Let the N-variate complex vector $\mathbf{x}_k(t)$ denote the antenna array snapshot received at the kth range cell and t^{th} repetition period. When the array is tuned to a desired or interfering HF signal source, the received array snapshots $\mathbf{x}_k(t)$ may be represented as a superposition of $m = 1, 2,.., M$ signal modes $\mathbf{s}_{k,m}(t)$, propagating from source to receiver along different ionospheric paths, and additive background noise $\mathbf{n}_k(t)$.

$$\mathbf{x}_k(t) = \mathbf{s}_k(t) + \mathbf{n}_k(t) = \sum_{m=1}^{M} \mathbf{s}_{k,m}(t) + \mathbf{n}_k(t) \begin{cases} t & = & 1,2,..,P \\ k & = & 1,2,..,K \end{cases}. \tag{1}$$

After reflection from a random and inhomogeneous ionospheric layer, each signal mode $\mathbf{s}_{k,m}(t)$ may be represented by the following multidimensional (N-variate) statistical process which incorporates space-time diffuse scattering causing Doppler and angular spread.

$$\mathbf{s}_{k,m}(t) = A_m \mathbf{S}(\theta_m)\mathbf{c}_m(t)g_k(t, \tau_m)e^{j2\pi\Delta f_m t}. \tag{2}$$

For a narrowband uniform linear array (ULA) such as Jindalee, the $N \times N$ diagonal matrix $\mathbf{S}(\theta_m)$ contains the array steering vector elements parameterised by the mean cone angle-of-arrival θ_m of the mth mode;

$$\mathbf{S}(\theta_m) = diag[e^{j2\pi\frac{n\Delta d}{\lambda}\sin\theta_m}]_{n=0}^{N-1} \tag{3}$$

where Δd is the distance between adjacent receivers (subarray centres) and λ is the carrier wavelength. Other deterministic quantities in Eqn.(2) include the regular component of ionospheric Doppler shift Δf_m, the group delay τ_m and the RMS amplitude A_m of the mth mode.

The complex scalar $g_k(t, \tau_m)$ represents the transmitted waveform after it has been delayed by a transit time τ_m, demodulated with a sweep-FMCW (in the case of Jindalee), filtered, digitised and range processed at a reference receiver ($n = 0$). When the source transmits *broadband* interference of bandwidth not less than the receiver bandwidth f_b, the term $g_k(t, \tau_m)$ can be modelled as white noise with the following correlation properties.

$$E\{g_k(t, \tau_m)g_{k'}^{*}(t', \tau_{m'})\} = \mathrm{sinc}(f_b[\tau_m - \tau_{m'}])\delta(k - k')\delta(t - t') \tag{4}$$

The sinc function ($\frac{\sin \pi x}{\pi x}$) used to describe the inter-mode correlation coefficient implicitly assumes that the power spectral density of the interference is flat over the receiver bandwidth. If the source transmits a synchronised replica of the demodulating sweep-FMCW, then $g_k(t, \tau_m)$ does not vary as function of PRI (t) and its form in range (k) is determined by the range processing window function.

Temporal fluctuations of the ionospheric channel causing Doppler spread are described by an N-dimensional scalar-type auto-regressive (AR) process;

$$\mathbf{c}_m(t) + \sum_{i=1}^{I} \alpha_{m,i}(\Delta t)\mathbf{c}_m(t - i\Delta t) = \mu_m \epsilon_m(t) \tag{5}$$

where the scalar AR coefficients $\alpha_{m,i}(\Delta t)$ and the normalising constant μ_m define the Doppler spectrum characteristics of the mth mode for a PRI of Δt seconds. As pointed out in [6], a first order AR(1) process with coefficients $\alpha_m(\Delta t) = e^{-B_t(m)\Delta t}$ and $\mu_m = \sqrt{1 - \alpha_m(\Delta t)^2}$ may be adopted to model the Doppler spectral density as a Lorentzian profile parameterised by the temporal bandwidth $B_t(m)$ of the mth propagation mode.

As shown in [1], [2], spatial fluctuations of the ionospheric channel causing angular spread may also be statistically described by a first order AR(1) process;

$$\epsilon_m^{[n]}(t) = \beta_m(\Delta d)\epsilon_m^{[n-1]}(t) + \nu_m \gamma_{m,n}(t) \tag{6}$$

where $\epsilon_m^{[n]}(t)$ denotes the nth element of the vector $\epsilon_m(t)$ and AR(1) normalising constant $\nu_m = \sqrt{1 - \beta_m(\Delta d)^2}$. The spatial AR(1) coefficient is analogously defined as $\beta_m(\Delta d) = e^{-B_s(m)(1-\sin\theta_m)\Delta d}$ where $B_s(m)$ is the spatial bandwidth of the mth propagation mode.

The driving white noise process $\gamma_{m,n}(t)$ is a zero-mean complex Gaussian process having independent identically distributed real and imaginary parts and the following correlation properties.

$$E\{\gamma_{m_1,n_1}(t_1)\gamma^*_{m_2,n_2}(t_2)\} = \delta(m_1, m_2)\delta(n_1, n_2)\delta(t_1, t_2). \tag{7}$$

This driving noise produces statistical processes in accordance with the assumptions of Rayleigh fading and mutually independent mode fluctuations, as experimentally verified by Watterson [7].

To summarise, the essential feature of this model is the complex N-dimensional random vector $\mathbf{c}_m(t)$ which models the statistical characteristics of the ionospheric channel propagating each mode $m = 1, 2,.., M$. Variation of the individual elements in $\mathbf{c}_m(t)$ with respect to slow-time (t) represents the complex (gain and phase) temporal modulation sequence induced on the mth mode by the ionosphere to produce Doppler spread in the different receivers. On the other hand, changes in the gain and phase relationship between the elements of $\mathbf{c}_m(t)$ with respect to slow-time (t) represent the time evolution of the spatial modulations imparted by the ionosphere to produce angular spread on the mth mode.

It is worth noting that over short time intervals (commensurate with the PRI) the model assumes that each mode wavefront has a fixed spatial modulation, in other words $\mathbf{c}_m(t)$ is range (k)-independent, which is justifiable in cases when the pulse repetition frequency f_p

is much greater than the mode Doppler bandwidths $B_t(m)$. This implies that a mode wavefront has a spatial rank of unity when observed over the "quasi-instantaneous" PRI, but the spatial modulations vary from one PRI to another in a manner determined by the two-dimensional (space-time) AR process and the resulting angular spread may be described in statistical terms by the space-time auto-correlation sequence.

If the background noise contribution is negligible compared with that of the signal, the two-dimensional (space-time) auto-correlation sequence (ACS) statistically expected at the ith temporal lag (at a sampling interval of Δt seconds) and l^{th} spatial lag (at a receiver spacing of Δd metres) may be derived as;

$$r(i\Delta t, l\Delta d) = E\left\{\mathbf{x}_k^{[n]}(t)\mathbf{x}_k^{[n-l]*}(t-i)\right\} = \sum_{m=1}^{M} h_m z_m^i w_m^l \begin{cases} i = 0,..,L_t - 1 < P \\ l = 0,..,L_s - 1 < N \end{cases} \tag{8}$$

where $h_m = A_m^2$ is the power of the mth distributed signal mode, $z_m = \alpha_m(\Delta t)e^{j2\pi\Delta f_m\Delta t}$ is the temporal AR(1) process pole incorporating the regular component of Doppler shift Δf_m and $w_m = \beta_m(\Delta d)e^{j2\pi\frac{\Delta d}{\lambda}\sin\theta_m}$ is the spatial AR(1) process pole incorporating the mean cone angle-of-arrival θ_m.

The values of positive scalars $\alpha_m(\Delta t)$ and $\beta_m(\Delta d)$ is not greater than one and determine the degree of Doppler and angular spread induced respectively by the ionosphere on the mth signal mode. Lower values imply that the magnitude of the ACS tends to zero more rapidly which is equivalent to an increase in the width of the associated power density function. The phase progression terms $e^{j2\pi\Delta f_m\Delta t}$ and $e^{j2\pi\frac{\Delta d}{\lambda}\sin\theta_m}$ simply centres the even power density function about the mean Doppler shift Δf_m and cone angle-of-arrival θ_m of the mth signal mode respectively.

3.2. *Parameter Estimation*

The objective of parameter estimation is to estimate the pole pairs (z_m, w_m) and the corresponding residues h_m for all the modes $m = 1, 2,.., M$. In [1], [2], a closed form least squares technique, based on the ideas originally developed in [10], [11], was proposed and used in [1], [2] to jointly estimate the required model parameters from the sample space-time ACS $\bar{r}(i\Delta t, l\Delta d)$ in a computationally attractive manner. The sample ACS $\bar{r}_k(i\Delta t, l\Delta d)$ used for parameter estimation is computed by first averaging the sum of lagged products within each dwell to form $\hat{r}(i\Delta t, l\Delta d)$,

$$\hat{r}_k(i\Delta t, l\Delta d) = \frac{1}{N_t N_s}\sum_{t=1}^{N_t}\sum_{d=1}^{N_s} \mathbf{x}_k^{[d]}(t)\mathbf{x}_k^{[d+l\Delta d]*}(t+i\Delta t) \begin{cases} i = 0,..,L_t - 1 \\ l = 0,..,L_s - 1 \end{cases} \tag{9}$$

where $N_t = P - i\Delta t + 1$, $N_s = N - l\Delta d + 1$ and then averaging these lags over different dwells to form the mean ACS, denoted by $\bar{r}_k(i\Delta t, l\Delta d)$, over the period of data collection.

To introduce the parameter estimation method we define an $L_t \times L_s$ matrix $\mathbf{T}$ such that its $(i, l)^{th}$ entry equals $r(i\Delta t, l\Delta d)$ and note that $\mathbf{T}$ may be factorised in the following way;

$$\mathbf{T} = \sum_{m=1}^{M} h_m \mathbf{z}_m \mathbf{w}_m^H = \mathbf{ZHW}^H \tag{10}$$

by setting the vector $\mathbf{z}_m = [z_m^0 \cdots z_m^{L_t-1}]^T$, the vector $\mathbf{w}_m = [w_m^0 \cdots w_m^{L_s-1}]^H$ and the matrices $\mathbf{Z} = [\mathbf{z}_1 \cdots \mathbf{z}_m]$, $\mathbf{W} = [\mathbf{w}_1 \cdots \mathbf{w}_m]$, $\mathbf{H} = diag[h_m]_{m=1}^M$.

In practice, we have an estimate $\hat{\mathbf{T}}$ containing the sample space-time lags evaluated for a particular range cell k known to contain one or more HF signal modes. Even in the absence of modelling errors and additive noise, the matrix $\hat{\mathbf{T}}$ will generally be of full rank even if $L_t > M$, $L_s > M$ due to the presence of estimation errors in the sample lags $\bar{r}(i\Delta t, l\Delta d)$. We can attempt to reduce the influence of estimation uncertainty on the parameter estimates by taking the reduced rank-M approximation of $\hat{\mathbf{T}}$ using a truncated singular value decomposition (SVD).

$$\hat{\mathbf{T}}_s = \sum_{m=1}^{M} \sigma_m \mathbf{u}_m \mathbf{v}_m = \mathbf{U}_s \mathbf{\Sigma}_s \mathbf{V}_s^H. \tag{11}$$

In Eqn.(11), σ_m are the principal singular values of $\hat{\mathbf{T}}$ and $\mathbf{u}_m$, $\mathbf{v}_m$ are the corresponding left and right singular vectors respectively. The proposed parameter estimation method is based on finding the model parameters h_m, z_m, w_m $\forall m = 1, 2,.., M$ which provides the best least-squares fit to $\hat{\mathbf{T}}_s$ by minimising the Frobenius norm $\| \cdot \|_F$ in the following criterion function.

$$f_{cr}(\mathbf{Z}, \mathbf{H}, \mathbf{W}) = \|\hat{\mathbf{T}}_s - \mathbf{ZHW}^H\|_F. \tag{12}$$

This multidimensional optimisation problem is separable and it may be shown that the argument $\mathbf{Z}$ which minimises $f_{cr}(\cdot)$ is given by Eqn.(13).

$$\mathbf{Z} = \text{argmin} \ \|(\mathbf{I} - \mathbf{Z}(\mathbf{Z}^H\mathbf{Z})^{-1}\mathbf{Z}^H)\mathbf{U}_s\mathbf{\Sigma}_s\|_F. \tag{13}$$

By defining $\mathbf{Z}^+ = (\mathbf{Z}^H\mathbf{Z})^{-1}\mathbf{Z}^H$ as the Moore-Penrose pseudo inverse of $\mathbf{Z}$ and the projection matrix $\mathbf{P}_z = (\mathbf{I} - \mathbf{ZZ}^+)$, it is possible to re-write Eqn.(13) in terms of a trace $\text{Tr}\{\cdot\}$ operator.

$$\mathbf{Z} = \text{argmin Tr}\{\mathbf{P}_z \hat{\mathbf{T}}_s \hat{\mathbf{T}}_s^H\} \tag{14}$$

Assuming the number of modes M is known or can be estimated by a suitable method beforehand, solving for $\mathbf{Z}$ is more tractable when the projection matrix is re-parameterised as $\boldsymbol{P}_z = \mathbf{A}(\mathbf{A}^H\mathbf{A})^{-1}\mathbf{A}^H$ where the $L_t \times (L_t - M)$ Toeplitz matrix $\mathbf{A}$ is defined as;

$$\mathbf{A} = \begin{bmatrix} a_M & a_{M-1} & \cdots & a_0 & & \cdots & 0 \\ & \ddots & \ddots & & & \ddots & \\ 0 & & a_M & a_{M-1} & & \cdots & a_0 \end{bmatrix}^H \tag{15}$$

such that its elements a_0, a_1,..,a_M are the coefficients of the following characteristic polynomial [10].

$$a_0 z^M + a_1 z^{M-1} + \cdots + a_M = \prod_{m=1}^{M} (z - z_m). \tag{16}$$

As pointed out in [10], the coefficient vector $\mathbf{a} = [a_0 \cdots a_M]^T$ which minimises Eqn.(14) may be estimated analytically by sequentially solving two quadratic optimisation problems;

$$\begin{aligned} 1)\ \tilde{\mathbf{a}} &= \text{argmin Tr}\{\mathbf{A}\mathbf{A}^H \hat{\mathbf{T}}_s \hat{\mathbf{T}}_s^H\} \\ 2)\ \hat{\mathbf{a}} &= \text{argmin Tr}\{\mathbf{A}(\tilde{\mathbf{A}}^H\tilde{\mathbf{A}})^{-1}\mathbf{A}^H \hat{\mathbf{T}}_s \hat{\mathbf{T}}_s^H\} \end{aligned} \tag{17}$$

subject to a linear constraint $\mathbf{a}^H\mathbf{e} = 1$, where $\mathbf{e} = [1, 0, \cdots, 0]^T$ ensures a non-trivial solution. The roots corresponding to the estimated polynomial coefficient vector $\hat{\mathbf{a}}$ represent the temporal pole estimates $\hat{z}_m$ for $m = 1, 2,.., M$. Note that to estimate (Doppler) spread the roots ($\hat{z}_m$) are not required to be constrained to the unit circle as they are for the DOA estimation problem treated in [10]. Similarly, the argument $\mathbf{W}$ minimising the criterion function $f_{cr}(\cdot)$ is the solution of Eqn.(18) which can be solved in analogous fashion to estimate the spatial poles $\hat{w}_m$ for $m = 1, 2,.., M$. Eqn.(18).

$$\mathbf{W} = \text{argmin Tr}\{\mathbf{P}_w \hat{\mathbf{T}}_s^H \hat{\mathbf{T}}_s\}. \tag{18}$$

The number of modes $M = 5$ and their associated time-of-arrivals were determined by oblique sounding in [1], [2], a procedure for pairing the temporal and spatial poles associated with each mode is also provided [1], [2] in order to estimate the corresponding residues (mode powers) h_1,..,h_M using the least squares criterion. Table 1 lists the parameters estimated for the five modes using the channel scattering function data. Two modes, denoted by $1E_s(a)$ and $1E_s(b)$ in Table 1, were propagated by single-hop reflections from the mid-latitude sporadic-E layer, the remaining three modes were propagated by single-hop reflections from the F_2-layer. The modes denoted by $1F_2(o)$ and $1F_2(x)$ correspond to the ordinary and extraordinary magneto-ionic components in the high angle rays while the mode denoted by $1F_2$ represents the low angle ray. The ratio of the mode powers estimated from experiment 1 were held fixed and scaled in Table 1 so that the total simulated interference power was equal to the total received interference power of 30.2 dB.

Table 1. Interference simulation parameters estimated from the channel scattering function data. The five ionsopheric propagation modes were identified by an oblique sounder which also measures their group delay τ_m (see [1] for more information). The sampling interval or PRI $\Delta t = 1/f_p$ with $f_p = 5$ Hz and the spacing between adjacent subarray centres $\Delta d = 84$ m. Note that Darwin is approximately 22.0 degrees off the array boresight and that modes incident from higher elevation angles (with greater group delays) appear closer to boresight due to the coning effect.

Mode	Power, dB	τ_m, ms	Δf_m, Hz	θ_m, deg	$\alpha_m(\Delta t)$	$\beta_m(\Delta d)$
$1E_s(a)$	20.5	4.30	0.46	22.2	0.998	0.920
$1E_s(b)$	17.9	4.33	0.39	21.7	0.997	0.963
$1F_2$	28.6	4.77	0.44	20.7	0.994	0.988
$1F_2(o)$	21.8	4.92	0.47	20.5	0.997	0.986
$1F_2(x)$	7.9	5.13	0.53	19.9	0.994	0.953

Real background noise, recorded in the absence of interference, was substituted for the additive noise in the simulation.

4. Adaptive Beamforming Methods for OTH Radar

The optimal weight vector $\mathbf{w}_{opt}$ that maximises the SINR at the output of an adaptive beamformer may be derived according to the minimum variance distortionless response (MVDR) criterion in cases where the desired signal (of known spatial structure) is uncorrelated with a (wide-sense stationary) interference-plus-noise process. Under the above-mentioned assumptions, the optimal adaptive beamforming weight vector is given by;

$$\mathbf{w}_{opt} = \frac{\mathbf{R}_n^{-1}\mathbf{s}(\theta)}{\mathbf{s}^H(\theta)\mathbf{R}_n^{-1}\mathbf{s}(\theta)} \tag{19}$$

where $\mathbf{R}_n$ is the statistically expected interference-plus-noise spatial covariance matrix and $\mathbf{s}(\theta)$ is the array response vector or steering vector matched to the desired signal. In practice, $\mathbf{R}_n$ is unknown and often substituted by the sample covariance matrix $\hat{\mathbf{R}}_n$ in Eqn.(19) to yield the sample matrix inverse (SMI) estimate of the optimal weight vector $\hat{\mathbf{w}}_{opt}$ [12]. However, during normal operation, OTH radar systems cannot receive desired signals without simultaneously receiving clutter, the presence of which significantly complicates the estimation of $\mathbf{R}_n$. To circumvent this difficulty, a number of operational adaptive beamforming schemes may be proposed. Four different schemes are illustrated in Figure 1 and explained below.

The first approach (scheme 1) is to turn the transmitters off for a period of Δ_p PRI prior to each CPI and to estimate $\mathbf{R}_n$ from the interference-plus-noise snapshots received over this interval. The "optimal" weight vector, denoted by $\hat{\mathbf{w}}_1$, may then be formed using the SMI technique and frozen to beamform the active (with transmitters on) CPI of data which

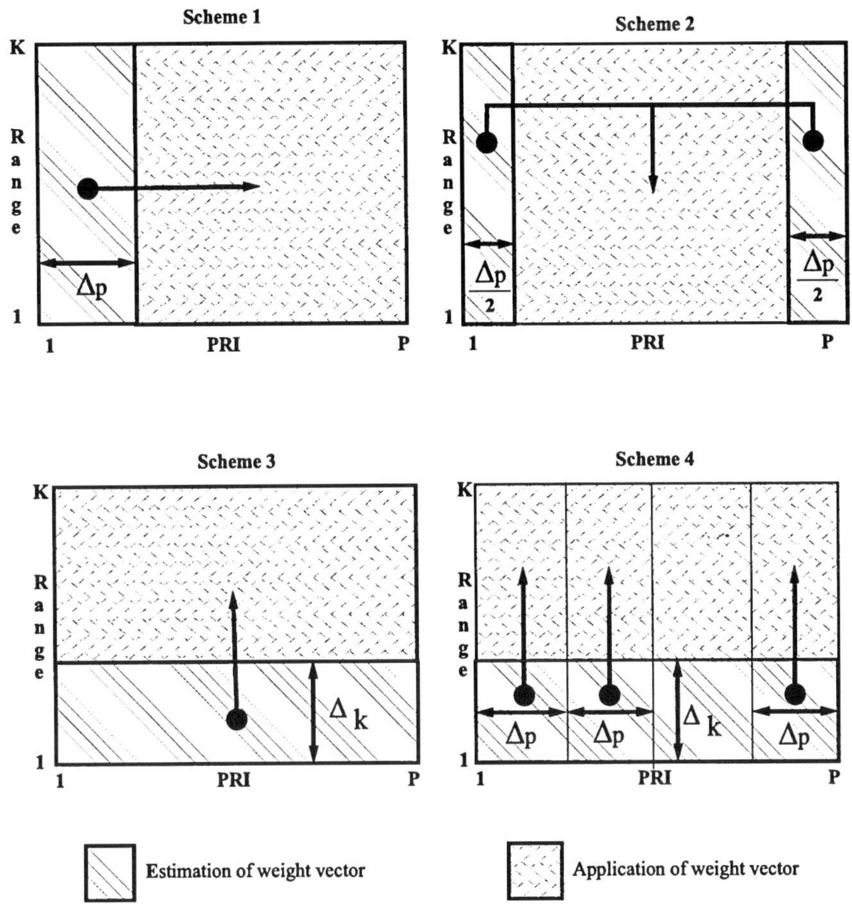

Figure 1. Diagrammatic illustration of the four adaptive beamforming schemes. Each diagram shows the dwell regions in which the recorded array snapshots are used for estimation of the adaptive weight vector and those where this weight vector is applied to beamform the data.

immediately follows the passively sampled data. In mathematical terms, the scalar beamformed output is given by $y_k^{[1]}(t) = \hat{\mathbf{w}}_1^H \mathbf{x}_k(t)$ for $t = \Delta_p + 1, \Delta_p + 2,.., P$ where $\hat{\mathbf{w}}_1$ is derived from the passively received interference-plus-noise process according to Eqn.(20).

$$\hat{\mathbf{w}}_1 = \frac{\hat{\mathbf{R}}_n^{-1}\mathbf{s}(\theta)}{\mathbf{s}^H(\theta)\hat{\mathbf{R}}_n^{-1}\mathbf{s}(\theta)}, \quad \hat{\mathbf{R}}_n = \frac{1}{K\Delta_p}\sum_{k=1}^{K}\sum_{t=1}^{\Delta_p} \mathbf{n}_k(t)\mathbf{n}_k^H(t). \tag{20}$$

A modification is to form the sample covariance matrix from the interference-plus-noise data received passively at both ends of the CPI (scheme 2) as any variation in the

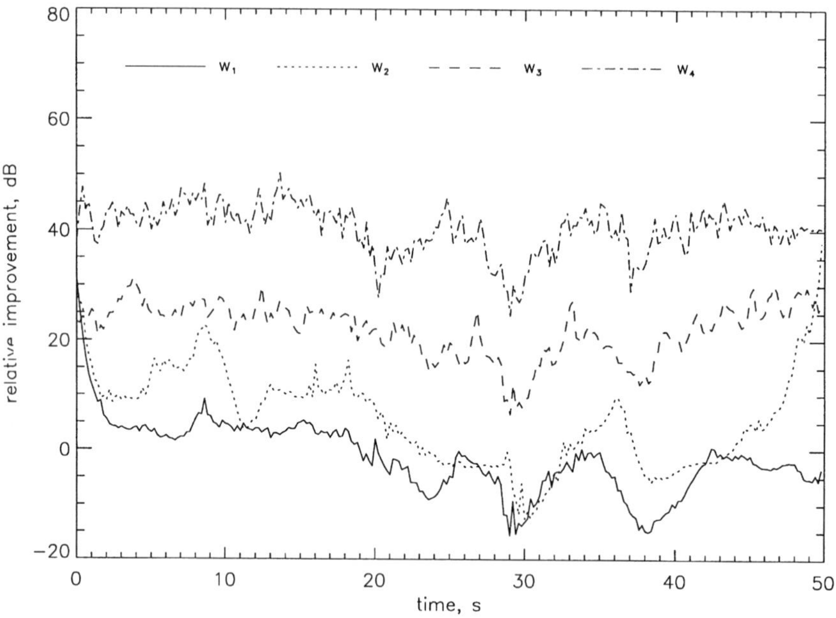

Figure 2. Quasi-instantaneous relative improvement in interference rejection performance obtained by four adaptive beamforming schemes as a function of PRI number over the CPI.

interference-plus-noise field over the CPI is better captured. In this case, the "optimal" weight vector $\hat{\mathbf{w}}_2$, formed according to Eqn.(21), is held fixed to beamform the active CPI which additionally contains clutter and potentially desired signals. (i.e., $y_k^{[2]}(t) = \hat{\mathbf{w}}_2^H \mathbf{x}_k(t)$ for $t = \Delta_p/2 + 1, \Delta_p + 2,.., P - \Delta_p/2$)

$$\hat{\mathbf{w}}_2 = \frac{\hat{\mathbf{R}}_n^{-1}\mathbf{s}(\theta)}{\mathbf{s}^H(\theta)\hat{\mathbf{R}}_n^{-1}\mathbf{s}(\theta)}, \quad \hat{\mathbf{R}}_n = \frac{\sum_{k=1}^{K}\sum_{t=1}^{\Delta_p/2}\mathbf{n}_k(t)\mathbf{n}_k^H(t) + \sum_{k=1}^{K}\sum_{t=P-\Delta_p/2+1}^{P}\mathbf{n}_k(t)\mathbf{n}_k^H(t)}{K\Delta_p}. \tag{21}$$

Another approach (scheme 3) is based on the observation that while interference and noise are present in all range cells, the clutter and desired signals may only be present over a certain range extent. In OTH radar applications, this can arise due to the appearance of a skip-zone between the receiver and region of the earth surface before which the ionosphere cannot effectively support propagation of the radar signal. Hence, the data received in the skip-zone range cells (say $k = 1, 2,.., \Delta_k < K$) is dominated by the interference and noise component and this natural phenomenon can be exploited to estimate $\mathbf{R}_n$ *within the CPI* rather than at its extremities. The "optimal" weight vector $\hat{\mathbf{w}}_3$, formed according to Eqn.(22), is then used to beamform the entire CPI in

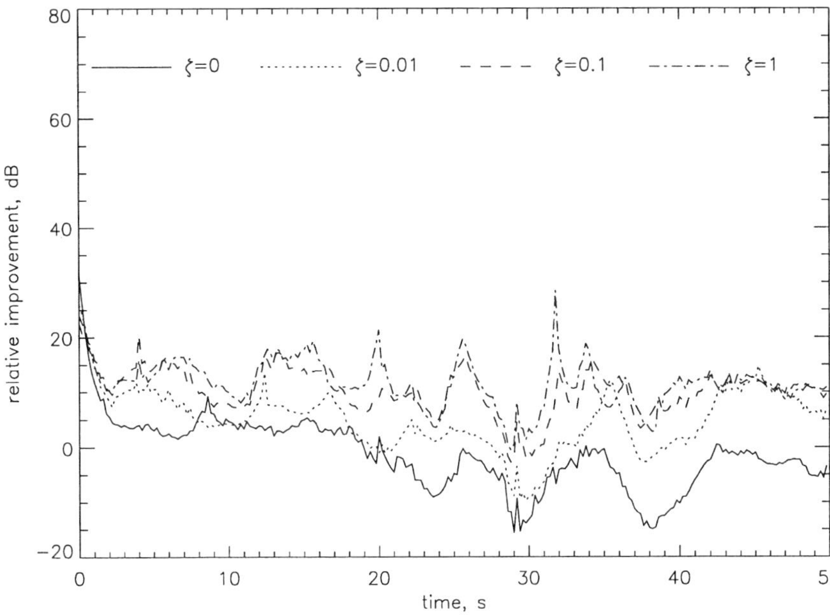

Figure 3. Quasi-instantaneous relative improvement in interference rejection performance obtained by the robust version of adaptive scheme 1 with different parameters $\zeta = 0, 0.01, 0.1, 1$.

the range cells beyond the skip-zone as these cells potentially contain desired signals. (i.e., $y_k^{[3]}(t) = \hat{\mathbf{w}}_3^H \mathbf{x}_k(t)$ for $k = \Delta_k + 1, \Delta_k + 2,.., K$)

$$\hat{\mathbf{w}}_3 = \frac{\hat{\mathbf{R}}_n^{-1}\mathbf{s}(\theta)}{\mathbf{s}^H(\theta)\hat{\mathbf{R}}_n^{-1}\mathbf{s}(\theta)}, \quad \hat{\mathbf{R}}_n = \frac{1}{P\Delta_k}\sum_{k=1}^{\Delta_k}\sum_{t=1}^{P} \mathbf{n}_k(t)\mathbf{n}_k^H(t). \tag{22}$$

When the spatial characteristics of interference-plus-noise field vary rapidly it may be necessary to re-adapt the beampattern over short time segments *within the CPI* (scheme 4). An apparently reasonable method is to form the slow-time (t) varying sample covariance matrix $\hat{\mathbf{R}}_n(t)$ using the skip-zone range cells in successive data frames of Δ_p PRI duration and to calculate the fluctuating weight vector $\hat{\mathbf{w}}_4(t)$ in accordance with the well known rule.

$$\hat{\mathbf{w}}_4(t) = \frac{\hat{\mathbf{R}}_n^{-1}(t)\mathbf{s}(\theta)}{\mathbf{s}^H(\theta)\hat{\mathbf{R}}_n^{-1}(t)\mathbf{s}(\theta)}, \quad \hat{\mathbf{R}}_n(t) = \frac{1}{\Delta_k\Delta_p}\sum_{k=1}^{\Delta_k}\sum_{t'=t}^{t+\Delta_p-1} \mathbf{n}_k(t')\mathbf{n}_k^H(t'). \tag{23}$$

As illustrated in Figure 1, this slow-time varying weight vector is applied to generate the beamformed output $y_k^{[4]}(t) = \hat{\mathbf{w}}_4^H(t)\mathbf{x}_k(t)$ in the range cells beyond the skip-zone $k = \Delta_k + 1$,

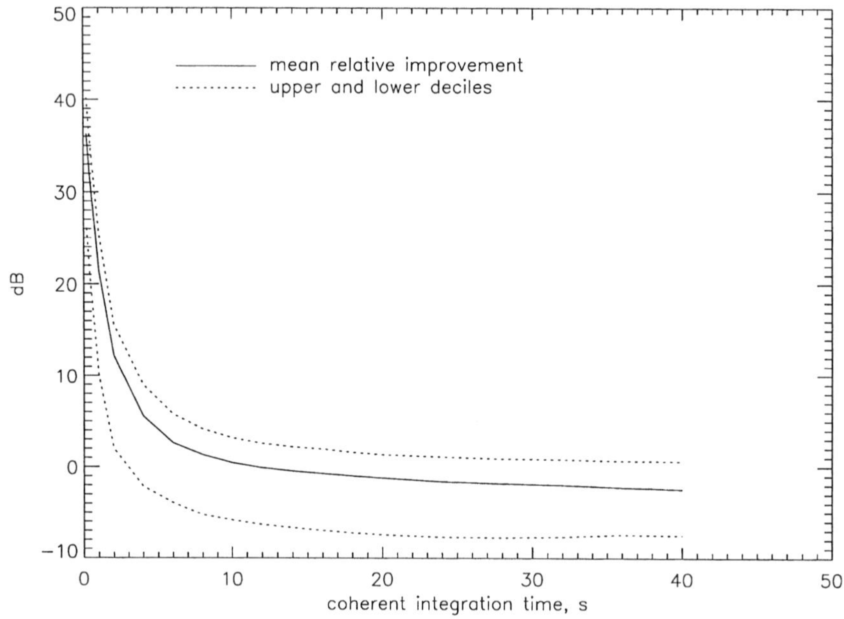

Figure 4. Mean and deciles of the improvement in interference rejection performance achieved by adaptive scheme 1 over the conventional beamformer as a function of CPI length.

$\Delta_k + 2,.., K$ within the current set of PRI $t, t+1,.., t+\Delta_p - 1$. While the re-estimation of weight vector from block to block is expected to enhance interference rejection, the problem encountered in practice is that the resulting changes in the beampattern within the CPI effectively destroy the temporal correlation properties of the clutter causing it to spread over the search space after Doppler processing and this can potentially obscure the presence of useful signals. Hence, the objective is to allow weight fluctuations within the CPI to ensure the interference is cancelled effectively but to prevent these fluctuations from broadening the Doppler spectrum of the clutter signals at the output of the time-varying beamformer. This problem has been addressed in [13] and will not be considered here as we are primarily concerned with the interference cancellation performance of this scheme.

A *fixed* weight vector method designed for robustness against interference with rapidly varying spatial characteristics or "jammer motion" was developed in [14], [15] and makes use of so-called "derivative snapshots" to widen the adaptive pattern nulls formed on the interference at a particular time in anticipation of interference nonstationary beyond the training interval. In the case of scheme 1, the robust adaptive beamformer which makes use of first derivatives only can be mathematically formulated as;

$$\tilde{\mathbf{w}}_1 = \frac{\tilde{\mathbf{R}}_n^{-1}\mathbf{s}(\theta)}{\mathbf{s}^H(\theta)\,\tilde{\mathbf{R}}_n^{-1}\mathbf{s}(\theta)}\,, \quad \tilde{\mathbf{R}}_n = \frac{1}{K\Delta_p}\sum_{k=1}^{K}\sum_{t=1}^{\Delta_p}\left[\,\mathbf{n}_k(t)\mathbf{n}_k^H(t) + \zeta\mathbf{B}\mathbf{n}_k(t)\mathbf{n}_k^H(t)\mathbf{B}\,\right] \quad (24)$$

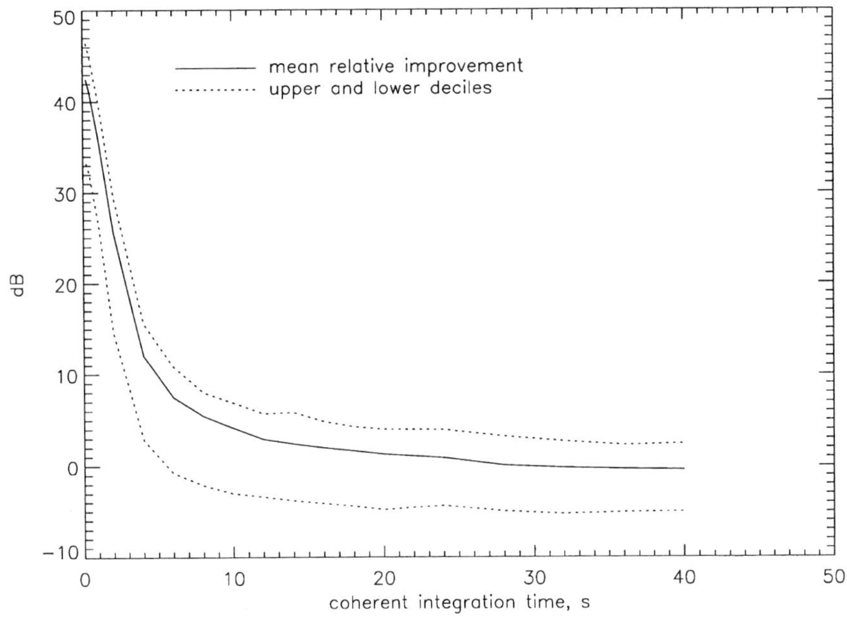

Figure 5. Mean and deciles of the improvement in interference rejection performance achieved by adaptive scheme 2 over the conventional beamformer as a function of CPI length.

where the parameter ζ controls the relative contribution of the "derivative data" and **B** is an $N \times N$ diagonal matrix of known sensor coordinates [14], [15].

5. Experimental Results

The experimental results are presented in two sections. The first section quantifies the interference cancellation improvement gained by adaptive relative to conventional beamforming. As both adaptive and conventional beamformers are normalised to provide the same (unit) response to ideal desired signals, the measured improvements correspond to SINR improvements in the case of ideal target signals. The second section compares the experimentally measured performance improvements with those derived from a multi-sensor signal processing model that simulates the received interference in a statistical sense.

5.1. Interference Cancellation Performance

To illustrate the HF interference cancellation performance of adaptive beamforming schemes $l = 1, 2, 3, 4$ relative to a conventional beamformer $\mathbf{v}(\theta)$, the quasi-instantaneous

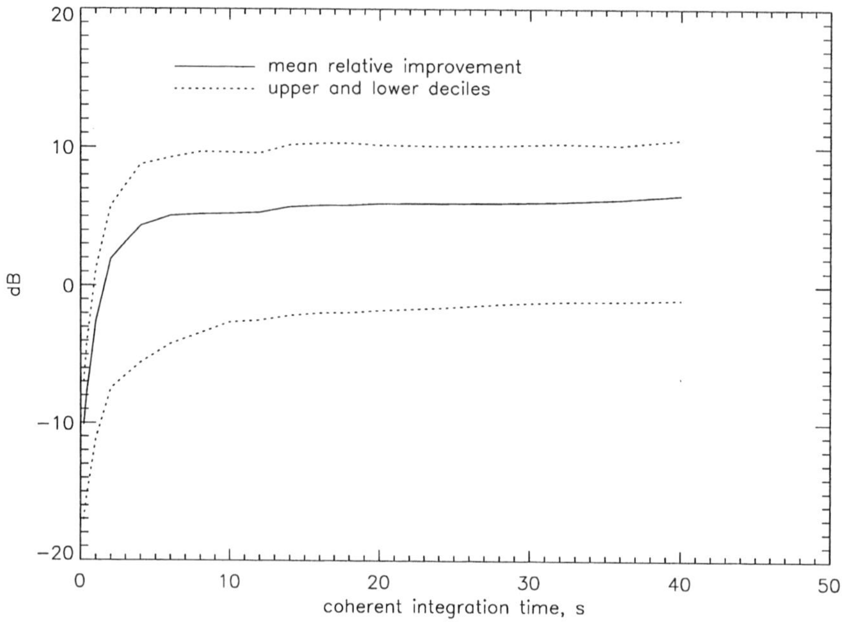

Figure 6. Mean and deciles of the improvement in interference rejection performance achieved by the robust version of adaptive scheme 1 (using $\zeta = 1$) over the standard SMI version of adaptive scheme 1 ($\zeta = 0$).

relative interference cancellation improvement $\hat{q}_{inst}^{(l)}(t)$ was calculated as;

$$\hat{q}_{inst}^{(l)}(t) = \frac{\mathbf{v}^H(\theta)\,\hat{\mathbf{R}}_n(t)\mathbf{v}(\theta)}{\hat{\mathbf{w}}_l^H\,\hat{\mathbf{R}}_n(t)\hat{\mathbf{w}}_l}\ , \quad \hat{\mathbf{R}}_n(t) = \frac{1}{K'}\sum_{k=\Delta_k+1}^{\Delta_k+K'} \mathbf{n}_k(t)\mathbf{n}_k^H(t) \tag{25}$$

where $\mathbf{v}(\theta) = \mathbf{Ts}(\theta)/\mathbf{s}^H(\theta)\mathbf{Ts}(\theta)$ ($\mathbf{T}$ is a diagonal matrix containing the Hamming window taper as its elements), $\Delta_k = 16$, $K' = 16$ and $t = \Delta_p + 1, \Delta_p + 2, \ldots, P = 256$. Note that $\hat{\mathbf{w}}_1$ and $\hat{\mathbf{w}}_2$ were formed according to Eqn.(20) and Eqn.(21) respectively using $K = 32$ and $\Delta_p = 6$ while $\hat{\mathbf{w}}_3$ and $\hat{\mathbf{w}}_4$ were formed according to Eqn.(22) and Eqn.(23) respectively with $\boldsymbol{\Delta}_k = 16$ and $\Delta_p = 4$. The first 10 ranges of the 42 recorded were not used as a calibration signal that would have biased the results was known to be present in some of these range cells, hence $K = 32$ rather than $K = 42$.

Curves 1 to 4 in Figure 2 show the relative improvement in interference rejection $\hat{q}_{inst}^{(l)}(t)$ for schemes $l = 1, 2, 3, 4$ as a function of slow-time (t) for a 50 second CPI of HF interference data. It is evident from Curve 1 in Figure 2 that $\hat{\mathbf{w}}_1$ has an initial effectiveness which is 30 dB better than the conventional beamformer, but this adaptive filter “ages” rapidly and the improvement is diminished to almost 0 dB in a matter of 2 seconds.

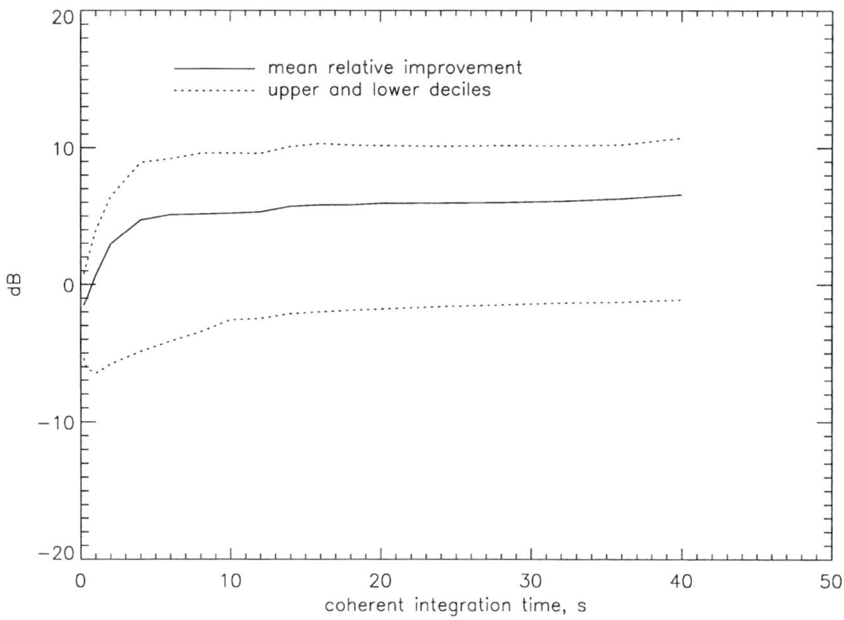

Figure 7. Mean and deciles of the improvement in interference rejection performance achieved by the robust version of adaptive scheme 1 with CPI-dependent values of ζ over the standard SMI version of adaptive scheme 1 ($\zeta = 0$).

Although the relative improvement offered by $\hat{\mathbf{w}}_2$ is superior to that of $\hat{\mathbf{w}}_1$ throughout almost all of the CPI, and 30–40 dB better than $\mathbf{v}(\theta)$ at the extremities near where the weights $\hat{\mathbf{w}}_2$ are trained, the improvement also fades quickly towards the central portion of the CPI since this adaptive filter has no "knowledge" of the intermediate variations in the interference spatial structure.

Unlike the first two schemes, adaptive scheme 3 uses an estimate of the interference spatial structure which is averaged over the *whole CPI*. The relative improvement offered by $\hat{\mathbf{w}}_3$ is 15–20 dB higher than that observed for scheme 2 over the great majority of the CPI. By far the best performance is obtained by re-adapting the weight vector $\hat{\mathbf{w}}_4(t)$ to match the time-evolution of the interference-plus-noise spatial properties over short time segments within the CPI. This leads to a fairly consistent improvement of about 40 dB relative to the conventional beamformer.

Using the same data, Figure 3 (in the same format as Figure 2) shows the results corresponding to the robust version of adaptive scheme 1 in Eqn.(24) for different values of the parameter ζ. Note that the curve labelled $\zeta = 0$ corresponds to the standard SMI adaptive filter $\tilde{\mathbf{w}}_1 = \hat{\mathbf{w}}_1$, so it is a replica of the first curve in Figure 1 and is included here for comparison only. The analysis demonstrates that the robust approach improves performance by 5–10 dB relative to the standard SMI technique after approximately 1 second has elapsed but a 5–10 dB performance loss is evident before 1 second has

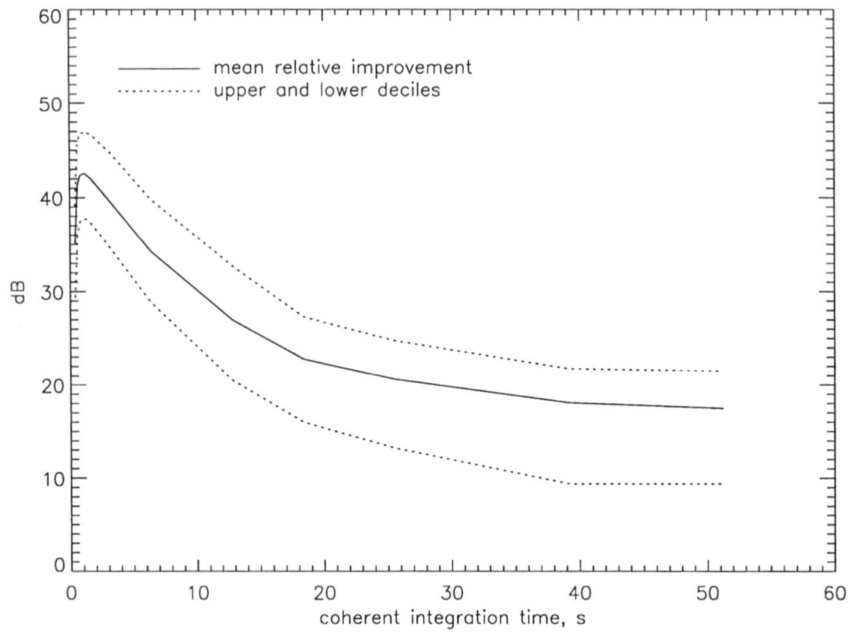

Figure 8. Mean and deciles of the improvement in interference rejection performance achieved by adaptive scheme 3 over the conventional beamformer as a function of CPI length.

elapsed. This occurs because the recorded interference may be considered effectively "stationary" over a time interval less than 1 second in which case better performance results from deeper rather than wider "nulls" in the adaptive beampattern.

However, this is only an illustration of performance for particular CPI. Of more interest is the expected performance of the schemes and its variability over different CPI sets and beam steer directions. To estimate the distributions of performance arising for a given CPI length, a total of 10 mutually orthogonal beams were processed in each of 40 different CPI and the relative improvement calculated to yield 400 samples at each investigated CPI length. Figures 4 and 5 show the mean and deciles of the relative improvement distributions obtained for schemes 1 and 2 respectively, both schemes are on average not much better than conventional beamforming when the CPI length is increased past 5–10 seconds.

The performance of scheme 1 with robustness included ($\zeta > 0$) is statistically compared against that of the standard SMI version ($\zeta = 0$) as a function of CPI length in Figures 6 and 7. When the parameter $\zeta = 1$ is held fixed for all CPI lengths (Figure 6), the tradeoff between "stationary and nonstationary" performance mentioned in [14], [15] is evidenced by a mean performance loss of 5–10 dB for CPI lengths less than 1 second and a relative performance gain of 5–10 dB for CPI lengths greater than about 5 seconds. Figure 7 shows the results when the parameter ζ is made CPI-dependent, more specifically, it is set to the reciprocal of the mean improvement of adaptive scheme 1 relative to the conventional beamformer depicted in Figure 4. As shown in Figure 7, this choice greatly reduces

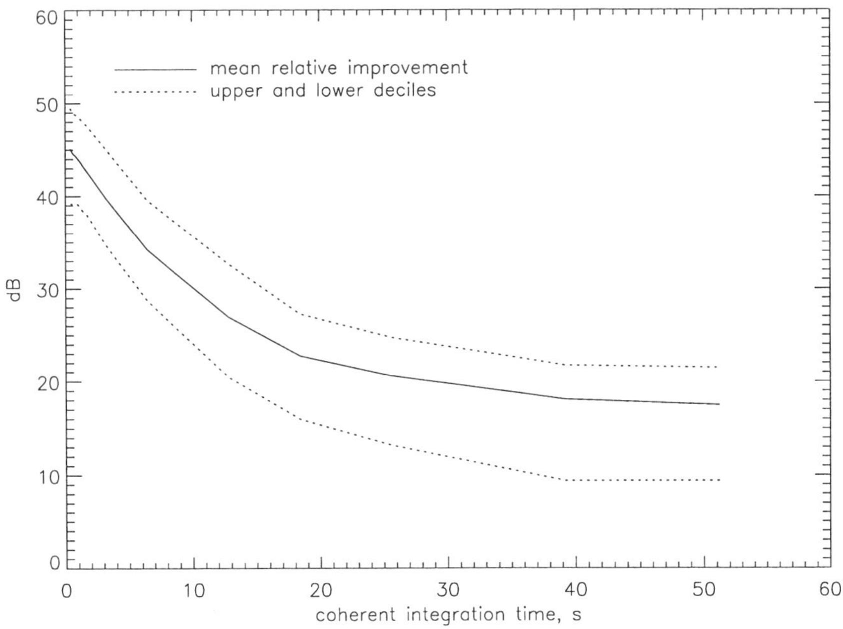

Figure 9. Mean and deciles of the improvement in interference rejection performance achieved by adaptive scheme 3 *after diagonal loading* relative to the conventional beamformer as a function of CPI length.

the performance losses observed for the robust approach relative to the standard SMI technique at CPI lengths less than 1 second (Figure 6) while retaining the relative performance benefits obtained at greater CPI lengths.

Figure 8 illustrates the mean and deciles of the relative improvement distributions resulting for adaptive scheme 3. The maximum mean relative improvement gained by this scheme is about 43 dB and occurs for a CPI length slightly over 1 second. A degradation of up to 8 dB is observed for smaller CPI lengths due to the reduced number of samples available for spatial covariance matrix estimation when the CPI is small. For CPI lengths longer than 1 second, the degradation is caused by the increased consumption of adaptive degrees of freedom as a result of the temporal variations in the interference spatial structure within the CPI. Diagonal loading [16] can be employed to improve the performance at very small CPI lengths if required, as illustrated by Figure 9 where a loading factor of -20 dB is applied (i.e., $0.01 \times \mathbf{I}$ is added to the sample spatial covariance matrix used to estimate the weight vector). This relatively simple operation improves the relative interference cancellation effectiveness from 35 dB to 45 dB at very small CPI lengths, a remarkable 10 dB increase.

As each CPI length in Figures 8 and 9 may be considered as a short data segment or sub-CPI of a longer (total) CPI, the results shown in these figures also provide information regarding the rate at which the adaptive beamformer in scheme 4 should be re-adjusted to maximise interference cancellation performance. For example, if the adaptive weight

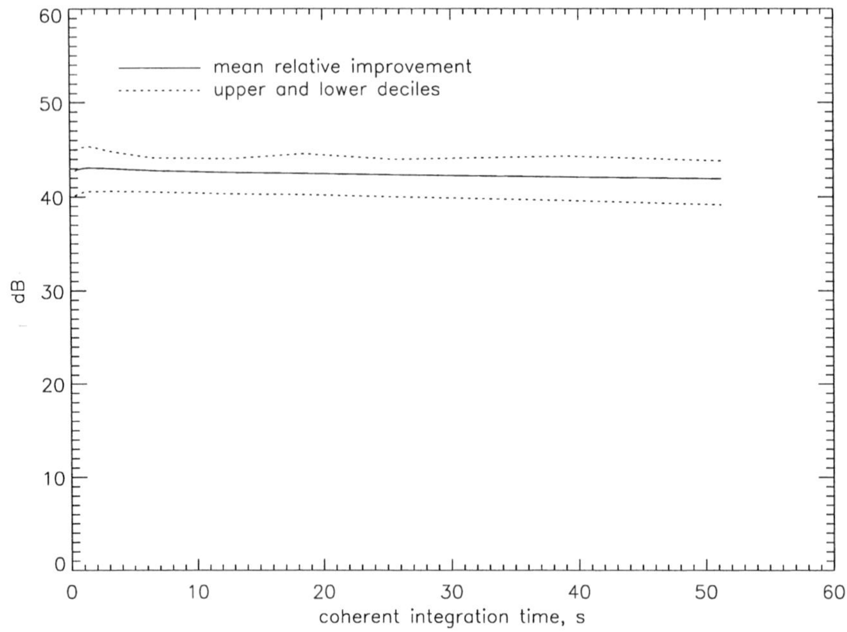

Figure 10. Relative improvement achieved by adaptive scheme 3 (after diagonal loading) for simulated HF interference with temporal distortions but in the *absence* of spatial distortions.

vector were re-adapted every second without diagonal loading then the mean relative improvement of about 43 dB indicated in Figure 8 is realisable for longer CPI lengths of say 30 seconds duration whereas adaptive scheme 3 only manages a mean improvement of about 20 dB.

5.2. *Simulated Interference Data*

Figure 10 shows the mean and deciles of the relative improvement distribution which results when the interference modes are simulated according to the previously described model but with *no spatial distortions* and processed in identical manner to that which produced Figure 9. No spatial distortions refers to spatial poles on the unit circle, this has been simulated by setting the *magnitude* of the spatial pole estimated for each mode in Table 1 to unity. In other words, the interference modes are modelled as plane waves with Doppler shifted and Doppler spread waveforms. Obviously, the traditional plane wave interference model does not provide an accurate representation of the distributions derived from experimental data processing and the substantial difference between Figure 9 and Figure 10 illustrates the need to interpret results obtained from such a model with caution.

Figure 11 shows the relative interference cancellation improvement distribution which results when the spatial distortions described by the first order AR model parameters in Table 1 are introduced. The mean relative improvement ranges from 46 dB at small CPI

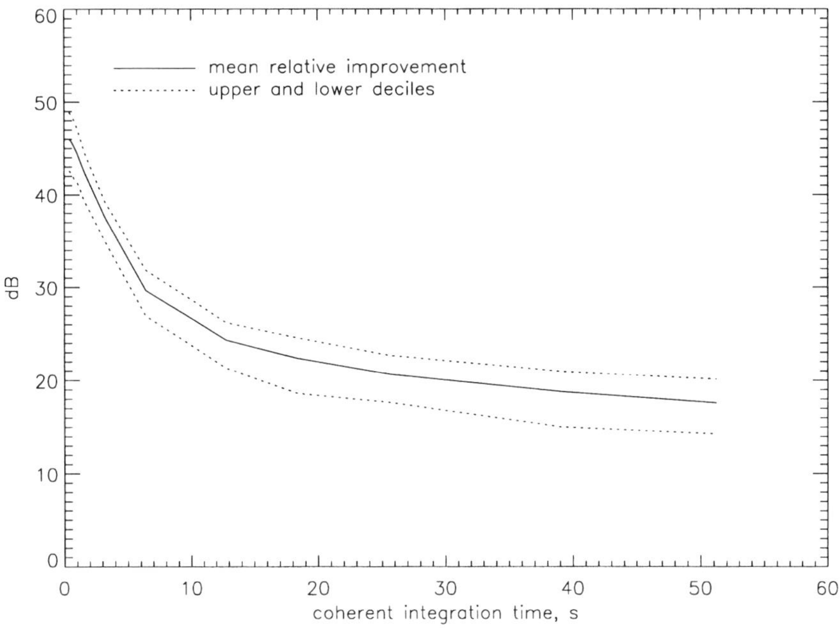

Figure 11. Relative improvement achieved by adaptive scheme 3 (after diagonal loading) for simulated HF interference with the inclusion of space-time distortions.

lengths (less than one second) to 18 dB at CPI lengths of about 50 seconds. This compares favourably with the experimentally observed reduction from 45 dB to 17 dB over the same CPI range. The precise shapes of the curves in Figure 9 are not exactly the same as those in Figure 11 but the estimated model parameters provide a fairly accurate representation of adaptive beamformer performance in the HF environment. A similar experiment that favourably compares the simulated and experimental performance of another adaptive beamforming scheme (scheme 1) over the same propagation path can be found in [17].

6. Summary and Conclusion

This paper has experimentally evaluated the interference cancellation performance of various adaptive beamforming schemes applicable to OTH radar systems. A statistical analysis of performance using real high frequency (HF) interference propagated by the ionosphere revealed that adaptive beamforming schemes based on the use of training data taken before and/or after the coherent processing interval (CPI) were not effective in the HF environment for CPI lengths greater than 1 second. The severe degradation in mean interference cancellation performance observed relative to the conventional beamformer with increasing CPI length is not consistent with the idea of fixed plane wave interference signals, and the large discrepancy between the experimental results and those expected for these relatively simple (but often used) theoretical models is due to intra-CPI variations in the spatial properties of the received interference being unaccounted for.

The robust implementation of the first adaptive beamforming scheme considered (scheme 1) significantly improved the interference cancellation performance at CPI lengths greater than 1 second relative to the standard SMI version, the mean relative improvement ranging between 5–10 dB for practical OTH radar CPI. The most promising adaptive beamforming scheme re-adapted to the changing spatial structure of the received interference *within the CPI*, and was on average up to 40 dB more effective than conventional beamforming in terms of interference cancellation. This effectiveness was critically dependent upon the update rate and it was demonstrated that a more sophisticated interference model, which included angle and Doppler spread terms estimated directly from real data, was able to accurately forecast the mean improvement in interference cancellation achieved by this scheme relative to a conventional beamformer as a function of the re-adaptation interval.

This experiment involved single-hop propagation on a mid-latitude ionsopheric path, and while the theoretical model is in good agreement with the experimental measurements made on this path, it would be of interest to determine whether such a model is suitable for describing HF channels involving multiple ionospheric reflections and (perhaps over shorter time-scales) propagation via the typically perturbed equatorial and auroral ionospheric regions. Another area which deserves further research is the generalisation of the multi-sensor HF channel model to include wave polarisation for general antenna systems and the frequency dispersive effects of the ionosphere for wideband HF applications.

Note

1. Correspondence author.

References

1. G.A. Fabrizio, "*Space-Time Characterisation and Adaptive Processing of Ionospherically-Propagated HF Signals*," Ph.D. thesis, Australia: Adelaide University, July 2000.
2. G.A. Fabrizio, D.A. Gray, and M.D. Turley, "Parametric Localisation of Space-Time Distributed Sources," *IEEE International Conference on Acoustics, Speech and Signal Processing (ICASSP 2000)*, Istanbul, Turkey, June 2000, pp. 3097–3100.
3. T.W. Washburn and L.E. Sweeney, "An On-Line Adaptive Beamforming Capability for HF Backscatter Radar," *IEEE Transactions on Antennas and Propagation*, vol. AP-24, no. 5, September 1976, pp. 721–732.
4. L.J. Griffiths, "Time-Domain Adaptive Beamforming of HF Backscatter Radar Signals," *IEEE Transactions on Antennas and Propagation*, vol. AP-24, no. 5, September 1976, pp. 707–720.
5. G.A. Fabrizio, Y.I. Abramovich, S.J. Anderson, D.A. Gray, and M.D. Turley, "Adaptive Cancellation of Nonstationary Interference in HF Antenna Arrays," *IEEE Proceedings - Radar, Sonar and Navigation: Special Issue on Antenna Array Processing Techniques*, vol. 145, no. 1, February 1998, pp. 19–26.
6. Y.I. Abramovich, A.Y. Gorokhov, and C. Demeure, "Experimental Verification of a General Multivariate Propagation Model for Ionospheric HF Signals," *Proc. EUSIPCO-96*, Trieste, Italy, 1996.

7. C.C. Watterson, J.R. Juroshek, and W.D. Bensema, "Experimental Confirmation of an HF Channel Model," *IEEE Transactions on Communication*, vol. 18, no. 6, 1970, pp. 792–803.
8. A. Paulraj and T. Kailath, "Direction of Arrival Estimation by Eigenstructure Methods with Imperfect Spatial Coherence of Wavefronts," *J. Acoust. Soc. Am*, vol. 83, March 1988, pp. 1034–1040.
9. J. Ringelstein, A. Gershman, and J. Bohme, "Sensor Array Processing for Random Inhomogeneous Media," *SPIE'99, Advanced Signal Processing: Algorithms, Architectures and Implementations IX*, vol. 3807, Denver, Colorado, 1999, pp. 267–276.
10. P. Stoica and K.C. Sharman, "Novel Eigenanalysis Method for Direction Estimation," *IEEE Proceedings, Pt. F*, vol. 137, no. 1, 1990, pp. 19–26.
11. J. Li, P. Stoica, and D. Zheng, "One-Dimensional MODE Algorithm for Two Dimensional Frequency Estimation," *Multidimensional Systems and Signal Processing*, vol. 8, 1997, pp. 449–468.
12. J.D. Mallet, I.S. Reed, and L.E. Brennan, "Rapid Convergence Rate in Adaptive Arrays," *IEEE Transactions on Aerospace and Electronic Systems*, vol. 10, no. 6, 1974, pp. 853–863.
13. Y.I. Abramovich, A.Y. Gorokhov, V.N. Mikhaylyuvov, and I.P. Malyavin, "Exterior Noise Adaptive Rejection for OTH Radar Implementations," *IEEE International Conference on Acoustics, Speech and Signal Processing (ICASSP 94)*, Adelaide, Australia, 1994, pp. 105–107.
14. A.B. Gershman, U. Nickel, and J.F. Bohme, "Adaptive Beamforming Algorithms with Robustness Against Jammer Motion," *IEEE Transactions on Signal Processing*, vol. 45, no. 7, July 1997, pp. 1878–1885.
15. A.B. Gershman, E. Nemeth, and J.F. Bohme, "Experimental Performance of Adaptive Beamforming in a Sonar Environment with a Towed Array and Moving Interference Sources," *IEEE Transactions on Signal Processing*, vol. 48, no. 1, January 2000, pp. 246–250.
16. B.D. Carlson, "Covariance Matrix Estimation Errors and Diagonal Loading in Adaptive Arrays," *IEEE Transactions on Aerospace and Electronic Systems*, AES-24, vol. 4, July 1988, pp. 397–401.
17. G.A. Fabrizio, D.A. Gray, M.D. Turley, and S.J. Anderson, "Modelling the Spatial Characteristics of Ionospherically Propagated HF Signals," *Proceedings of the International Radar Symposium*, Munich, Germany, September 1998, pp. 1187–1197.

Multidimensional Systems and Signal Processing, 14, 265–279, 2003

Contributing Authors

Stanley C. Ahalt received his BSEE and MSEE degrees from the Virginia Polytechnic Institute and State University in 1978 and 1980 respectively. He obtained his Ph.D. in Electrical Engineering from Clemson University in 1986. Since 1987, he has been with the department of Electrical Engineering where he is currently a Professor in the Department of Electrical Engineering. During 1980 and 1981, Dr. Ahalt worked at Bell Telephone Laboratories where he developed industrial data products. Dr. Ahalt's research interests include high performance computing, pattern recognition and data compression, with applications to automatic target recognition, image compression, and video annotation. He has published over 90 archival journal papers, conference papers, and book chapters in these areas. Dr. Ahalt is a member of the International Neural Network Society and the Institute of Electrical and Electronics Engineers–Circuits and Systems, and Signal Processing societies. During 1996 he was on sabbatical leave from OSU as a visiting professor at the Universidad Polytecnica de Madrid. He is a former editor of the IEEE Transactions on Neural Networks, and he was the recipient of the 1997 OSU Lumley Research Award and the 2000 OSU College of Engineering Research Award.

Guoan Bi received a B.Sc. degree in Radio communications, Dalian University of Technology, PRC, 1982, M.Sc. degree in Telecommunication Systems and Ph.D. degree in Electronics Systems, Essex University, UK, 1985 and 1988, respectively. Since 1991, he has been with the school of Electrical and Electronic Engineering, Nanyang Technological University, Singapore. His current research interests include DSP algorithms and hardware structures and digital signal processing for communications.

Martin Biancalana received the B.S. degree from the University of Illinois in 1969 and the MSEE degree from the Air Force Institute of Technology in 1978. He is Research Program Manager for Veridian Systems Ann Arbor Research and Development Center. In the Air Force, he worked on the PAVE MOVER advanced development radar and managed the TACIT BLUE stealth research project. For ERIM, he worked on foliage penetrating radar and managed the Gold Pan '93 project. For Veridian, he has worked on applications of imaging radars to moving targets since 1995. He is an active member of the IEEE Engineering Management Society.

Faio Bovenga received the Laurea Degree from the Dipartimento di Fisica, University of Bari, Italy, in 1997. He was at Consiglio Nazionale delle Ricerche – IESI (Istituto per l'Elaborazione dei Segnali e delle Immagini) for one year as grant Researcher and since 2000 he is at University of Bari as a contract Researcher. His main research interests concern advanced processing techniques for SAR imaging ad radar interferometry.

Alejandro E. Brito received the B.S. degree in Computer Systems Engineering (with Honors) from the Universidad de las Americas, Puebla, Mexico in 1989. He received the M.S. and Ph.D. degrees in Computer Engineering from the University of Texas at El Paso, in 1994 and 2001, respectively.

He is currently a member of the technical staff in the Image Processing and Advanced Computer Technology Group at XEROX Corporation.

His research interests include superresolution in SAR imaging, signal recovery and reconstruction, optimal signal representation, and high performance computer architectures and ASIC design for signal and image processing applications.

Sergio D. Cabrera was born in Cananea, Sonora, Mexico. He obtained the B.S. and M.S. degrees in Electrical Engineering from MIT and the University of Arizona in 1977 and 1979 respectively. He obtained the Ph.D. degree in Electrical Engineering from Rice University in 1985 specializing in the area of Digital Signal Processing. From 1985–87 he worked at Anadrill/Schlumberger as a Senior Engineer. From 1987–1991 he was Assistant Professor of Electrical and Computer Engineering at the Pennsylvania State University. He joined the University of Texas at El Paso in January 1992 where he is currently Associate Professor of Electrical and Computer Engineering and serves as the Graduate Advisor. He is also the current holder of the Texas Instruments Foundation Professorship in Digital Signal Processing.

From 1995–2001 he was part of the Army Research Office sponsored Center for Imaging Science. Summer faculty fellowships participations include: the Jet Propulsion Lab. (1991), the NRaD Navy Lab. in San Diego (1992), Sandia National Labs. (1994), and the Army Research Lab. at White Sands Missile Range (2001). His research interests are in Signal and Image Processing including: multirate and time-frequency signal processing; digital image compression; super-resolution for SAR imaging and telecommunications applications; and applied computer vision. Throughout his career, he has co-authored nearly 100 publications, supervised 5 doctoral dissertations and more than 15 masters theses. He has received extramural funding for research and educational projects from NASA, the Army Research Office, the Office of Naval Research, the Texas Advanced Technology Prog., the National Science Foundation (NSF), and the Texas Instruments Foundation.

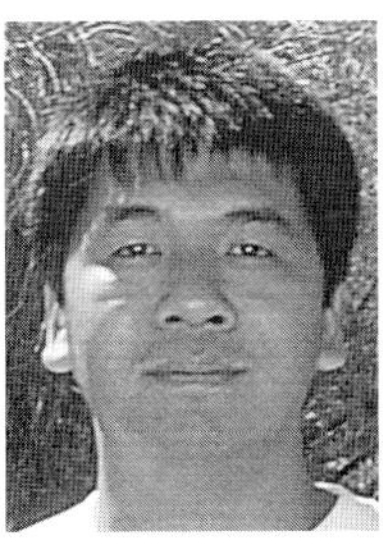

Shiu Hung Chan obtained his B.S. and M.S. degrees in Electrical Engineering from The University of Texas at El Paso in 1997 and 2000, respectively. His research was focused on the study of the effects of signal extrapolation in SAR imaging. His research interests include signal processing, communications and operating systems. Currently, he is working for Ericsson Wireless Communications on the CDMA2000 infra-structure design and development.

Victor C. Chen received his Ph.D. degree in electrical engineering from Case Western Reserve University. Since 1990 he has been with Radar Division, the U.S. Naval Research Laboratory in Washington DC. His research interests include radar imaging, moving target detection, and time-frequency applications to radar. He has more than 90 publications in books, journals and proceedings including a recent book: *Time-Frequency Transforms for Radar Imaging and Signal Analysis* (V. C. Chen and Hao Ling), Artech House, Boston, MA, January 2002.

Rui J.P. de Figueiredo holds the B.S. and M.S. degrees in electrical engineering from MIT and the Ph.D. degree in Applied Mathematics from Harvard University. He served on the faculty of Rice University, Houston, Texas, for several years until he moved in 1991 to the University of California, Irvine, where he holds the position of Professor of Electrical and Computer Engineering and of Mathematics. He is also a co-founder and member of Board of Directors of Neural Computing Systems, LLC.

Professor de Figueiredo has authored or co-authored several books and book chapters and more than 350 papers published in archival journals and reviewed conference proceedings. He has served IEEE (Institute of Electrical and Electronic Engineers) in various positions including that of the President of the IEEE Circuits and Systems (CAS) Society in 1998 and a Distinguished Lecturer of the Society. He has also served on several national and international committees and panels and editorial boards of journals.

For all these contributions Dr. de Figueiredo has received several awards including the IEEE Fellow Award, the IEEE CAS Society Technical Achievement Award (1994), the IEEE CAS Society Golden Jubilee Medal (1999), the Third Millennium Medal (2000), the IEEE Neural Networks Council Best Paper Award (2000), and the IEEE CAS Society Mac E. Van Valkenburg Society Award (2002).

Bradley S. Denney is the Chief Technologist of Neural Computing Systems, LLC. He received his B.S. degree (summa cum laude) at Polytechnic University, New York, his M.S. degree at Rice University, Houston, and his Ph.D. at the University of California, Irvine in Electrical and Computer Engineering. He was formerly an employee of McDonnell Douglas Aerospace where he researched autonomous space-based robot vision and teleoperated planetary exploration vehicles.

Dr. Denney served on the 1999 and 2000 award committee for the IEEE Dennis J. Picard Medal for Radar Technologies. His research interests include hyperspectral image analysis, medical imaging, robot vision, target recognition, HRR radar and SAR modeling and prediction, business intelligence and data mining, and automated network problem detection and declaration.

Michael D. DeVore received B.S. degrees in Electrical Engineering, Computer Engineering, and Mathematics from the University of Missouri at Columbia in 1991 and an M.S. degree in Electrical Engineering in 1993. In 2001 he received the DSc degree at Washington University in St. Louis doing research in the area of automatic target recognition from synthetic aperture radar data. Prior to attending Washington University, he led the Operations Support and System Integrity and Integration teams Amdocs, Inc. in St. Louis. He is currently an Assistant Professor in the Systems and Information Engineering department of the University of Virginia. His research interests include target recognition, statistical model assessment, model representation and storage, and performance of inference systems subject to complexity constraints.

Xun Du received his MSEE degree from The Ohio State University in 1997 and his BSEEE from Shanghai Jiaotong University, China, in 1994. He was an Assistant Engineer at the China Telecom from 1994 to 1996. Mr. Du is currently a Ph.D. Candidate in the Department Electrical Engineering at The Ohio State University. His current research interests include image and video compression, image understanding, and pattern recognition.

D. David Ensley received the Bachelor of Electrical Engineering degree in 1989 and the Master of Science degree in 1992, both from Auburn University. From 1992 to 1998 he researched computer intelligence and automatic target recognition algorithms with the U.S. Air Force Research Laboratory. Currently with the U.S. Air Force Special Operations Forces System Program Office, he manages software and avionics projects on the most technologically advanced helicopters in the world.

Giuseppe A. Fabrizio received his electrical and electronic B.E. degree from Adelaide University (1992) where he also completed his Ph.D. in the field of high frequency signal modelling and adaptive signal processing for over-the-horizon radar (2000). He is currently working as a research scientist in the radar signal processing group of the Intelligence, Surveillance, and Reconnaissance Division at DSTO Edinburgh (Australia), where he applies adaptive signal processing techniques to enhance the signal detection and parameter estimation performance of sky-wave and surface-wave over-the-horizon radar systems.

no photo
available

John W. Fisher III

No bio available.

Douglas A. Gray holds the Chair of Sensor Signal Processing at Adelaide University, South Australia. He is also Deputy Director of the Cooperative Research Centre for Sensor Signal and Information Processing, leading the Sensor Signal Processing Program. He received the B.Sc. (Hons) and Ph.D. degrees in mathematical physics from the University of Adelaide, South Australia, in 1969 and 1973 respectively. From 1973 to 1993 he was with the Australian Defence Science and Technology Organisation applying signal processing to sonar and electronic warfare. From 1977 to 1979 he was visiting a scientist at the Royal Aircraft Establishment, U.K. and in 1985 was a visiting fellow at the Dept of Systems Engineering of the Australian National University. His research interests are in the application of signal processing to sonar, radar, GPS and electronic warfare, particularly in adaptive processess, beamforming, signal sorting and classification techniques, data fusion and estimation.

Andrew J. Kim received the degree of Bach. of Elec. Eng. from the Georgia Institute of Technology, Atlanta, GA in 1995. He received the degrees of S.M. and Ph.D. in electrical engineering from the Massachusetts Institute of Technology (MIT) in 1997 and 2001 while a research assistant in the Stochastic Systems Group in the Laboratory for Information and Decision Systems at MIT. In 2001, Dr. Kim joined the research staff of Quellan, Inc. in Atlanta, GA where he develops signal processing algorithms for multi-gigabit communications systems. Dr. Kim is a member of Sigma Xi, Eta Kappa Nu and Tau Beta Pi. His research interest is in the field of stochastic signal processing with particular emphasis on SAR image processing, multiscale processing, and high-speed communications.

Junfei Li was born in Wuhan, China in 1971. He received the B.S. and M.S. degrees from Northwestern Polytechnical University, Xi'an, China, in 1992 and 1995, respectively, and the Ph.D. degree from the University of Texas at Austin in 2002, all in Electrical Engineering. He worked at the Institute of Remote Sensing Applications, Chinese Academy of Sciences as a Research Engineer from 1995 to 1998. He is currently an Assistant Professor at the University of Texas – Pan American. His research interests include radar imaging, digital signal processing, RF and microwave engineering.

Zhiping Lin received the B.Eng. degree from South China Institute of Technology, China in 1982, and the Ph.D. degree from the University of Cambridge, England in 1987. Subsequently, he worked as a postdoctoral researcher at the University of Calgary, Canada. He was an associate professor at Shantou University, China from 1988 to 1993, and a senior engineer at DSO National Laboratories, Singapore from 1993 to 1999. Since Feb. 1999, he has been an associate professor at Nanyang Technological University, Singapore.

Dr. Lin has been an editorial board member for the International Journal of "Multidimensional Systems and Signal Processing" since 1993, and is a guest editor for the special issue on "Applications of Grobner bases in multidimensional systems and signal processing" in the same journal in 2001. He has been an associate editor for the Journal of *Circuits, Systems and Signal Processing* since 2000. He is the Program Co-Chair of the Third International Conference on Information, Communications and Signal Processing, 2001. His research interests include multidimensional systems and signal processing, wavelets and applications, neural network applications, array signal processing and biomedical signal processing.

Hao Ling was born in Taichung, Taiwan, on September 26, 1959. He received the B.S. degrees in electrical engineering and physics from the Massachusetts Institute of Technology, in 1982, and the M.S. and Ph.D. degrees in electrical engineering from the University of Illinois at Urbana-Champaign, in 1983 and 1986, respectively. He joined the faculty of the University of Texas at Austin in September 1986 and is currently a Professor in the Department of Electrical and Computer Engineering and holder of the L. B. Meaders Professorship in Engineering.

During 1982, he was associated with the IBM Thomas J. Watson Research Center, Yorktown Heights, NY, where he conducted low temperature experiments in the Josephson Department. While in graduate school at the University of Illinois, he held a research assistantship in the Electromagnetics Laboratory as well as a Schlumberger fellowship. He participated in the Summer Visiting Faculty Program in 1987 at the Lawrence Livermore National Laboratory. In 1990, he was an Air Force Summer Fellow at Rome Air Development Center, Hanscom Air Force Base. Dr. Ling's principal area of research is in computational electromagnetics. During the decade, he has actively contributed to the development and validation of numerical and asymptotic methods for characterizing the radar cross section from complex targets. His recent research interests also include radar signal processing, automatic target identification and wireless channel modeling.

Dr. Ling is a recipient of the National Science Foundation Presidential Young Investigator Award in 1987, the NASA Certificate of Appreciation in 1991, as well as several teaching awards from the University of Texas. He was elected to IEEE Fellow in 1999.

Junshui Ma received the M.S. degree from Xi'an Jiaotong University, China, in 1997, and the Ph.D. Degree from the Ohio State University, Columbus, USA, in 2001, both in Electrical Engineering. He is currently a postdoctoral researcher in the Los Alamos National Lab, USA.

His research interests include pattern recognition, machine learning, and signal/image processing.

Dale E. Nelson received the B.S. degree in aeronautical engineering from Tri-State University, Angola, IN in 1969. He received the M.S. degree in computer science from Wright State University in 1981. The Ph.D. degree in electrical engineering was awarded in 2001 by Ohio University.

From 1969 to 1972 he was a structural fatigue engineer at Cessna Aircraft in Wichita KS. In 1972 he joined the Air Force Flight Dynamics laboratory where he worked in the preliminary aircraft design group. In 1986 he was promoted into the Aeronautical Systems Division (ASD) Deputy for Development Planning where he was the database administrator responsible for the budgeting submission of ASD to HQ AFMC. In 1989 he became the supervisor of the Advanced Systems Research Group responsible for performing machine intelligence research. He is currently chief of the ATR and Fusion Algorithms Branch of the Sensors Directorate. He is leading a group doing basic research into advanced algorithms for target identification. He is also serving as an adjunct faculty member at Wright State University. His current research interests are in the area of rough set theory, automatic target recognition, and machine intelligence.

no photo available

Joseph A. O'Sullivan was born in St. Louis, MO, on January 7, 1960. He received the B.S., M.S., and Ph.D. all in Electrical Engineering from the University of Notre Dame in 1982, 1984, and 1986, respectively. In 1986, he joined the faculty in the Department of Electrical Engineering at Washington University, where he is now a Professor. He has joint appointments in the Department of Radiology and the Department of Biomedical Engineering. He was a founding member and is now Director of the Electronic Systems and Signals Research Laboratory. He is a member of the Magnetics and Information Science Center and the Center for Imaging Science at Washington University.

Prof. O'Sullivan is chair of the Faculty Senate and Chair of the Faculty Senate Council, and Faculty Representative to the Board of Trustees at Washington University. He is a consultant for and on the Board of Directors of Abacus Controls, Inc.

He was the Publications Editor for the IEEE Transactions on Information Theory from 1992 to 1995, was Associate Editor for Detection and Estimation from 1998 to 2001, and was a Guest Associate Editor for the 2000 Special Issue on Information Theoretic Imaging. He has served on the organizing and program committees for several conferences and workshops. He was co-chair of the 1999 Information Theory Workshop on Detection, Estimation, Classification, and Imaging, and was in charge of travel grants and registration for the 1995 Information Theory Workshop on Information Theory, Multiple Access, and Queueing. His research interests include information-theoretic imaging, automatic target recognition, systems integration issues in magnetic recording, CT imaging in the presence of known high density attenuators, information hiding, and hyperspectral imaging.

Prof. O'Sullivan is active in local IEEE activities as well, including being chair of the St. Louis Section of the IEEE in 1994. He is a Senior Member of the IEEE, a member of Eta Kappa Nu, and a member of SPIE. He was awarded an IEEE Third Millenium Medal.

Lee C. Potter received the B.E. degree from Vanderbilt University and M.S. and Ph.D. degrees from the University of Illinois, Urbana, all in electrical engineering. Since 1991 he has been with the Department of Electrical Engineering at The Ohio State University where he is currently Associate Professor. His research interests include statistical signal processing, inverse problems, detection, and estimation, with applications in radar target identification and ultra wide-band systems. Dr. Potter is a 1993 recipient of the Ohio State College of Engineering MacQuigg Award for Outstanding Teaching.

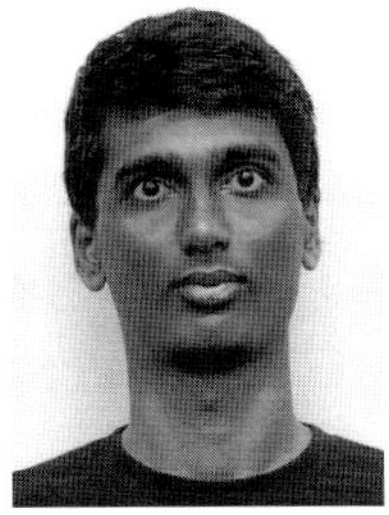

Ravikiran Rajagopal received the B.E. degree from the University of Madras and the M.S. degree from the Ohio State University, where he is currently a doctoral candidate. His research interests include algebraic systems theory, multivariate MIMO equalization, blind estimation, and fast algorithms for signal processing and coding theory.

Alberto Refice received the Laurea Degree and the Ph.D. in Physics from the Dipartimento di Fisica, University of Bari, Italy, in 1994 and 1998, respectively.

He was at the same University as a Post-Doc researcher for two years, and then as a contract Researcher for the next two years.

He is presently with Consiglio Nazionale delle Ricerche - IRA (Istituto di Radioastronomia) as a Researcher.

His main research interests concern advanced processing techniques for radar interferometry, and their applications to the retrieval of geo-biophysical parameters, with emphasis on environmental hazard modeling and management.

Pedro Sanchez received a Ph.D. in mathematics from the University of Michigan in 1972. He has taught a variety of courses in mathematics and information systems at Eastern Michigan University, the George Washington University and the University of Michigan. In addition to his professional work for Veridian Corporation, he has served as consultant to EDS's Center for Advanced Research and ERIM International (in Ann Arbor, Michigan). His present areas of research include Linear Systems and Knowledge Databases.

Janusz A. Starzyk received the M.S. degree in applied mathematics and Ph.D. degree in electrical engineering from Warsaw University of Technology, Warsaw, Poland, in 1971 and 1976 respectively.

From 1977 to 1981 he was an Assistant Professor at the Institute of Electronics Fundamentals, Warsaw University of Technology, Warsaw, Poland. From 1981 to 1983 he was a Post-Doctorate Fellow and research engineer at McMaster University, Hamilton, Canada. In 1983 he joined the Department of Electrical and Computer Engineering, Ohio University, Athens, Ohio, where he is currently a professor of EECS. He has cooperated with the National Institute of Standards and Technology in the area of testing and mixed signal fault diagnosis. He has been a consultant to ATT Bell Laboratories, Sverdrup Technology, and Magnetek Corp. In 1991 he was a visiting professor at University of Florence, Italy. He was a visiting researcher at Redstone Arsenal-U.S. Army Test, Measurement, and Diagnostic Activity and at the Air Force Research Laboratory-Advanced Systems Research and Sensor ATR Technology Development. His current research is in the areas of neural networks, rough sets, VLSI design and test of mixed signal MOS circuits, and reconfigurable design for wireless communication systems.

Mark A. Stuff received a master's degree in mathematics from Western Illinois University in 1976, and a master's degree in applied mathematics from the University of Colorado in 1980. He is currently a Ph.D. candidate in statistics, at the University of Michigan. From 1980 to 1985 he worked as a geophysicist for Shell Oil Company, where he developed new methods for seismic signal processing. From 1985 to the present he has worked for ERIM (which has now become the Ann Arbor Research and Development Center for Veridian Systems). There, he has developed methods for thermal and electromagnetic simulations and new signal processing methods for radar signals.

Mike D. Turley completed his Ph.D. in plasma physics at The Flinders University of South Australia in 1986. Since 1986, he has been employed by DSTO. He presently leads an EW team as a Senior Research Scientist in the Intelligence, Surveillance & Reconnaissance Division. His research activities focus on signal processing techniques including signal detection, CFAR, adaptive beamforming and spectral analysis.

Nicola Veneziani received the Laurea Degree in Physics from the University of Bari, Italy, in 1977. In 1978 he joined Ansaldo S.p.A., where he worked in the field of industrial automation. From 1979 to 1982 he was researcher at the Italian National Council of Research (CNR) for a research program concerning computerized cartography and Remote Sensing data processing for Geology. Since 1983 he has had a permanent position as staff of the Institute for Signal and Image Processing (IESI - CNR), in Bari. Since then he has collaborated with the Italian national institutions for space activities in the field of Remote Sensing. He contributed to the implementation of the first Italian digital SAR-processor, coordinating also the design of an advanced ultrasound scanner for Ophthalmology. From 1987 to 1992 he collaborated with the Caltech C3P Group (Caltech Concurrent Computation Program) at the project "SAR-processor on Hypercube", and with the Caltech Concurrent Supercomputing Facilities (CCSF), in order to evaluate the suitability of multiprocessor systems with different architectures to perform fast processing of remotely sensed data. Beginning from the early years '90, his main research concerns advanced techniques for data processing in radar-interferometry, in order to get soil morphology as well as the survey of Geodynamic Hazard. Currently his research effort is directed towards the compensation of atmospheric effects on the phase terms exploited in differential interferometry, in order to allow the detection of subtle movements between borders of a tectonic fault. At the press of this paper he will be with the Institute of Intelligent Systems for Automation (ISSIA-CNR), which replaces the previous IESI-CNR at the same address in Bari.

Alan S. Willsky received both the S.B. degree and the Ph.D. degree from the Massachusetts Institute of Technology in 1969 and 1973 respectively. He joined the M.I.T. faculty in 1973 and his present position is as the Edwin S. Webster Professor of Electrical Engineering. From 1974 to 1981 Dr. Willsky served as Assistant Director of the M.I.T. Laboratory for Information and Decision Systems. He is also a founder and member of the board of directors of Alphatech, Inc. and is also a member of the US Air Force Scientific Advisory Board. In 1975 he received the Donald P. Eckman Award from the American Automatic Control Council. Dr. Willsky has held visiting positions at Imperial College, London, L'Université de Paris-Sud, and the Institut de Recherche en

Informatique et Systèmes Aléatoires in Rennes, France. He was program chairman for the 17th IEEE Conference on Decision and Control, has been an associate editor of several journals including the *IEEE Transactions on Automatic Control*, has served as a member of the Board of Governors and Vice President for Technical Affairs of the IEEE Control Systems Society, was program chairman for the 1981 Bilateral Seminar on Control Systems held in the People's Republic of China, was special guest editor of the 1992 special issue of the *IEEE Transactions on Information Theory* on wavelet transforms and multiresolution signal analysis, and served as co-chair of the 1998 ONR Working Group on the Role of Probability and Statistics in Command and Control. Also in 1988 he was made a Distinguished Member of the IEEE Control Systems Society. Dr. Willsky has given several plenary lectures at major scientific meetings including the 20th IEEE Conference on Decision and Control, the 1991 IEEE International Conference on Systems Engineering, the SIAM Conf. on Applied Linear Algebra, 1991, the 1992 Inaugural Workshop for the National Centre for Robust and Adaptive Systems, Canberra, Australia, the 1993 IEEE Symposium on Image and Multidimensional Signal Processing, Cannes, France; and the 1997 SPIE Wavelets and Applications Symposium, San Diego.

Dr. Willsky is the author of the research monograph *Digital Signal Processing and Control and Estimation Theory* and is co-author of the undergraduate text *Signals and Systems*. He was awarded the 1979 Alfred Noble Prize by the ASCE and the 1980 Browder J. Thompson Memorial Prize Award by the IEEE for a paper excerpted from his monograph. Dr. Willsky's present research interests are in problems involving multi-dimensional and multiresolution estimation and imaging, statistical image and signal processing, data fusion and estimation of complex systems, image reconstruction, and computer vision.

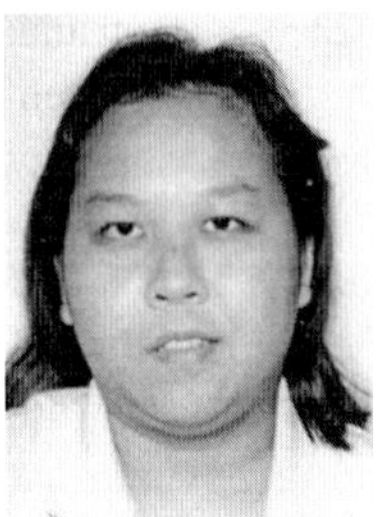

Jocelyn Yeo received BSE degrees in Electrical Engineering (1998) and Computer Engineering (1998), and an MSE degree in Electrical Engineering: Systems (1999) from the University of Michigan in Ann Arbor, Michigan, USA. She is a member of Eta Kappa Nu, IEEE and ACM. Her current professional interests include areas in signal processing and information theory. She is presently with DSO National Laboratories, Singapore.

Yonghong Zeng received the B.S. degree in mathematics from the Beijing University in 1983, the M.S. degree in applied mathematics and Ph.D. degree in computer science and technology from the National University of Defense Technology in 1986 and 1998 respectively, all in P.R. China.

Since 1986, he has been with the department of system engineering and mathematics, National University of Defense Technology in P.R. China. He has served as an associate professor since 1993. He joined the School of Electrical and Electronic Engineering, Nanyang Technological University, Singapore, to work as a research fellow in 1999. His main research interests include algorithms and transforms for signal processing, parallel computing, wavelets and filter banks, radar and array signal processing. He received ministry level Scientific and Technological Development Awards in P.R. China three times. He has published five books and a number of papers.